New ingredients in food processing

Selected titles from Woodhead's food science, technology and nutrition list:

Food technology

Tamime and Robinson, *Yoghurt: science and technology* Second edition
ISBN: 1 85573 399 4

Widely accepted as the standard work in the industry, the second edition has been completely rewritten and greatly expanded, incorporating major new developments such as 'bio-yoghurts'.

Greensmith, *Practical dehydration* Second edition ISBN: 1 85573 394 3

'This book is primarily addressed to the food industry professional. The text is written with great care and provides a large amount of valuable, practical information.' *Lebensmittel Wissenschaft und Technologie*

Grandison and Lewis (eds), *Separation processes in the food and biotechnology industries* ISBN: 1 85573 287 4

'. . . a comprehensive and practical view of the separation technologies that are currently available.' *Trends in Food Science and Technology*

Food quality

Singhal, Kulkarni and Rege, *Handbook of indices of food quality and authenticity*
ISBN: 1 85573 299 8

'. . . this book will be an essential reference work for industry and an indispensable guide for technologists, analysts, microbiologists and food chemists.' *Food Science and Technology*

Kress-Rogers (ed.), *Instrumentation and sensors for the food industry*
ISBN: 1 85573 363 3

'. . . the most comprehensive workbook for practitioners ever written.' *Food Science and Technology*

Details of these books and a complete list of Woodhead's food science, technology and nutrition titles can be obtained by:

- visiting our web site at www.woodhead-publishing.com
- contacting Customer Services (e-mail: sales@woodhead-publishing.com; fax: +44 (0) 1223 893694; tel.: +44 (0) 1223 891358; address: Woodhead Publishing Ltd, Abington Hall, Abington, Cambridge CB1 6AH, England)

New ingredients in food processing

Biochemistry and agriculture

Guy Linden
Denis Lorient

Translation by Maggie Rosengarten, Translate Language Services, Bury St Edmunds, UK
UK advisory editor: Dr M J Lewis, Dept of Food Science and Technology, University of Reading

CRC Press
Boca Raton Boston New York Washington, DC

WOODHEAD PUBLISHING LIMITED
Cambridge England

Published by Woodhead Publishing Limited, Abington Hall, Abington
Cambridge CB1 6AH, England
http://www.woodhead-publishing.com

Published in North and South America by CRC Press LLC, 2000 Corporate Blvd, NW, Boca Raton FL 33431, USA

British Library Cataloguing in Publication Data
A catalogue record for this book is available from the British Library.

Library of Congress Cataloging in Publication Data
A catalog record for this book is available from the Library of Congress.

Woodhead Publishing ISBN 1 85573 443 5
CRC Press ISBN 0-8493-0631-0
CRC Press order number: WP0631

Cover design by The ColourStudio
Typeset by Best-set Typesetter Ltd, Hong Kong
Printed by TJ International, Cornwall, England

Contents

Part Two – Extraction and modification of biomolecules

Abbreviations

ADF: anhydrous dairy fat
AMP: adenosine monophospate
ATP: adenosine triphosphate
a_w: water activity
BHA: butyl hydroxyanisol
BHT: butyl hydroxytoluen
BV: biological value
CMC: critical micelle concentration; carboxymethylcellulose
DA: degree of amidation
DE: dextrose (glucose) equivalent
DEAE: diethyl amino ethyl cellulose
DEX: dry extract
DH: degree of hydrolysis
DM: dry matter
DP: degree of polymerisation
DS: degree of substitution
DSC: differential scanning calorimetry
DUC: digestive utilisation coefficient
ERH: equilibrium relative humidity
FPC: fish protein concentrates
GMP: guanosine monophosphate
HDL: high-density lipoproteins
HLB: hydrophilic lipophilic balance
HPLC: high-performance liquid chromatography
HT: hydrotimetric titre
IC_{50}: concentration of the molecule being tested for which 50% of the maximum effect is achieved

ID_{50}: mass of the molecule being tested for which 50% of the maximum effect is achieved
IMP: inosine monophosphate
IU: international units
LD 50: lethal half-dose
LDL: low-density lipoprotein
MM: molecular mass
NCE: new chemical entity
NMR: nuclear magnetic resonance
NPN: non-protein nitrogen
NUT: nephelometer unit of turbidity
PDI: protein dispersibility index
PEC: protein efficiency coefficient (also known as protein efficiency ratio (PER))
PPS: plant protein substances
RDA: recommended daily amount
SDS: sodium dodecyl sulphate
SED: storage expiry date
SP: sweetening power
T_c: Kraft temperature
TMAO: trimethylamino oxalate
TNS: total nitrogenous substances
TOS: transgalacto-oligosaccharides
UF: ultrafiltration
UHT: ultra high temperature
WHC: water-holding capacity
WPNI: whey protein nitrogen index

Foreword

Training and innovation are the major assets which ought to accompany and, if possible, anticipate, the profound changes that are a feature of modern life. The farming and food industries, in particular, are experiencing a revolution that is all the more significant since together they form one of the world's oldest industries, which has been around since the neolithic age.

The sudden appearance of science and technology within manufacturing processes which up to then had been based on skills accumulated over the years, the appearance, all over the world, of products which had been very local right up to the Second World War, and recourse to modern management techniques have completely overturned the financial and industrial structures as well as storage, production and marketing techniques within this industrial sector.

One of the most significant phenomena of this amazing development is the appearance, about twenty years ago, of intermediate food products (IFPs) particularly associated with a progressive distancing from the places in which they were originally produced, and with conversion and consumption.

Formerly the food factory was in direct contact with its suppliers and its consumers. It was capable of both preventing the biochemical phenomena responsible for the degradation of the quality of the agricultural products and of constantly adapting the characteristics of its products to changes in consumers' tastes.

Today the various food industries have expanded considerably. The first stage involved a split between the agricultural industries close to the areas of production (such as sugar refineries and malting, for example) and the food industries located within large built-up areas (such as breweries, for example). However, this phenomenon became progressively more complex, resulting in products designed to meet the consumer's changing demands (satisfaction, convenience, health, safety) at the best possible cost.

The response of the food industry was to refine the agricultural raw materials, once they were stabilised, to produce ingredients of a standardised quality, and then to recombine them in factories close to the areas of consumption which now fulfil functions similar to those of car-assembly plants.

This development led to the exploration of new fields of research aimed at analysing the various components of agricultural and food products, in order to understand how they interact, and to move towards an increasingly comprehensive mastery of all the biochemical and physico-chemical phenomena on which the quality of food depends.

This work has been covered in numerous specialist publications and many conferences and symposia. However, to date there has been no general text describing in ordinary terms an activity whose principal characteristic is the fact that it interacts closely with the farming and the food industries.

Professors Guy Linden and Denis Lorient are to be congratulated for having written this book, which is intended for laboratory researchers in the field of public and private research and development, more generally for farmers and manufacturers and for final year and postgraduate university students and those from the '*Grandes Ecoles*'.

This book offers a comprehensive summary of research into the biochemistry of food products and of the experience the authors have acquired in the course of their long professional careers. Readers will find the style clear and readable, offering a widely accessible and attractive style of learning. The format is lively, well organised and concise, recording the results of research work carried out in many public and private laboratories, both in France and abroad.

Finally, this book, as well as offering a summary of the work carried out on IFPs over the past twenty years or so, is also preparing for the future by laying the biochemical foundations for commercial exploitation of agricultural products which we know will be important in the development of the European agro-industrial system.

I hope this work will quickly reach the wide audience awaiting it.

Hervé Bichat
Director General of Teaching and Research
Ministry of Forestry and Agriculture
France

* Translator's note: '*Grandes Ecoles*' – non-university establishments, awarding diplomas of comparable prestige to an Oxbridge degree in Britain.

Preface

. . . we have given the name of plastic foods to albumin, fibrin and plant casein, as well as to animal casein and fibrin, because these are the only substances from the plant and animal kingdom, which, in nutrition, can produce the essential parts of animal blood and organs.

Justus Liebig,
Nouvelles Lettres sur la Chimie,
Victor Masson, 1852, p. 116

The fractionation of agricultural products and the substitution of a biochemical ingredient of animal origin with an ingredient of plant origin are long-standing concerns, but for many years they were the prerogative of a small band of pioneers.

Nowadays, in order to take into account all the requirements of health, convenience, hygiene and consumer satisfaction, the agro-food industry has broken free of the conventional chains of food technology. This is why, at the end of the twentieth century, it is undergoing a revolution comparable to that previously experienced in other industrial sectors.

Basically, the primary conversion industry markets products called intermediate products which can be used as raw materials, providing ingredients for the secondary conversion industry which processes foods. Using farming produce, and implementing techniques and processes based on the physicochemical properties of these ingredients, this primary conversion industry isolates fractions in order to produce ingredients or intermediate food products (IFPs) with well-defined nutritional and functional qualities which have special uses. In addition, this industry not only extracts major biochemical ingredients but is also able to produce texturising agents, sweeteners, colorants, aromas, and so on. In the development of end-products, IFPs have become the kit parts of the food industry. The same is true for the catering trade which has become a kitchen for assembling these ingredients.

The advantages of this revolution in the logistics of the food industry are associated with the requirements for manufacturing and marketing products whose quality is constant (in time and space), and which can be processed where they are eaten, thus avoiding most of the problems of storing end-products (ready-made meals, semi-preserves, etc.).

This book aims to offer a concise explanation of the two branches of this industrial activity:

- Manufacture and functional properties of intermediate food products.
- Extraction and modification of biomolecules.

The authors had a problem, if they were to keep the book to a manageable length, in terms of choice of subjects and methods of presentation. They solved this by basing the text on biochemistry alone. Microbiology has not been included, and metabolic aspects have only been touched upon.

The growing number of specialist works, the organisation of numerous conferences and symposia on this subject have produced voluminous and costly reference material. Given the way this type of teaching is developing within universities and 'Grandes Ecoles', this work aims to offer students a biochemical approach to these new industrial developments. Finally, professionals working in the food industry will find that this book summarises and complements their own scientific and technological knowledge.

Acknowledgements

Our grateful thanks to the kind, patient and long-suffering Denise who worked with us on the book. We would like to express our warm thanks to the open-minded and talented Laurent Diez who was responsible for the iconography. We also thank our colleagues from the Nancy I University and the ENSBANA who were kind enough to read certain chapters.

Since we know that a text can never be perfect, and that, of course, there is always room for improvement, we will be pleased to hear any criticisms, alongside any praise.

Guy Linden
Denis Lorient

Part One – Manufacture and properties of intermediate food products

1

Intermediate food product strategy

1.1 Introduction

Over recent years the food industry has been undergoing major change. To meet growing variations in demand and increasingly specific requirements from consumers, the food industry needs to display a huge capacity for innovation. Nowadays food products must always be safe, must meet nutritional and sensory requirements and must offer more and more benefits to satisfy the needs created by our changing lifestyles. In other words, they must offer four essential elements: health, taste, safety and convenience.

Although traditional foodstuffs (bread, wine, cheese, beer, meat, etc.) are the outcome of processing agricultural raw materials, this is far from being the case with new products which are the result of combining a more complex range of ingredients. Over the last few years we have seen a new processing industry come into being in the fields of carbohydrates, lipids, proteins, colorants, flavourings, etc. Its purpose is to provide a wider range of tailor-made ingredients, or 'intermediate' food products (IFPs) for the secondary processing industries (Fig. 1.1).

The IFP supplier operates between the food manufacturer and the agricultural supply stage. This has important consequences from both a cultural and an economic point of view. By turning the relationship with the raw material upside down in this way, this sector is using new methods of operation closer to those in other industries. IFPs are types of processed material with distinct characteristics. They can be used with increasing flexibility, and they have increasingly refined functional properties. The challenge for the industry lies in making such improvements to the constituents

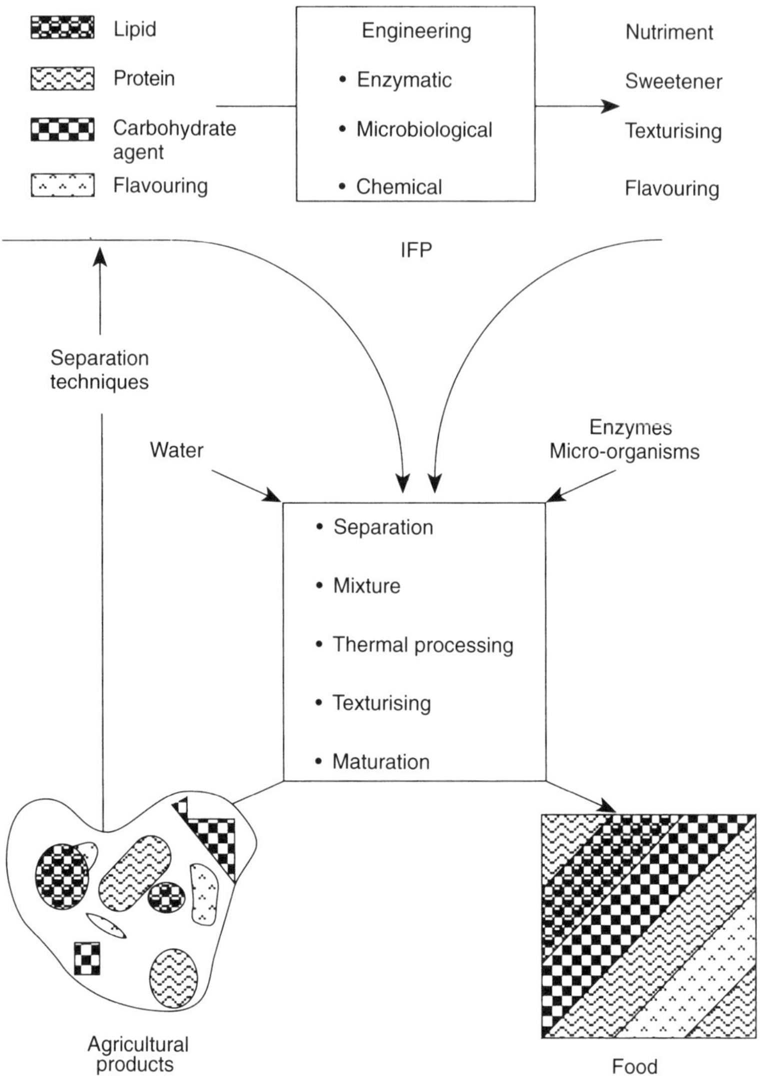

Fig. 1.1 General diagram showing the development of new food products.

of agricultural raw materials which can, for example, vary over time and which are often difficult to store. As an example, products and waste from the food industry could become the basis for ingredients themselves, whose constituents, such as casein and whey, blood, abattoir waste, etc., need to be refined for further use.

The IFP industry has frequently succeeded in improving all these ingredients and even in introducing some of the compounds obtained from the

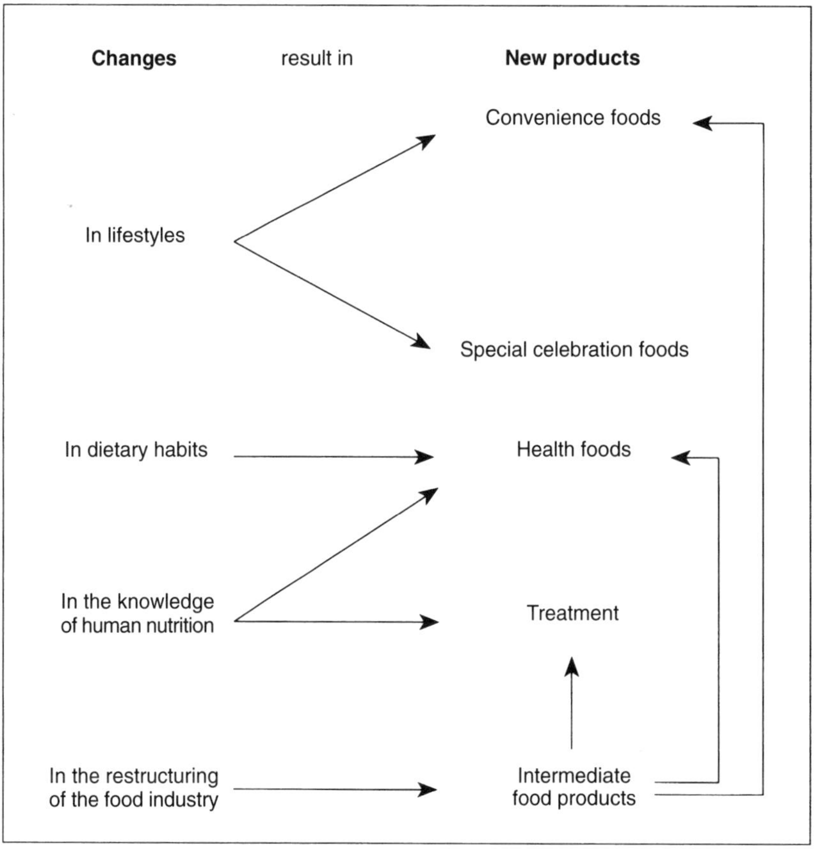

Fig. 1.2 Diversification of dietary types.

food sector into non-food sectors such as the pharmaceutical or other industrial sectors. We need to bear in mind that even the most useful plant products contain only 50% of usable components. These consist of carbohydrates, lipids or proteins, depending on whether we are dealing with cereals or legumes, half of which is concentrated in the outer skin. These 'packaging' materials constitute an extremely important source, but at the moment it is still very difficult to exploit them economically.

1.1.1 Consumers' viewpoint

Consumers want to obtain ready-to-use products at the best possible price, whether they are purchasing catering services or simply buying for themselves (Fig. 1.2). The health value of these various foods must also be guaranteed, by reconciling technological dictates (making products attractive to consumers) with good nutritional value. An example of this is the production of appealing and palatable fibre (see Section 12.5.2) or the production of low-calorie products (see Section 1.3).

1.1.2 Manufacturers' viewpoint

We are witnessing a gradual development within IFP manufacturing companies. They are extending the use of all the constituents of a raw material, whether food or non-food. Each purified fraction derived from these constituents makes up a new raw material. Fractions obtained in this way from these raw materials are first-generation IFPs. However, in order to encourage the use of these ingredients by secondary processors, IFP companies must create new compounds from these fractions, and also provide them in a balanced proportion, providing each user with the technology for exploiting these mixtures. These companies now provide a comprehensive mix of products and services for customers, tailored to their individual requirements. This occurs, for example, in companies that fractionate the components in milk in order to obtain products used in the manufacture of ice-cream, biscuits, cooked meats or chocolate. The same is true of firms that make gelatin for the food industry. The new compounds obtained from these fractions are second-generation IFPs, also called 'mixes', and they are sold together with their technologies and methods of application.

One well-known instance which illustrates how IFP manufacturing companies and users have worked together is that of lipids and lipid chemistry. They have co-operated to bridge the gap between the properties of raw animal and plant lipids and the technological and nutritional properties required in the final product. The technological requirements can be illustrated by the various uses of fats. These uses include oils for deep-fat frying that are stable at 180 °C, fats that are spreadable at 5 °C, products for ice-cream, cake-making, biscuit-making (shortening), cooked meats, etc. Nutritionally, these products must satisfy the needs of the average consumer. Ideally fats should constitute a maximum of 30% of energy intake (such fat products must include a balanced mix of essential fatty acids), and must combine unsaturated fatty acids, adapted to different physiological ages. As an example, elderly people need arachidonic acid as they can no longer convert linoleic acid into arachidonic acid.

IFP manufacturers have succeeded in using developments in lipid chemistry to satisfy these various requirements, using methods such as dry or surfactant fractionation, trans-esterification, selective or total hydrogenation. For example, it is possible to obtain various products from palm oil which are either liquid at 20 °C or solid up to 45–50 °C. In the same way, one of the key developments in the dairy industry recently has been to develop butters which can be spread at temperatures of between 5 and 30 °C. Another example is the preparation of emulsifiable fats ('shortenings') for the biscuit industry. In this case the requirement was to obtain a brittle structure by combining fats with flour and sugar. Its melting point needed to be between 33 and 35 °C, yet it must not stick to the tongue. After controlled trans-esterification the density of the 'shortenings' is brought to around 0.70 by introducing inert gas. This suspension of lipid crystals in a liquid phase is stable between 15 and 45 °C (see Chapter 13).

1.2 Scientific and economic essentials

1.2.1 Scientific and technical criteria

By definition an IFP is a standardised product, whose composition and functional properties are constant. The example of flours and bread illustrates this particularly well. The manufacturer has responded to the diversity in the composition of wheat, which results from the different varieties cultivated and the various conditions under which the plant develops (for example, location, seasonal variations, farming practice), by adding corrective agents (e.g. enzymes or chemical products) in order to supply the baking industry with flour whose quality never varies. This basic requirement is sometimes forgotten, when we wish to improve a by-product in order to extend its range of applications. Improving the quality of syrup produced from soaking maize illustrates this. Soaking is the first operation in the wet starch industry: among other things it enables soluble substances to be extracted (carbohydrates, amino acids, peptides, minerals, vitamins, etc.). This syrup has been used for a long time in cattle foodstuffs without any attempt to improve its quality, although it has a very high biological value. If the quality was standardised it would have a high added value as a raw material in the fermentation industry.

A high-performance IFP is polyvalent: it can be used in many food products. This quality is obviously not an essential requirement but it does permit significant economies of scale. The widespread use of sorbitol illustrates this point. It is used not only in the cake and confectionery industry, but also in ice-creams, sauces and dressings, and in the manufacture of drugs, resins and cosmetic products (see Section 10.6.1.1).

To fulfil this condition of polyvalency, an IFP:

- must be compatible with many other ingredients and with many technological and preparation processes;
- must be easy to store in a dehydrated form and be easy to use (powder or liquid).

The IFP industry has been able to expand to such an extent over the past few decades because of the considerable technological progress that has been made. This includes the development and improvement of extraction, purification, thermal and texturisation processes (cooking-extrusion, spinning, etc.) and preservation methods (see Chapter 3).

1.2.2 Economic criteria

Between the conception and birth of an IFP lies its gestation period and it is during this time that an examination of the economic profitability of the product is a key requirement. The IFP manufacturer must get as close as possible to the technological development and industrial requirements of its potential customers. The use of starch in the brewing industry illustrates very clearly the type of approach needed. In most European countries, apart from Germany, the raw materials of beer are made up of the following: 70% malt and 30% of other starchy substances known as 'raw grains', which in the past were essentially constituted by maize grits. Approximately fifteen years of technical discussions between brewers and starch producers have led to the conclusion that starches could be more attractive to brewers if they were delivered in the form of 'starch milk'. In addition, this working relationship has enabled manufacturers to discover that the ideal IFP is hydrolysed starch. The interests of the two manufacturing partners are converging on glucose syrup.

In addition, the IFP must be an improvement on the raw material from which it originates, in terms of the price and superior qualities it offers, even if the raw material is already the reference material in terms of quality, as is the case for example with egg and egg by-products (see Chapter 6).

This far-sighted attitude becomes the basic rule for adapting to the specific requirements of the market. Apart from the obvious need to produce for a market that is as large as possible, the IFP manufacturer must anticipate the ways in which legislation will develop and must be able to adapt to a country's food traditions. Two examples can be given:

- A French cyclodextrin manufacturer (see Chapter 10) gained early market entry by anticipating government authorisation to use these products within the food field.
- For a French or an Italian person pasta is a unique product, based on a single raw material: durum wheat flour. In Asia and America, however, dozens of different types of pasta are manufactured, using several raw materials: wheat flour, cereal or leguminous starch, dairy proteins, etc. Some European manufacturers are now taking advantage of these new raw materials in manufacturing for their own markets.

1.3 Illustrating the IFP strategy: low-calorie foods

The enormous popularity of low-calorie foods is a response to the growing demand from consumers who are increasingly concerned about their weight, and about eating a balanced diet. The combined progress made in dietary studies, food engineering and food science has enabled us to reconcile pleasure with a balanced diet and to reduce the average caloric intake of food eaten.

Reducing calories in food involves 'lightening' or reducing those ingredients that are now considered 'suspect' for health (e.g. animal fats, cholesterol, carbohydrates, sodium chloride) and adding substances that are considered beneficial to health (e.g. vitamins, oligo-elements, fibre and sweeteners), as well as novel future ingredients such as algae, plankton and synthetic fats.

As a consequence, food manufacturers are expanding most of their product ranges to include low-calorie versions, the sales of which have risen an average of 8% per year (as opposed to 3% for standard versions). Modifying traditional recipes means introducing new ingredients, within the new recipe formulations, which ensure that original organoleptic qualities are matched or even improved.

Texture is provided by the following:

- Thickeners and gelling agents whose concentration has been readjusted: gelatins, carrageenan, xanthan, guar, carob, pectin, alginates, modified starch.
- Dairy fermentation products: bifidus and acidophilus types.
- Bulking agents: soluble fibres, crystalline cellulose, polyhydric alcohols, maltodextrins, proteins.

Taste is provided by the following:

- A careful selection of flavours and taste enhancers.
- Sweeteners, to recreate a sweet taste, modifying the intensity of flavours ('booster' or 'masker' effect) or those that cause 'special effects' such as the 'cold in the mouth' sensation produced by polyol alcohol (xylitol).
- Spices and seasonings that bring out flavour.
- Preservatives (dependent on legislation) which prevent deterioration and the emergence of off-flavours.

As far as is possible, the manufacturer will try to interfere as little as possible with the core process, and will try to adapt to new problems. We can quote several examples:

- In low-calorie ice-creams, the excess of free water increases the risk of crystallisation which will adversely affect the texture of the product; more stabilising agents therefore have to be used.
- Gelatin must be added to several products to ensure that the soft, melting sensation of fats is retained.
- It is essential to add flavours that recreate the flavour of the products that would have resulted from the Maillard reaction to foods in which the carbohydrates have been replaced by sweeteners.

The use of new ingredients has to involve studies relating dosage levels to final product quality. It is particularly important to study the interactions between all these ingredients, their stability during the industrial processes and their behaviour with water. Some of these matters are covered in Chapter 2 which deals with the functional properties of IFPs.

To sum up, in the face of the rapid changes in eating habits, and owing to IFP techniques, foods of the future could all become balanced from a caloric point of view. The mix of ingredients in such foods will also be more perfectly adjusted to the nutritional aspirations of consumers. Optimisation of nutritional intake in any one food product will therefore result in more variations in what constitutes a healthy diet. It will therefore become easier to implement successful nutritional programmes among consumers.

In conclusion, this restructuring of the food industry must be put into its socio-economic context (Fig. 1.2). Thus manufacturers have a great interest in developing IFPs in order to obtain, from primary agricultural raw materials, other products whose uses can be extremely varied. This essentially involves meeting the widely divergent needs of consumers who want balanced products that are relatively attractive, but that ensure that all the elements needed for a low-calorie diet are provided. There must also be a wide range of products to suit a variety of lifestyles, and plenty of choice for consumers.

2

Functional properties

2.1 Definition and classification: role of functional properties of food components within sensory quality

2.1.1 Definition and factors of variation

Traditionally processes within the food industry have involved transposing culinary techniques. Certain familiar ingredients used as additives (vinegar, lemon juice, etc.) are willingly accepted by the consumer, while others with chemical names (magnesium carbonate, for example) are less so. The same is true for the traditional processes of cooking or grinding which are better accepted as they are more familiar on a domestic level than certain new thermal (cooking-extrusion) or sterilisation (ionisation) processes. The truth is, we like what we know, but we do not necessarily know what we like. Today's consumers require much more information about their health and what they eat since they want to select their food in accordance with a wide number of criteria (origin of ingredients, processing and preservation treatments, etc.) and to understand the reason why such and such an ingredient has been added or a particular process used. Food scientists also want to know more about the physico-chemical behaviour of the ingredients they use, so that they have better control over the nutritional and sensory qualities of the food they are producing.

However, the progress that has been made in the fundamental knowledge of macromolecules (proteins, polysaccharides, etc.) in terms of their structure and the interaction they are likely to establish with the small molecules that are naturally present (water, lipids, flavours, minerals, etc.) is making the formulation of foods less and less empirical; for this we need to

Table 2.1 Relationship between functional and sensory properties

Functional properties		Physical conditions	Sensory properties
Adsorption	Flavour holding Lipid holding Water adsorption	Gas	Flavour
Interfacial	Bulking Emulsification	Foam Liquid	Taste
Hydration	Water holding Solubility Viscosity	Paste	Kinaesthetic properties
Texture	Porosity Aggregation Gelling Coagulation	Dispersed solid	Touch Rheology
	Elasticity Microstructure (cellular)	Compact solid	Hearing

develop raw products and ingredients with constant characteristics, often because of increasing automation of processes.

In addition to the nutritional characteristics that need to be conserved while the food is being processed, it must be possible to define the sensory properties, either directly through tasting (which is often very risky when used as a routine method) or by linking them with functional properties which are easier to measure, but only give an incomplete idea of the sensory characteristics (Table 2.1). For example, the texture of a yoghurt can be characterised sensorially by its firmness, its elasticity, its cohesion, its creaminess, its visual appearance (shiny, matt, etc.) and how quickly it can release its flavour.

What functional properties, which can be measured using instruments, are able to represent all these characteristics: rheological measurements of the gel (penetrability, viscosimetry), microstructure of the fatty globules in the emulsion or retention of flavours? The most representative functional properties are global properties that simultaneously associate different but interdependent physico-chemical properties. These are closely dependent on the spatial structure of the molecules (more or less unfolded conformation, for example) and their state of association (between each other or with other molecules). The factors that intervene are principally those shown in Fig. 2.1:

- The composition of the medium: water, presence of other molecules, pH, ionic strength.
- Physical or chemical processes which alter the medium (concentration, drying, mechanical processes).

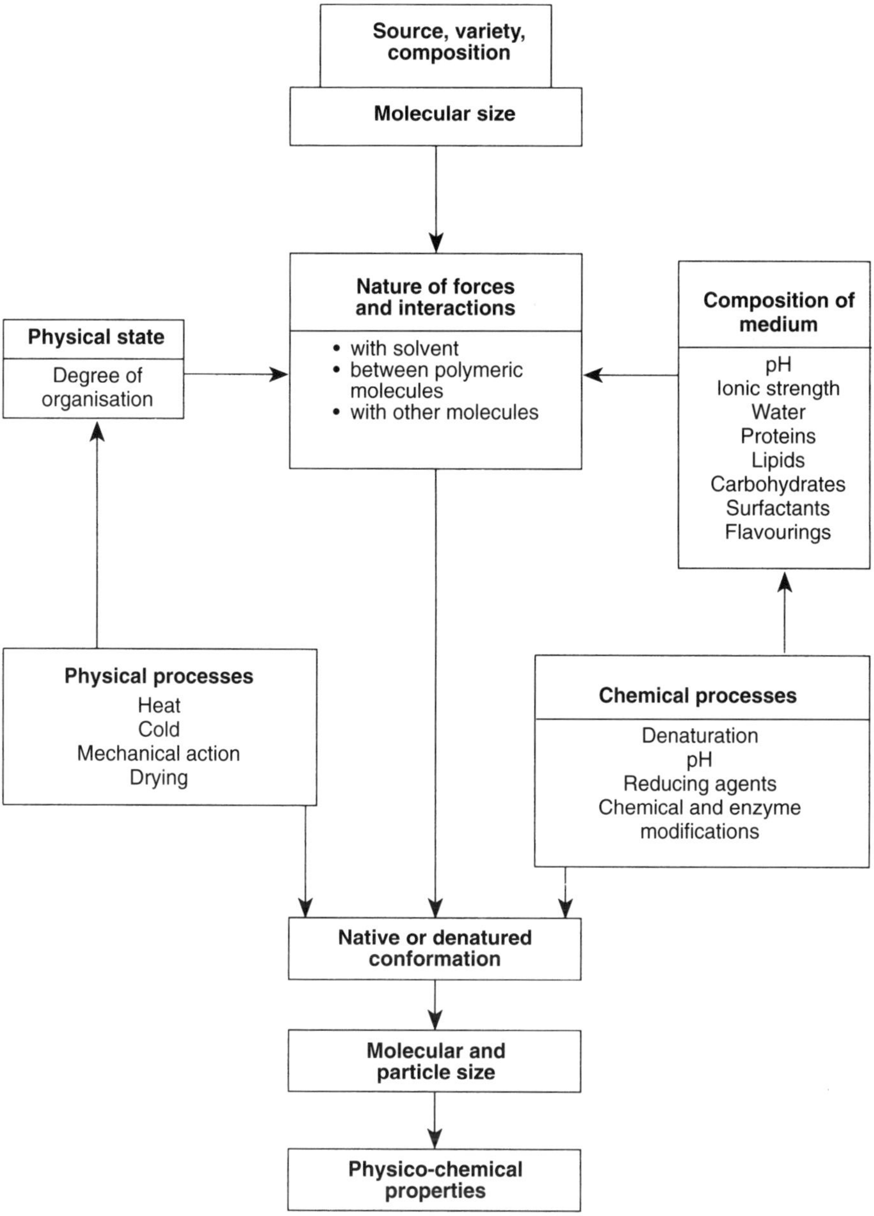

Fig. 2.1 Factors affecting the structure and functional properties of the macromolecules.

Table 2.2 Chemical groups that produce interactions with water

Ionised polar groups that can be dissolved	
Negatively charged	
• Carboxylic (COO^-): (if pH > p*K*a)	Polysaccharides, proteins (Asp, Glu), organic acids
• Phosphates (HPO_4^{2-}):	Phosphates, polyphosphates, phosphoproteins, nucleotides, phospholipids
• Sialic acids	Glycoproteins
• Sulphates (HSO_4^-):	Polysaccharides, sulphates (galactan sulphates)
• Other mineral anions:	Cl^-, NO_2^-, NO_3^- . . .
Positively charged	
• Amines, imines . . . —NH_3^+—NH_2^+— (if pH < p*K*a)	Proteins (Lys, His, Arg, Trp, Pro), osamines, nitrogenous bases, polyamines (cadaverin, spermin)
• Divalent mineral cations	Ca^{2+}, Mg^{2+} free or associated with proteins and polysaccharides
• Monovalent mineral cations	Na^+, K^+
Non-ionised polar groups (H bonds)	
• Hydroxyl (—OH):	All carbohydrates, proteins (Ser, Thr, Tyr), polyols
• Carboxyl (—COOH): (if pH < p*K*a)	Polysaccharides, proteins (Asp, Glu)
• Amine (—NH_2):	Proteins (Lys)
• Amide (—$CONH_2$)	Proteins (peptide bond, Asn, Gln)
• Thiol (—SH):	Proteins (Cys)
Non-polar groups	
• Aliphatic hydrocarbon group —$(CH2)_n$—:	Lipids, fatty acids, proteins (side chains of Ala, Val, Leu, Ile, Met, Pro, part of the side chain of Lys, Arg, peptide chain), carotenoid pigments, terpenes (unsaturated groups)
• Cyclic hydrocarbon group, either aromatic or not: 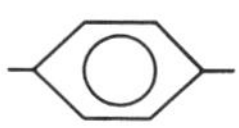	Proteins (side chains of Phe, Tyr, Trp), haem pigments (Hb, Mb, chlorophyll) and anthocyanins, polyphenols and tannins . . . , cyclic dextrins (cyclodextrins with hydrophobic centre) . . .

2.1.2 Nature of links and forces occurring within functional properties: classification

These properties (Table 2.2) are generally classified in accordance with the following criteria:

- The **behaviour of the ingredient towards water** which above all depends on hydrogen and van der Waals interactions: ionic interactions are also involved during the **solvation** of the ionisable groups.

- The **interactions of macromolecules with one another** (properties of polymerisation through intermolecular ionic, hydrophobic or covalent associations).
- The **interactions with molecules having little polarity** or those that are in a gaseous phase (properties at the interfaces, formation of polydispersed systems).

The interactions and forces that are involved therefore occur between solute and solvent (water) or between solutes, depending on the level of concentration of the solute in the solvent. So we can see that the balance between these two types of interaction will be influenced by the spatial requirements of the molecules of solute which itself depends on the concentration of their state of unfolding and their state of association (microstructure).

2.1.3 Influence of different phases of dispersion on functional properties

- **Mono and polydispersed systems.** A **monodispersed** system where all the particles are of the same size, can become, under certain circumstances, a **polydispersed** system consisting of particles of different sizes. In fact, in spite of the Brownian movement which maintains dispersion, particles can group together in **flocs** and separate from the continuous phase by gravity (**sedimentation** if the particles are denser than the medium, or by **creaming** if they are less dense). The process of **flocculation** can be reversed, unlike **coagulation** which cannot.
- **Different phases of dispersion and interdependency of different functional properties.** Because of the small size of the particles in colloidal substances, the surface/volume ratio is high and a significant proportion of molecules from these systems are located on the surface of heterogeneous parts. The molecules have different properties (energy, state of unfolding or association) from those of the continuous phase since they are a long way from the interface. Figure 2.2 shows that for a substance whose molecular volume is $30 cm^3/mol$, the proportion of molecules located on the surface of aggregates increases rapidly when the diameter becomes less than 1 μm (for example one molecule out of four is on the surface for particles of 10 nm diameter).

  This observation shows the importance of the interfacial properties in micro-heterogeneous systems characterised by particle sizes varying from 0.01 to 1 μm. This is precisely the case in food systems that contain dispersed phases such as **emulsions** (liquid in liquid), **foams** (gas in liquid), **suspensions** and aggregates (solids in liquids) or **poorly hydrated products** (liquids in solids).

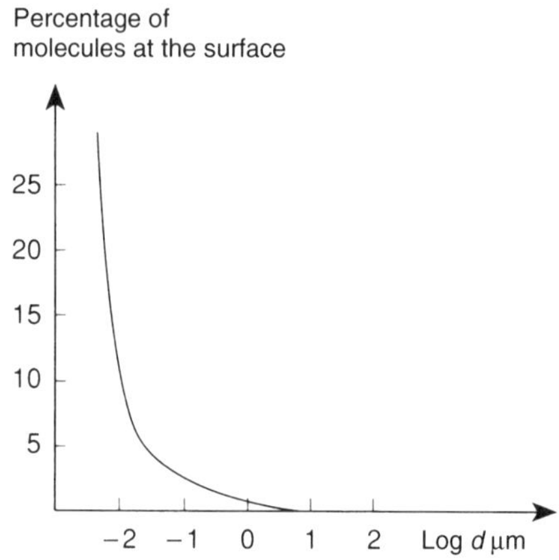

Fig. 2.2 Variation of percentage of molecules at the surface depending on the particle size for a substance with a molar volume of 30 cm^3/mol.

The separate study of the different properties simply aims to make their presentation clearer: the properties of texture, resulting from the behaviour of the molecules both within the medium and at the interfaces, should simultaneously integrate data on the different properties.

2.1.4 Methodology of functional properties

This requires the use of model systems and suitable tests that represent as closely as possible standard experimental conditions which are fairly close to technological reality.

Model food systems represent a simplification of the foodstuff which is intended to provide a better understanding of all the parameters that affect the measurement and to take these into consideration (for example, synthetic system with casein micelles being used as a dispersion medium without all the protein components of whey). The **physical properties** chosen should correspond to those that seem to be the most closely correlated with a sensory and representative property of the food. **Experimental conditions** must take into consideration the conditions of the medium (pH, ionic strength, composition, etc.) and the use of processes for shaping which often take place at the final stage of the process (cooking, drying, etc.). Sometimes, by juggling with several parameters, it will be possible to carry out accelerated preservation tests.

For several years we have been trying to standardise the methods of evaluating functional properties but these remain very controversial because of the great variability of the food systems involved: for example, should we use the same test of emulsifying capacity whether the emulsifier is incorporated in mayonnaises, homogenised recombined creams, cooked meats or ice-cream?

2.2 Properties of hydration

The properties of a macromolecular component depend on its interactions with water just as these depend on its conformational structure. Various states of water in a dispersed medium have been described (free, bound, immobilised, retained, freezable or non-freezable, solvent or non-solvent) but they often relate to the measurement technique used.

The water activity (a_w) allows us to describe the interaction between solute and water reasonably well if the system is in equilibrium; as this is not the case for foods that are often made up of several phases, kinetic parameters such as diffusivity must be used in order to obtain a better understanding of system dynamics.

As water has multiple roles (diffusion and reaction solvent medium, structure agent for the macromolecules), its effects will depend more on its interaction with the solutes than on its quantity within the medium.

2.2.1 Interaction between water and the components

These interactions occur because of ionisable groups capable of solvating or because of uncharged polar groups which establish hydrogen bonds with the water (Table 2.2). During hydration of protein powders, charged polar groups are hydrated before uncharged polar groups: the levels of hydration vary considerably, depending on the nature of the groups (Table 2.3) and on their position within the molecule, e.g. a polar residue located on the surface can be hydrated more rapidly. In addition, it is accepted that close to hydrophobic groups, polymerisation of the water can be found.

According to this information we might think that the amino acid composition of a protein would enable its behaviour towards water to be predicted. By calculating the average hydrophobicity of various proteins (average hydrophobicity of the various side chains), we can see that generally speaking it is not linked to the properties of hydration; in fact only

Table 2.3 Hydration capacity of side chains of amino acids in synthetic polypeptides (H_2O/mole residue) a_w 0.9

Polar groups				Non-ionised, non-polar groups		Non-ionisable polar groups	
Ionised		Non-ionised					
Asp^-	6	Asp (pH 4)	2	Ala	1.5	Asn	2
Glu^-	7.5	Glu (pH 4)	2	Val	1	Gln	2
Tyr^- (pH 12)	7.5	Tyr	3	Leu	1	Pro	3
Lys^+	4.5	Lys (pH 11)	4.5	Phe	0	Ser	2
His^+	4	Arg (pH 10)	3	Ile	1	Trp	2
				Met	1		

the exposure of polar groups on the molecule surface confers properties of hydration; this positioning depends on the polar/non-polar group ratio and above all on the three-dimensional structure essentially imposed by the disulphide bridges or by the medium conditions (pH, ionic strength).

2.2.1.1 Influence of solutes on the properties of water

- **Water activity (a_w):** this is lowered by any solute. At equal concentrations, solutes that are very hygroscopic (solvated, for example: as in the case of salts) and those of large size are the best depressors of a_w. If the interaction between solutes is significant (in the case of high concentrations of macromolecules), water activity can be increased.
- **Water sorption:** in the presence of increasing relative humidity, food components bind quantities of water which can be quantified using sorption isotherms (Fig. 2.3). According to numerous authors, section 1 of the isotherm corresponds to the binding of water on the most hydratable groups, while section 2 involves the uncharged polar groups. In section 3 the water is retained by capillary forces.
- **Freezable water:** only water which has not been immobilised by solutes can be frozen and measured using differential scanning calorimetry (DSC) or nuclear magnetic resonance (NMR). Water that cannot be frozen represents between 0.3 and 0.5 g/g of dry matter; it corresponds to **strongly bound** water or water that is not sufficiently mobile to access the ice network at the time the measurement is carried out (kinetic aspect).

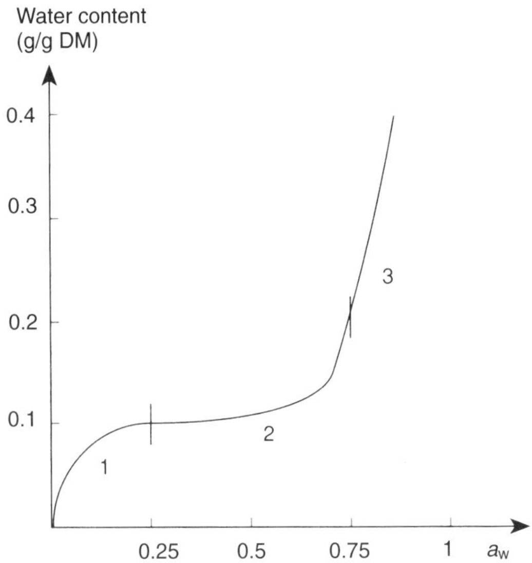

Fig. 2.3 Isotherm of water sorption by a food product.

- **Solvent water:** the quantity of water needed to dissolve solutes increases as the size of the molecule increases and its polarity decreases. This solubilisation of solutes is poor if they have reduced mobility.

2.2.1.2 Influence of hydration on the structure of macromolecular solutes
The various types of interaction among polar groups and/or among non-polar groups of the molecule or between these groups and water contribute to stabilisation of the conformation of the macromolecules.

Hydration alters the balance between intra- and intermolecular interaction, makes the polymeric chains more mobile and flexible and leads to structural reorganisation; the structures in a random coil structure would be favoured in the absence of water.

2.2.1.3 Influence of hydration on the mobility of chains of macromolecules
During the progressive hydration of proteins, we observe a mobilisation of the molecules; so, the internal movement of the chains of lysozyme are 1000 times slower at $0.04\,g^{-1}$ than at $0.2\,g^{-1}$ of water/g of dry material.

This mobilisation is accompanied by a swelling of the network, a reduction in the rigidity of the molecule and results from setting up macromolecule–water interactions; water plays a plasticising role. It is also after the minimum water content for mobilising the protein chains has been achieved that the appearance of activity in the enzymes is seen.

2.2.2 Hydration properties: influence of principal factors

2.2.2.1 Solid foods
Sorption of water and swelling
Instantaneisation is an industrial process intended to improve the dispersion of powders in water. It starts with humidification up to 10–12% water (for milk) which promotes the partial dissolution of certain components. This facilitates particle adhesion and aggregates with high porosity and a diameter of approximately 200 µm is obtained, which are then dried to improve preservation. This dispersion process can be improved by surfactant substances such as lecithin.

Water holding
Water holding by powders can be explained by the low quantities of strongly structured bound water (0.2–0.5 g/g dry matter). It can also be due to the following factors:

- The osmotic pressure created by the presence of solutes in the cellular systems which have a semipermeable membrane.
- The capillarity forces due to the organisation of the solute molecules or the microstructure and whose intensity increases as the size of the links

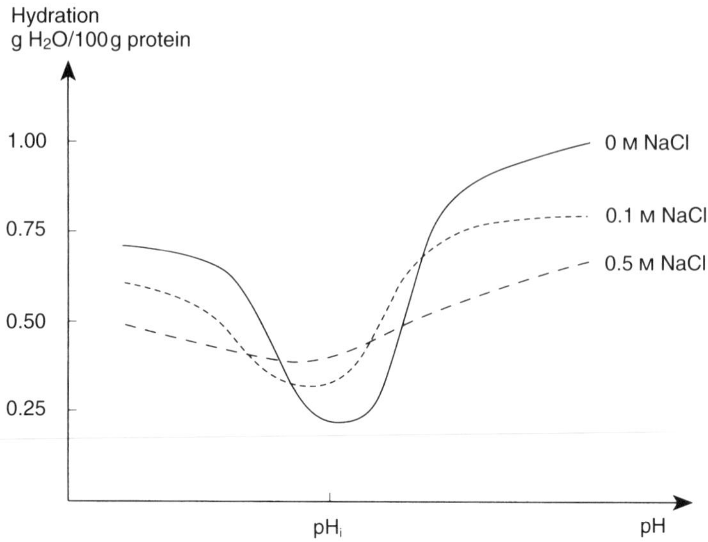

Fig. 2.4 Hydration of proteins according to pH and presence of NaCl.

decreases. We see this type of water holding in curds from cheese-making (difficult to drain fine-grained curds), in fine emulsions and in concentrated suspensions of polysaccharides. In food products it is mainly the capillary forces that are involved in water holding.

The pH plays a fundamental role in water holding by proteins (Fig. 2.4). At isoelectric pH (pH 5–6) water holding is minimal because of the increase in electrostatic attraction (between COO^- and NH_3^+) and because of the resulting network contraction. At acid and alkaline pHs, water holding increases, because electrostatic repulsion appears, either between NH_3^+ groups (in acid mediums) or between COO^- or PO_4^- groups (in neutral or alkaline mediums). In the case of negatively charged polysaccharides, the high pH also increases hydration. These variations in hydration depending on the pH are reduced by the presence of salts (Na^+Cl^-) whose ions, by neutralising the charges on the proteins, reduce attraction and repulsion.

The use of calcium complexing agents (citrates, polyphosphates, etc.) in meat salting and when manufacturing processed cheeses enables the calcium bridges of the myofibrillary (of meat) or micellar (of cheese) proteins to be dissociated thus increasing the absorption of water by opening the peptide chains (Fig. 2.5).

The denaturation of a globular protein in an unfolded structure leads to demasking of the side chains and the peptide chain which produces an increase in water binding (serum albumin binds between 33 and 46g H_2O/100g of protein in native and denatured states respectively). However, we often see a drop in hydration after denaturation because of the increase

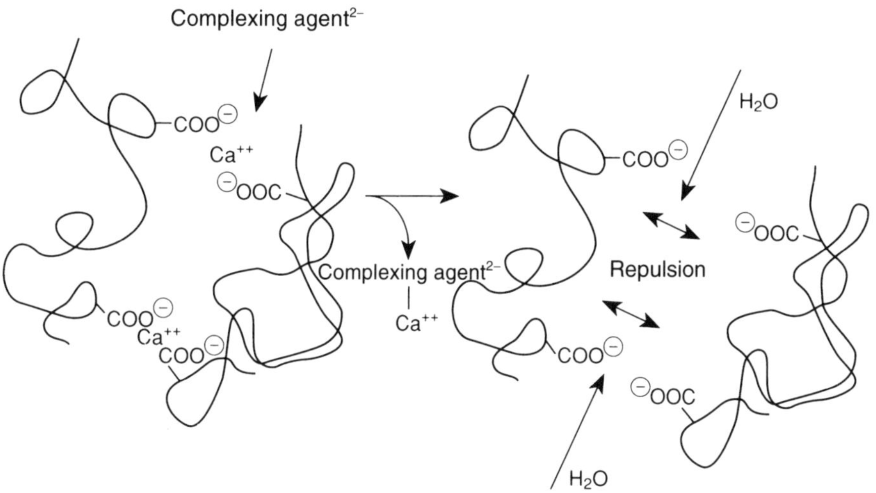

Fig. 2.5 Action of calcium complexing agents on protein hydration.

in protein–protein interaction: this is what occurs in moderate (20–60 °C) heat treatments.

Water binding is also influenced by the **particle interfacial area**, the number of surface binding sites and the porosity of the particles. The number and size of the pores of the protein matrix determine the area of total sorption, while the size and the interfacial properties of the pore influence the speed and the extent of the hydration. The diameter of each pore affects the speed of water entry or exit.

The increase of the surface area through denaturation of globular proteins means that more groups of ionised polar groups and amide groups of peptide links can be exposed and hydration can be increased by approximately 10%.

The measurement of the water-holding capacity remains empirical and is not suitable for most methods (draining, compression, centrifugation, etc.) unless the powder is insoluble; for example it is impossible to use these methods to measure the water-holding capacity of caseinates.

In addition the force required to extract the water from the network depends on the texture and the size of the pores, with the result that the methods used (Bauman apparatus, ultracentrifugation, etc.) are only useful for comparing several samples with each other.

2.2.2.2 Liquid foodstuffs

The solubility of a macromolecule depends on the dispersion solvent; the latter is very suitable if the attraction between molecules of solute and solvent predominate over solute–solute interactions. The **solubility of a protein** provides elements of information that enable its functional potential within foodstuffs to be predicted. As in the case of hydration, the solubility of proteins depends on numerous factors (pH, ionic strength, temperature, protein concentration, etc.). At the isoelectric pH (pH_i), only the denatured proteins precipitate, since they are no longer stabilised within their spatial structure by disulphide bridges. This feature was moreover the first test of protein denaturation. Obviously the denaturing action of high temperatures increases the insolubility at the pH_i.

The effect of the salts reduces the effect of the pH; at neutral pH in low concentrations they have a 'solubilising effect' whereas at higher concentrations they have a precipitating effect (salting out). The denaturing agents, such as urea, dissociate the hydrogen bonds and alter the native conformation of the proteins.

As far as polysaccharides are concerned, interactions with water take place mainly through hydrogen bonds and have little effect on the very numerous intermolecular bonds that stabilise the structure (crystalline cellulose) and maintain its solubility. Only the carboxyl groups, present naturally or chemically bound, permit solubilisation. These ionisable groups make the polysaccharides 'sensitive' to calcium (as in the case of pectins). If the concentration of macromolecular solute exceeds the saturation threshold, a balance is created between isolated molecules and aggregates which depends on the nature of the macromolecule, its concentration and the medium conditions (pH, salt, etc.).

As the intermolecular interactions and the degree of interlocking within the solutions of macromolecules increase (for example if the concentration increases or if the molecules unfold), the forces of friction increase as does the viscosity and, more particularly, the reduced viscosity at zero concentration (intrinsic viscosity) which represents the degree of spatial requirement of the molecule.

As shown in Fig. 2.6, the rheological characteristics of the solutions depend on numerous factors. Depending on the nature and conformation of the molecules, various general rules can be formulated:

- At equivalent concentrations, polysaccharides give solutions of higher viscosity than proteins.
- Globular proteins give less viscous solutions than fibrous or denatured proteins.
- When the molecules are polyelectrolytes, their viscosity can be controlled, by means of electrostatic repulsion, through ionic strength or the addition of di- and polyvalent cations (possible formation of gel through ionic cross-linkage).

The presence of macromolecules in solution disturbs the formation of the smaller ice crystals; this is reflected in ice-cream, for example, in improved creaminess.

Finally, we must not forget that the properties of hydration play an important role in the interfacial properties (foaming and emulsifying) at the

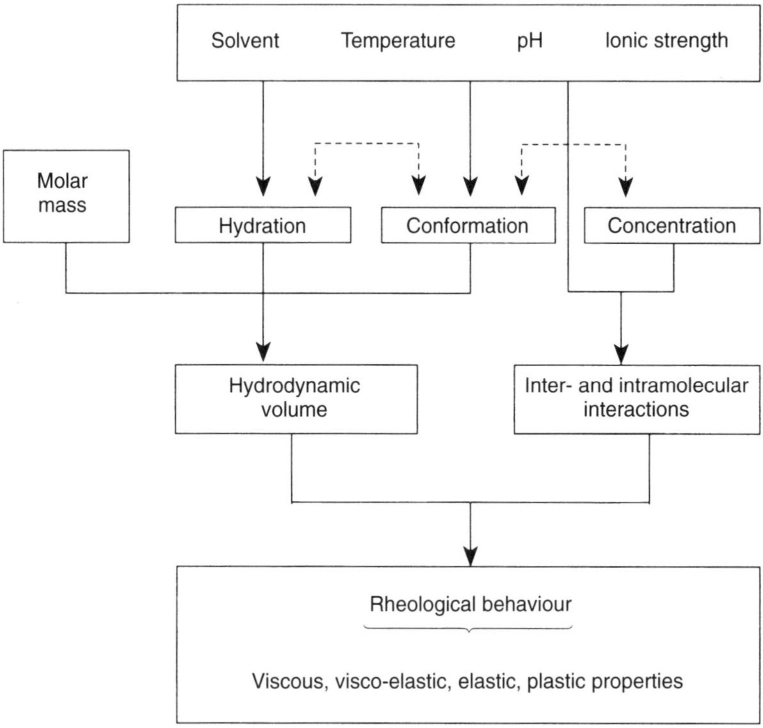

Fig. 2.6 Factors determining the rheological properties of macromolecule solutions.

level of the diffusion of the molecules to the interfaces and the formation of viscous interfacial films.

2.3 Properties of association and polymerisation

The trend of the molecules to associate when solutions are destabilised is due to a break in the balance between forces of attraction and repulsion, because of different medium parameters (pH, ionic strength, temperature, etc.); the progress of balance towards a stable system can sometimes be very slow (from several hours to several days).

2.3.1 Forces that are involved in molecular and interparticle interactions

The difference in free energy ΔG between particles that are wide apart and those that are very close to each other is obtained by adding these contributions:

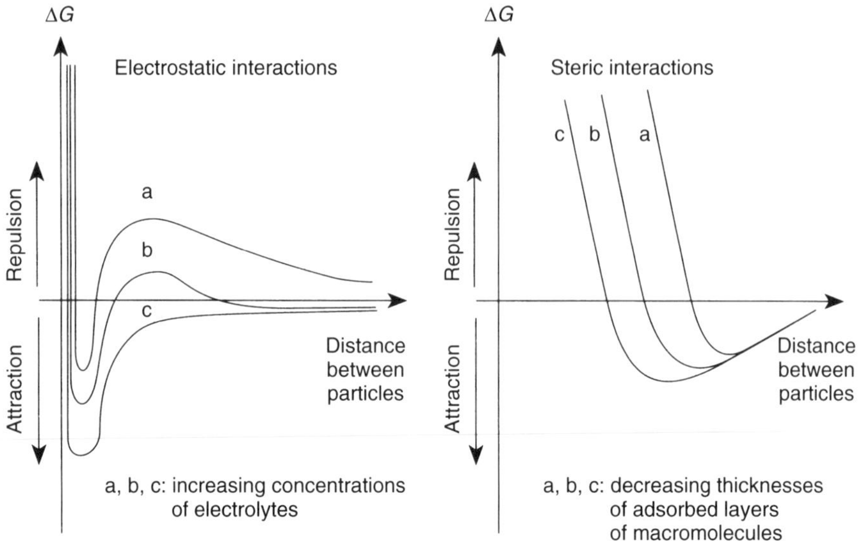

Fig. 2.7 Characteristic shapes of interactions between particles or molecules.

$$\Delta G = \Delta G^{att}(\text{van der Waals}) + \Delta G^{rep}(\text{short distance}) + \Delta G^{rep}(\text{electrostatic})$$

$$\text{or}$$

$$\Delta G^{att} + \Delta G^{rep}(\text{steric}) + \Delta G(\text{other effects})$$

(where rep = repulsion and att = attraction).

In accordance with Fig. 2.7, we can see that the electrostatic repulsion due to the double layer of ions on the particle surface depends on ionic strength (strong repulsion at low ionic strength) and that the steric repulsion is determined by the nature of the interaction between chains of macromolecules adsorbed on the particles and the solvent.

2.3.2 Motions of particles

The trend to association also depends to a great extent on the mobility of particles. This is due to the Brownian movement which is itself accelerated at high temperatures (increased probability of meeting) and to the forces of gravity owing to differences of density between particles and solvent (separation by creaming) and limited by the forces of friction imposed by the medium.

Depending on the size of the particles and their concentration, one or other of the phenomena predominate (creaming predominates for large particles).

2.3.3 Process of dispersion destabilisation

2.3.3.1 Flocculation and coagulation of dispersions stabilised electrostatically and sterically

The ions of salt solutions neutralise the repulsive surface charges of particles; the effectiveness of this flocculating effect can be appreciated by determining the critical coagulating concentration of counter ions; this depends on their valency (relative concentrations 1, 0.013 and 0.0016 respectively for valencies 1, 2, 3). For the same valency the order of effectiveness is that of the atomic number:

- For monovalent ions: $Cs^+ > Rb^+ > K^+ > Na^+ > Li^+$.
- For divalent ions: $Ba^{2+} > Sr^{2+} > Ca^{2+} > Mg^{2+}$ (lyotropic series).

The disappearance of repulsive charges on particle surfaces can also result from elimination by enzyme; this is what happens in the elimination of negatively charged glycomacropeptide, released from the surface of the casein micelle by the chymosin. The coagulation is therefore a secondary phenomenon of the enzymatic action, whose speed is greatly dependent on temperature (as with any hydrophobic interaction).

When there is no stability of the system (as is the case with flocculation), the phenomenon is reversible and often depends on temperature. Irreversible coagulation can be produced in particular when the steric repulsion due to macromolecules adsorbed on the surface has become too weak (depletion-flocculation). In fact, we know that the adsorption of macromolecules on the surface of particles creates steric stabilisation of colloidal suspensions (as is the case with gelatine or gums). The layers adsorbed can influence the van der Waals forces and create interparticle repulsion through interpenetration of polymeric chains: there is also a local increase of the concentration in polymers and the interparticle osmotic pressure, with a tendency for the water to infiltrate the particles.

2.3.3.2 Flocculation and coagulation by bridging

Flocculation and coagulation occur because of the many possibilities for association of polymeric chains that are absorbed on different particles. The phenomenon can be further increased by the presence of divalent cations capable of forming ionic bridges between negative charges (—COO^- in particular) of proteins or polysaccharides (pectins, alginates, etc.). This effect is closely linked to the pH and the pK_a of the ionised groups (marked coupling effect at neutral and alkaline pH). If the polymer concentration is sufficient, a gel will form.

Another type of bridging is currently encountered during thermal gelation of proteins. When the intramolecular disulphide bridges rupture, the structure is altered (partial unfolding); the reactive sites (thiol, hydrophobic, ionised, etc.) can group together in an intermolecular way especially if

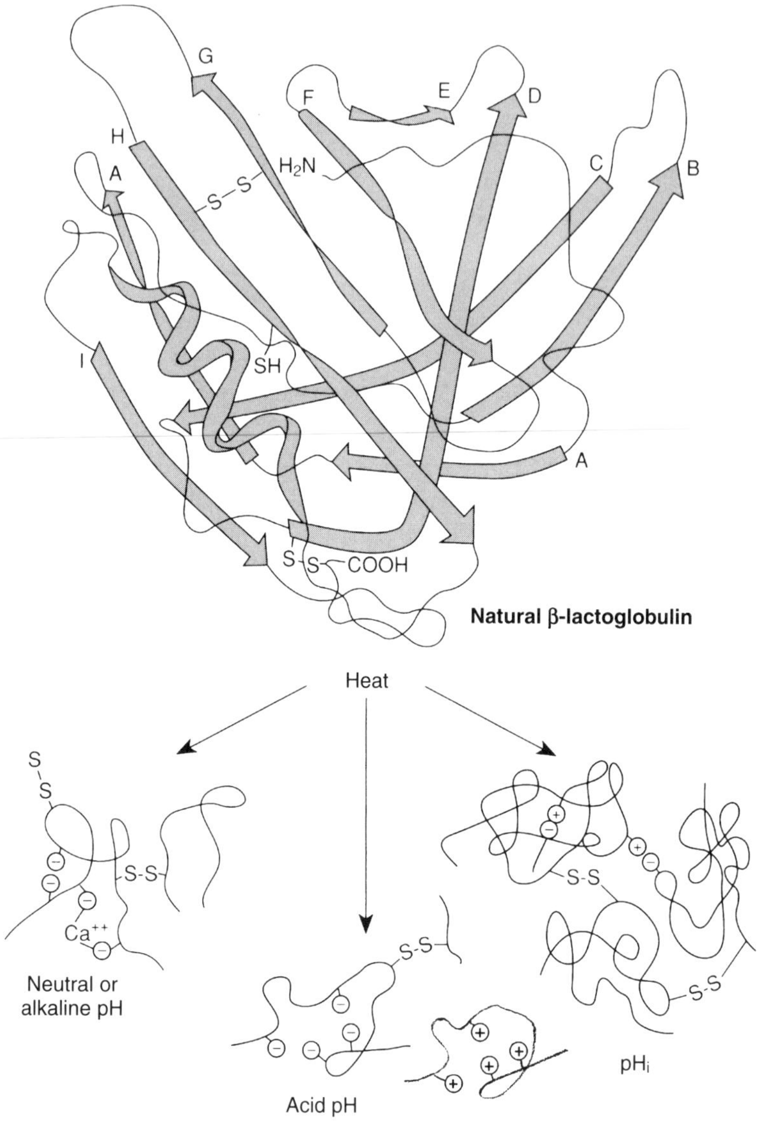

Fig. 2.8 Structural modification of β-lactoglobulin after thermal processing.

the concentration in polymers is sufficient to allow the chains to become entangled (Fig. 2.8).

Figure 2.8 shows that the gelation will take place more successfully if unfolding is stabilised by electrostatic repulsion occurs in neutral or alkaline mediums (firm, elastic gel) and if the protein concentration is high. At pH_i, on the other hand, precipitation is more likely (granular, brittle, opaque gel). We can see that in the latter case, gelation involves several mechanisms: disulphide, ionic and hydrophobic interactions.

2.3.4 Various gels obtained from macromolecules

2.3.4.1 Gels obtained by sol–gel transition

The macromolecules that form these gels (proteins, polysaccharides) form loose helicoidal chains at high temperature and in dilute solution. At high concentrations (>0.5% for polysaccharides with very extended chains, >6–7% for proteins) and low temperature, sections of chains group together to form double and triple helixes through H bonds and constitute a gel network (as is the case for gelatin or polysaccharides). If these sections are polyelectrolytes (alginates, pectins), coupling through divalent cations also takes place. This gelation process is very dependent on temperature and also on pH (for the polyelectrolytes).

2.3.4.2 Gels obtained through covalent cross-linkage

The macromolecule reacts with a bi- or polyfunctional cross-linking agent which forms covalent bridges between the chains. In the food domain, only exchanges of S–S intermolecular bridges involved in the thermal gelling process of proteins can exhibit this type of gelation. These gels can, after dehydration, form a solid, porous structure.

These gels swell in the presence of water. The swelling stops when the internal pressure is equal to the osmotic pressure of the 'internal solution': the gel network behaves like a semipermeable membrane. If the network is weak, the internal pressure of the water breaks up the gel; any aggregates that remain in suspension then influence the rheological behaviour of the solution.

2.3.5 Coagulation kinetics

In the absence of an energy barrier, the speed of the coagulation is entirely governed by the diffusion kinetics that enable the particles to come into collision; every impact between particle and floc allows the latter to grow.

According to Muller's equation, the speed at which the monomers (or single particles) disappear is given by:

$$\frac{dC}{dt} = -8\pi RDC^2$$

where C = concentration of monomers;
R = collision diameter, which is twice the particle radius;
D = monomer diffusion coefficient.

From this we can deduce the half-life:

$$t_{1/2} = \frac{1}{8\pi RDC_0}$$

where C_0 = initial concentration of monomers.

If the collision of the particles is halted by an energy barrier, the process resembles a chemical reaction involving activation energy. In this case the equation becomes:

$$\frac{dC}{dt} = -8\pi RDC^2 e^{-V\max/kT}$$

where V_{max} = level of energy barrier preceding coagulation.

Another factor can slow down coagulation: the interparticle liquid must be excluded from the collision zone; the speed is reduced by approximately half. Brownian movement can also govern the speed of coagulation; this can be increased by heating, by rapid stirring or by strong shearing. These factors become predominant as soon as the particle size exceeds a few micrometres.

2.4 Interfacial properties

2.4.1 Surface tension

Unlike what happens within a pure liquid where all the molecules are found in an attractive forcefield (van der Waals) and are in equilibrium, at the surface of this liquid one molecule is in an unbalanced field; in the case of a gas/liquid interface the molecules of liquid situated on the surface are drawn inside the liquid by free surface energy. The same phenomenon can be seen at the interface of two non-miscible liquids, with the interfacial energy coming from the inequality of the forces of attraction within the two liquids. This interfacial energy corresponds to work per surface unit and is expressed in J/m^2. From a dimensional point of view it is equivalent to a force per unit of length and can also be expressed in N/m.

As the intensity of the forces of attraction varies according to the solvent and the solutes present, the molecules that are the least attracted to the centre of the solution will be eliminated to a lesser extent than the others on the surface and will be preferentially adsorbed on the interface.

Two kinds of behaviour can be observed:

- Negative adsorption of mineral salts eliminated from the surface of the solution.
- Positive adsorption of organic components of the surface agents (bipolarised surfactants).

In equilibrium, the interfacial tension γ depends on the density of the molecules at the interface and therefore on the concentration (C) within the solution.

In accordance with Fig. 2.9, the surface tension is reduced as the concentration of the solute in solution increases, until a stage is obtained which corresponds to the formation of micelles of surfactant molecules; the minimum concentration is then called 'critical micelle concentration' (CMC).

Diffusion and adsorption of the molecules onto the interface are often slow, since the first molecules adsorbed oppose the arrival of new ones;

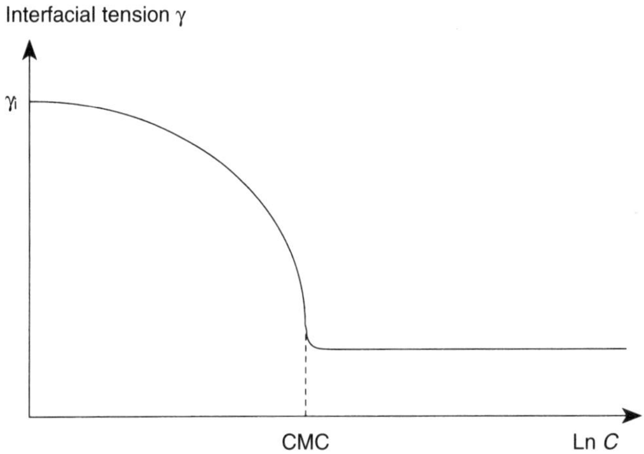

Fig. 2.9 Influence of the solute concentration on interfacial tension.

multiple rearrangements follow before the interface becomes saturated. As water molecules are small and very polar (very strong intermolecular links dependent on temperature) they have the highest surface tension when in liquid form (73 mN/m at 20 °C). The surface tension decreases very rapidly as the temperature increases.

2.4.2 Interfacial adsorption of the surfactant molecules

The interfacial concentration of adsorbed solute can be obtained using Gibbs' law:

$$C_{\text{interfacial}} = \frac{C}{RT} \cdot \frac{d\gamma}{dC}$$

where C = Concentration of the solute in the solution;
$d\gamma/dC$ = variation of the surface tension with concentration.

This law assumes that the temperature is constant (isothermal) and that the solute substances that lower the surface tension have a higher interfacial concentration than in the solution (as is the case with organic substances). On the other hand, the mineral ions preferentially bind with water in order to form hydrates and therefore move quickly away from the interface where the aqueous environment is less favourable to solvation.

2.4.2.1 Adsorption of the surfactants

Two types of surfactant substances are concentrated at the interfaces: liposoluble surfactants and water-soluble surfactants (ionic and non-ionic).

- **Liposoluble surfactants.** These are molecules that contain a hydrophobic pole (tail) made up from a hydrocarbon chain of C_8 to C_{18} and a hydrophilic (head) such as alcohol (OH), acids (COOH) or amino (NH_2) groups. Example: lauryl alcohol.

- **Water-soluble surfactants.** Although these molecules contain a short fatty chain, they are soluble in water because of groups such as COO^- (these are Na^+ soaps), sulphate ($—SO_4^-$) or sulphonates ($—SO_3^-$) groups. There are also cationic surfactants which contain quaternary ammonium (R_4N^+) groups.
- **Non-ionic surfactants.** These substances are soluble in water or in oil; they contain $—O—CH_2—CH_2$ groups in a variable number fixed to a hydrocarbonated skeleton. The hydrophilic group contains oxygen; so the solubility in water will increase as the number of oxyethylene groups increases. The most representative substances are the Tweens (20 or 80).

These surfactants can be defined by the ratio between hydrophilic and hydrophobic groups, that is to say by an empirical number placed on a scale from 0 to 30. This number is designated by the HLB (hydrophilic lipophilic balance). A low HLB, for example 3, means that the surfactant tends to be more soluble in oil than in water. For example: glycerol monostearate HLB = 3.8; polyoxyethylene monostearate of sorbitane HLB = 14.4; if HLB < 7, soluble in oil; if HLB > 7, soluble in water.

2.4.2.2 *Interfacial adsorption of polymers*

As polysaccharides and more particularly proteins are made up of peptide sections which are hydrophobic or hydrophilic depending on the sequence, they are equally well adsorbed from the hydrophobic side as from the hydrophilic side of the interface and are capable of lowering interfacial tension.

Adsorption improves as the molecule unfolds (and so denatures) at the interface (Fig. 2.10) which results in a loss of secondary structure and an increase in the rotational mobility of the peptide chain. The speed of adsorption depends on this mobility and on the speed of diffusion. We can calculate the interfacial concentration (C_i) in proportion to the time (t) using the following equation:

$$C_i = 2C_s\left(\frac{Dt}{3.142}\right)^{0.5}$$

where C_s = protein concentration in the solution;
D = diffusion coefficient.

The speed of diffusion and adsorption depends on the protein concentration in the solution (C_s); beyond a critical concentration the molecules begin to compete at the interface and must produce work $P\Delta A$ in order to compress the molecules which have already been adsorbed (pressure P) and create an adsorption area ΔA for themselves.

The speed of adsorption can be represented by:

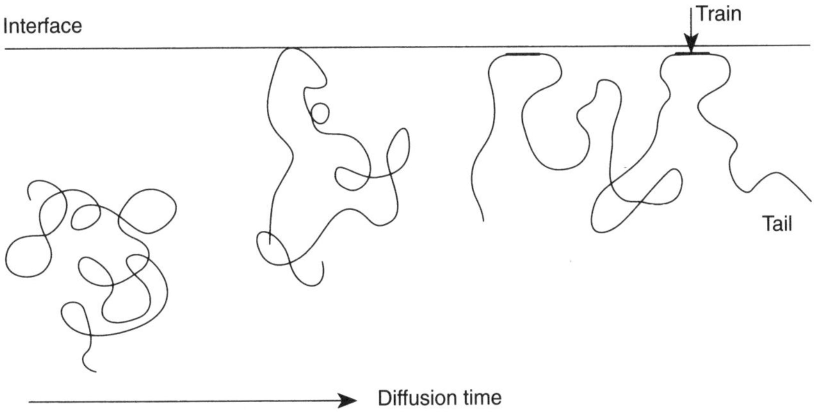

Fig. 2.10 Interfacial denaturation of a protein.

$$\mathrm{Log}\frac{\mathrm{d}C_\mathrm{i}}{\mathrm{d}t} = \mathrm{Log}kC_\mathrm{s} - \frac{P\Delta A + qE}{kT}$$

where q = charge of the molecule;
E = electrostatic potential of charges onto the interface;
qE = energy that opposes adsorption.

This equation enables us to calculate ΔA using a graphical representation: $\mathrm{Log}(\mathrm{d}C_\mathrm{i}/\mathrm{d}t)$ versus P. For neutral solutions where the electrostatic repulsions are weak, ΔA is between 1 and 1.75 nm^2. In the case of globular proteins, the protein is only adsorbed by a hydrophobic loop. At alkaline pHs, the molecule is more unfolded and ΔA is higher but the interfacial concentration or rather the interfacial protein layer is reduced.

2.4.2.3 Effects of molecular conformation on phenomena of adsorption of polymers

The capacity of proteins and polysaccharides for interfacial adsorption depends on the flexibility of the polymeric chain and its unfolding possibilities; thus β-casein, a protein without a tertiary structure, is more rapidly adsorbed than lysozyme, because of its greater flexibility and its ability to unfold, depending on the characteristics of the medium.

Various factors are involved in the capacity for interfacial adsorption:

- The pH plays a significant role in the case of molecules in 'random coil structure'; the migration of the casein towards the interface is greatest at $\mathrm{pH_i}$ (in spite of its poor solubility) because in the absence of any electrostatic repulsion the molecule is adsorbed in a compact structure having a low spatial requirement. An increase in ionic strength weakens the effects of the pH.
- The prior denaturation of proteins by means of heating or a denaturing reactant also increases the speed of adsorption.

2.4.2.4 Structure and properties of the interfacial films
The thickness of the films varies according to the interfacial concentration; this is higher for the globular proteins than for the unfolded proteins; the result is that even if the unfolded molecules greatly lower surface tension, they form films having little viscosity; on the other hand the globular proteins favour the formation of viscoelastic films. The polysaccharides improve the interfacial viscosity of the films, either by forming a gel film or by forming a gel complex with the proteins already adsorbed.

Surface viscosity is measured using a special rheometer equipped with a rotating disk positioned at the interface.

2.4.3 Emulsifying properties

2.4.3.1 Definition of emulsions
These are dispersions of a liquid phase (dispersed) in the form of droplets (0.1–10 μm) in another non-miscible phase (dispersant or continuous phase).

Two types of emulsion exist in the case of an oil/water mixture:

- Oil droplets in water: O/W (oil in water).
- Water droplets in oil: W/O (water in oil).

If the phase volume ratio is low, the least abundant phase is often the **dispersed** phase. If the ratio is close to 1, other factors determine the type of emulsion. We can identify the type of emulsion by adding, for example, water to an O/W emulsion: in this case we note a simple dilution of the emulsion. On the other hand if we add oil, this forms a separate phase. Milk can be diluted with water (O/W emulsion) whereas mayonnaise can be mixed with oil. Oil/water emulsions have a creamy appearance whereas water/oil emulsions look 'fatty'.

Movement from one type of emulsion to another can take place easily by measuring electrical conductivity; we move from an aqueous dispersed phase containing ions and having high electrical conductivity to a dispersed lipid phase which is a very poor conductor (Fig. 2.11).

2.4.3.2 Formation of emulsions
The formation of an emulsion involves an increase in the interfacial area, and is accompanied by an increase in free energy. The facility of forming an emulsion can be evaluated by measuring the mechanical effort required for emulsification. The lower the interfacial tension, the more easily the emulsion is obtained; the emulsifying role of the surfactant agents is precisely that of lowering the interfacial tension by being adsorbed on the interface; in the case of polymers they also form a coherent and rigid film around the droplets; if this film is made up of proteins or charged molecules, it can play a part, as a result of electrostatic repulsion, in stabilising the emulsion. Various types of homogenisers, whose effectiveness depends

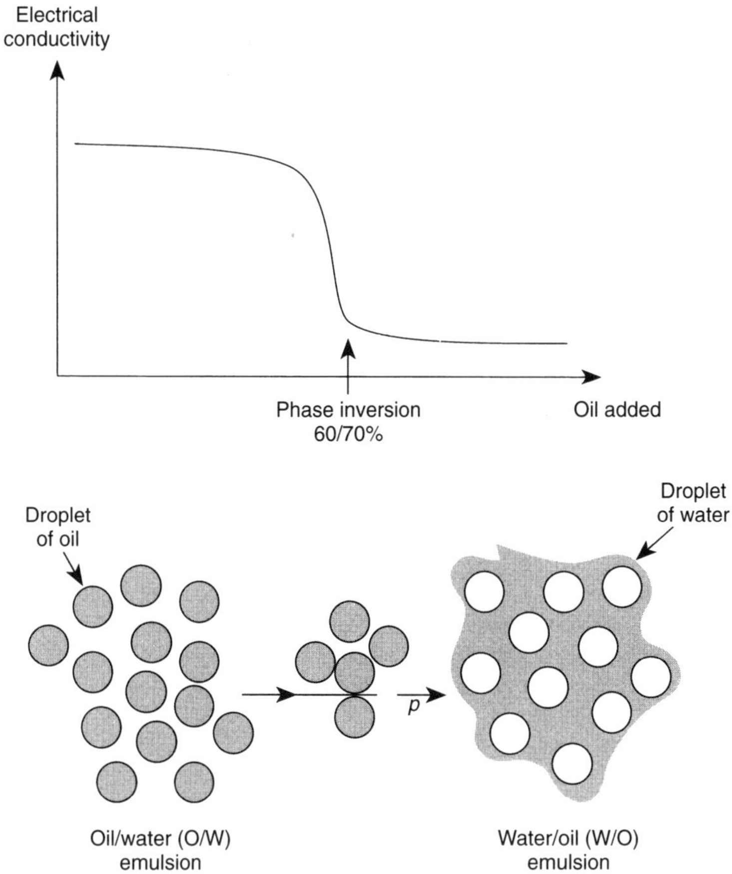

Fig. 2.11 Changes in electrical conductivity of an emulsion during phase inversion.

on the energy supply, the system of fragmentation of the droplets (helix, valve, percussion) and the temperature (increased fluidity at high temperature) are used on a laboratory or industrial level.

2.4.3.3 Stabilisation and destabilisation

Destabilisation of emulsions takes place by means of the same mechanisms used in the destabilisation of the colloidal dispersions.

Creaming is a separation of droplets of the dispersed phase due to the difference in density whereas **flocculation** is a **reversible** phenomenon of association of droplets; the flocs obtained undergo creaming once they have reached a certain size.

If separation is turbulent (centrifuging) or if the interfacial layer of surfactant destabilises, the droplets fuse by **coalescence** (Fig. 2.12). We can slow

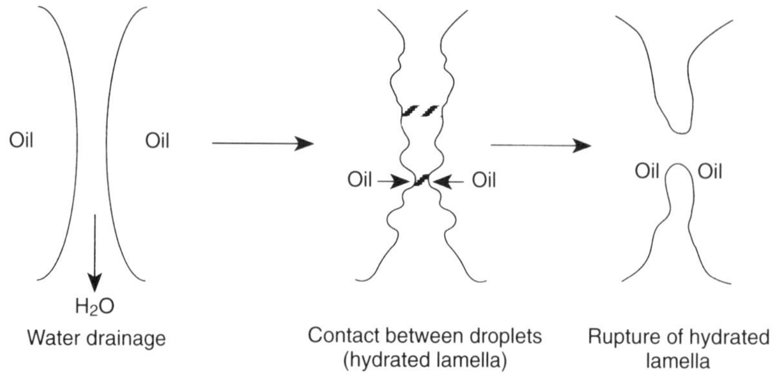

Fig. 2.12 Process of coalescence.

down the process of coalescence by adding molecules in the continuous phase. The molecules act in the following ways:

- **Increasing viscosity** – this is the case with water-soluble polysaccharides of a high molecular mass (10^6 Da) such as guar gum, xanthan gum, carrageenan, alginates; gelatin acts in the same way by creating a gel network in the continuous phase which stops creaming and flocculation (increased viscosity of the lamella).
- **Electrostatic** or **polymeric stabilisation** – this involves producing a thick and stable viscoelastic interfacial protein layer which prevents coalescence.

The addition of surfactants stabilises the interface but their action depends on the chemical nature of the surfactant (HLB value), the competition they exert towards the adsorption of proteins and polysaccharides. Stabilisation by proteins depends very much on concentration; it is often optimum for protein concentrations of over 3%.

The destabilisation of emulsions can be accelerated by:

- **centrifuging**, which accelerates creaming and, by compressing the droplets into each other, causes flocculation and then coalescence;
- **thermal shock** – freezing followed by thawing and abrupt changes in temperature destabilise emulsions; ice and lipid crystals grow and penetrate the droplets by crossing and damaging the interfacial layers.

2.4.3.4 Methodology of emulsifying properties

- **Emulsifying capacity:** represented by the quantity of oil emulsified per gram of emulsifier at the point of phase separation. This can be detected visually or by measuring the reduction in viscosity or electrical conductivity (Fig. 2.11).

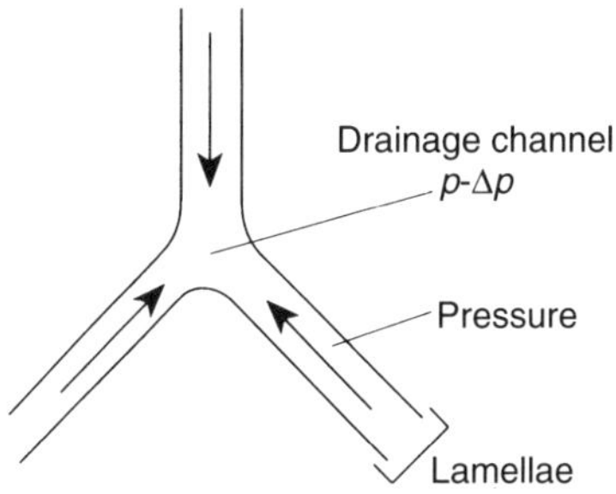

Fig. 2.13 Junction point (plateau border) of three bubbles where a depression is causing drainage.

- **Emulsifying activity:** defined as the area of the interface stabilised by a given concentration of emulsifier (m^2/g); this is measured using a turbidimetric method where it is possible to link the interfacial area with the actual size of the droplets in relation to the diffusion of light caused by the particles in suspension.
- **Stability of the emulsion:** defined as the capacity of the emulsion to retain its structure over time, its measurement can be **direct** (development of the size of the droplets over time evaluated by a light-scattering instrument) or **indirect** (development of quantities of oil or aqueous phase separated after a destabilisation treatment (centrifuging, heating).

2.4.4 Foaming properties

2.4.4.1 Definition of a foam: formation and structure

A foam is produced when a gas is introduced into a liquid, either through one or several small holes or through sintered glass (**bubbling through**) or when gas and liquid are shaken up together (**beating**, **shaking**).

The volume of gas is generally much higher than that in the liquid phase; the size of the bubbles is much larger than that of the colloidal particles; initially spherical, the bubbles take on a more compact structure in which they are separated by a very fine film. As the difference in pressure between various parts of the interface is slight, the bubbles become polyhedric when they are joined together by a flat lamella.

Foams are generally less stable than emulsions and the films that form them, because under the influence of gravity the liquid between the two lamellae (plateau border) runs all along a central channel (phenomenon of **drainage**) (Fig. 2.13). At this point, the reduction of pressure soaks up the liquid from the lamellae and can inhibit the stabilising forces of the film: a dry foam is obtained which is not very homogeneous and no longer forms a continuous network, and when the thickness of the film drops from 100 μm to less than 10 μm the film breaks. The drainage can be limited by increasing the viscosity of the lamellar liquid, by adding proteins or polysaccharides.

2.4.4.2 Formation of foams

Foaming agents such as soaps or surfactants are the most effective in forming foams that have a content close to the CMC; they form compact monolayers at the liquid/gas interface. Their effectiveness can be increased by using a second component. Non-ionic surfactants and water-soluble polymers are often good foaming agents. As far as proteins are concerned, it often happens that the interfacial adsorption denatures them and makes them irreversibly insoluble in the form of rigid and elastic interfacial layers. This is what happens with food foams produced from egg whites or milk cream.

Various systems are used to make foam:

- Bubbling through or incorporating air by means of low pressure (aerosol spray) allows foams with low concentrations of surfactants to be obtained (up to 0.1% for proteins).
- Beating requires higher concentrations of surfactants; this can be used as a **discontinuous** system (food mixer, industrial planetary beater) or in a **continuous industrial** system (gas pressure expansion, scraped surface exchanger, etc.).

2.4.4.3 Stabilisation and destabilisation of foams

The stability of foams depends on the following:

- The rigidity of the interfacial film – this is improved in the case of a protein foam, if the denaturation is accompanied by gelation or if the interfacial concentration is high (case of proteins at pH_i which are more compact). Adding polysaccharides enables stabilising complexes to be formed.
- The viscosity of the liquid phase – this is improved by adding polysaccharides or in the case of proteins by an increase in concentration and a pH well away from pH_i (rarely the case);
- The presence of solid particles which stabilise the interfacial film.
- The presence, in the case of protein foams, of basic proteins (lysozyme, for example, in egg white), and by glycosylated proteins (ovomucin and ovomucoid of egg white). The latter act by increasing viscosity.

Sometimes we wish to destabilise foams (as in the case of fermentation media). Mechanical means can be used, which apply, for example, a current of hot air to the foam. More often anti-foaming agents are used, which can act in various ways:

- Destabilisation through desorption of the foaming agent from the liquid/gas interface by an agent that is non-foaming but more strongly adsorbed (as in the case of ether, *n*-butanol, capric acid, etc.).

- Destabilisation by adding fine droplets of insoluble liquids, such as silicon oil, or solid hydrophobic particles to the water (these act by coupling the two films with a lamella).
- Destabilisation of soap foams using hard water; the calcium forms insoluble compounds with the soap which break the interfacial films.

2.4.4.4 Methodology of foaming properties

- Foaming capacity: defined as the quantity of foam formed per unit of volume of solution (or per unit of mass of solute). This is measured after expansion by determining the maximum volume of the foam or the minimum conductivity. It does not take into account the structure (size and shape of bubbles, thickness of the lamella).
- Foam stability: this is the capacity of the foam to retain its structure over time. It is evaluated by measuring either the volume of the foam or the volume of the separated liquid (drainage) over time. If bubbling-through equipment is used (column equipped at its base with sintered glass), the rate of increase of the conductivity can be measured.

2.4.5 Mixed systems: expanded emulsions or emulsified fatty foams

These systems are represented by whipped creams, ice-creams, meat and fish terrines, sponge cakes, etc.: certain types of confectionery are both emulsions and foams. The complexity of these systems comes from the simultaneous presence of several phases (liquid, solid, gaseous), of several interfaces (gas/liquid, liquid/liquid, solid/gas, solid/liquid) and several physical states (crystalline, amorphous, solution, etc.).

The model that has been most closely studied but is still poorly understood is that of ice-cream. This is an emulsion that is expanded and then cooled, where the water and the lipids are in both liquid and solid form and where the bubbles of air are surrounded by fat globules which the addition of emulsifier tends to partially destabilise. It is this example that shows us most clearly the importance of a large number of ingredients: emulsifiers (mono- and diglycerides, lecithin), expanders (milk proteins), stabilisers (hydrocolloids such as carrageenan, guar gum and carob gum); by acting simultaneously on the formation and the stabilisation of the interfaces as well as on the viscosity of the continuous phases, they contribute to the development of textures that only sensory evaluation is capable of testing.

3

Extraction and texturisation processes

3.1 Extraction and purification

3.1.1 Proteins

3.1.1.1 Aims and objectives of protein extraction
Until recently it was unusual to extract proteins for use in food, and only molecules having pharmacological properties or those demonstrating enzymatic or hormonal activity underwent extraction and purification processes. Although certain carbohydrate or lipid food ingredients obtained in their pure state for everyday use (sucrose, oil, margarine, etc.) have long been used by consumers, there is still reluctance to eat protein concentrates. However, it seems that recent progress made in fractionating proteins has removed the technical obstacles linked with the complexity and fragility of these molecules so that excellent quality protein extracts can be developed. In addition, controlling the functional properties of these protein extracts enables them to be used with increasing frequency in formulating many new foodstuffs.

The objectives in extracting proteins can be of a nutritional, functional, organoleptic or economic type:

- **Improvements in nutritional value** can be obtained by eliminating toxic substances or by extracting plant proteins which are strongly bound to substances difficult for humans to digest (cellulose, polyphenols, etc.), and by eliminating anti-nutritional substances (the trypsin inhibitor of the soya bean, goitrogenous substances from rape seed cake, gossypol from cotton seed cake, etc.; see Section 4.4.2).
- **Improvements in organoleptic characteristics** are achieved by removing pigments and flavoured components (deodorising, de-bittering, etc.)

during extraction. Functional properties are often improved as a result of eliminating undesirable ingredients (salts, lipids, etc.) and from actual protein enrichment. This is particularly important since new requirements for using intermediate food products have emerged recently: e.g. meat substitutes developed from plant proteins, imitation products.

- **The improved profitability of traditional food processes** which have become unprofitable because of the levels of investment required compared with returns in the market-place. It is not essential but it is more economic to make better use of certain fish and plant proteins to substitute for animal proteins which depend on costly and lengthy stock-rearing practices (Fig. 3.1).
- **The enhanced profitability of by-products from the food industry** (dairies, abattoirs, starch mills and plants, breweries, sugar refineries, distilleries) provides an answer to the problem of recovering valuable proteins (whey, for example) from existing proteins and of reducing the risks of polluting the environment.

3.1.1.2 Extraction methods

In comparison with other organic food components, proteins are very complex as far as their structures, heterogeneity and associations with other cell biopolymers are concerned, and this limits recovery by extraction. In addition, once isolated their structure may break down, losing their nutritional and functional properties; so the solubility (the first criterion of denaturation) of a particular protein varies according to the method of extraction.

Suitability of proteins for extraction: state of the protein
Several factors are involved:

- **Heterogeneity.** Soluble forms which can easily be dissolved in water (albumins, globulins, etc.) and strongly structured insoluble forms (myofibrillar proteins) can co-exist in the same raw material, as can insoluble forms strongly bound to polysaccharide (hemicellulose), nitrogen (tannins, etc.), oxidised lipid or pigments, so we need to use some dissociating processes (mechanical, chemical or enzymatic).
- **Susceptibility to denaturation:** alteration of spatial structurc. This is manifested by a loss of solubility and is caused by three factors: modification of the pH, high temperatures and change in ionic strength. Precipitation is often irreversible.
- **Molecular size and shape.** This criterion can be used in microfiltration, ultrafiltration and gel permeation, but only when proteins are the only macromolecules present: this process is particularly applicable to animal proteins present in physiological fluids (whey, blood, etc.) and less so to plant proteins which are often mixed with polysaccharides or other polymers (lignin, pectins, etc.).

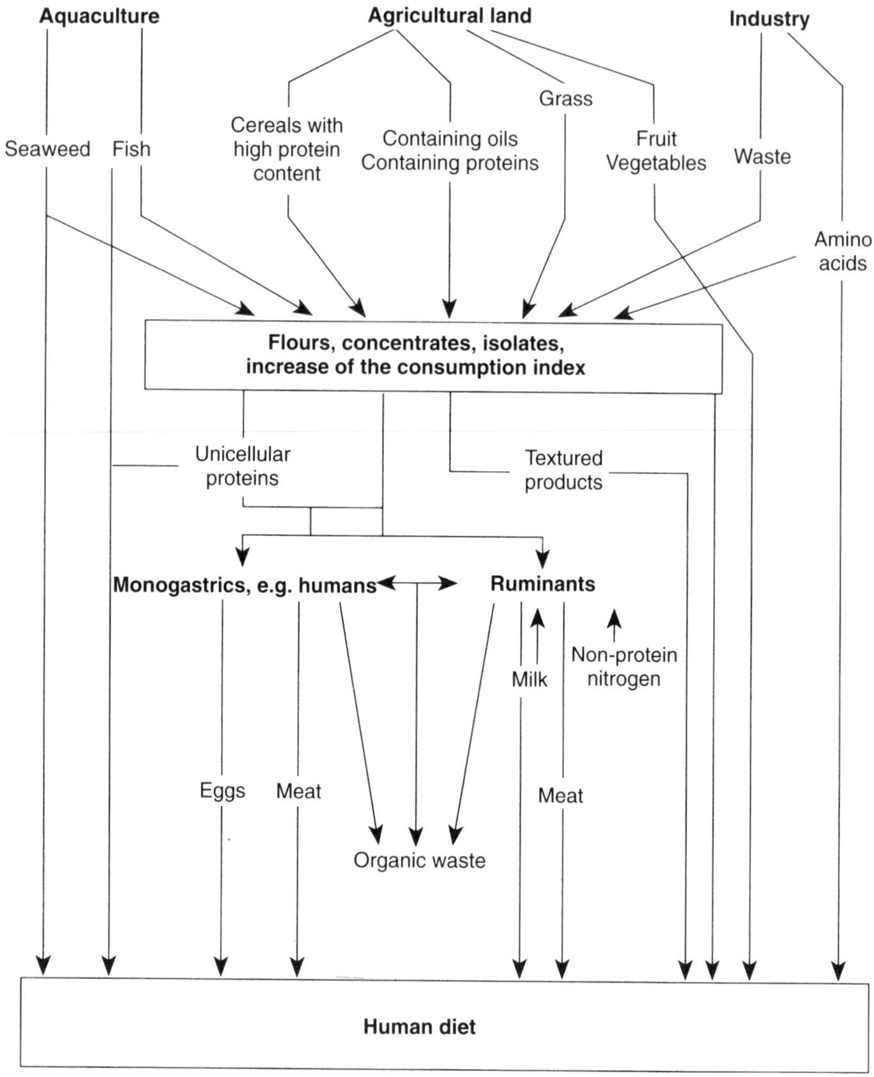

Fig. 3.1 Diagram showing how protein extraction can short-circuit animal rearing.

- **Polarity.** This depends on the proportion of both ionised (COO^-, $NH3^+$, etc.) and non-ionised polar groups (–OH, SH, etc.) compared with the hydrophobic non-polar groups (aliphatic or cyclic hydrocarbon chains, etc.); their availability also depends on spatial structure (globular proteins which are often hydrophobic in the centre of the molecule). Separation based on this criterion is effected by precipitation at various ionic strengths or at different pHs.

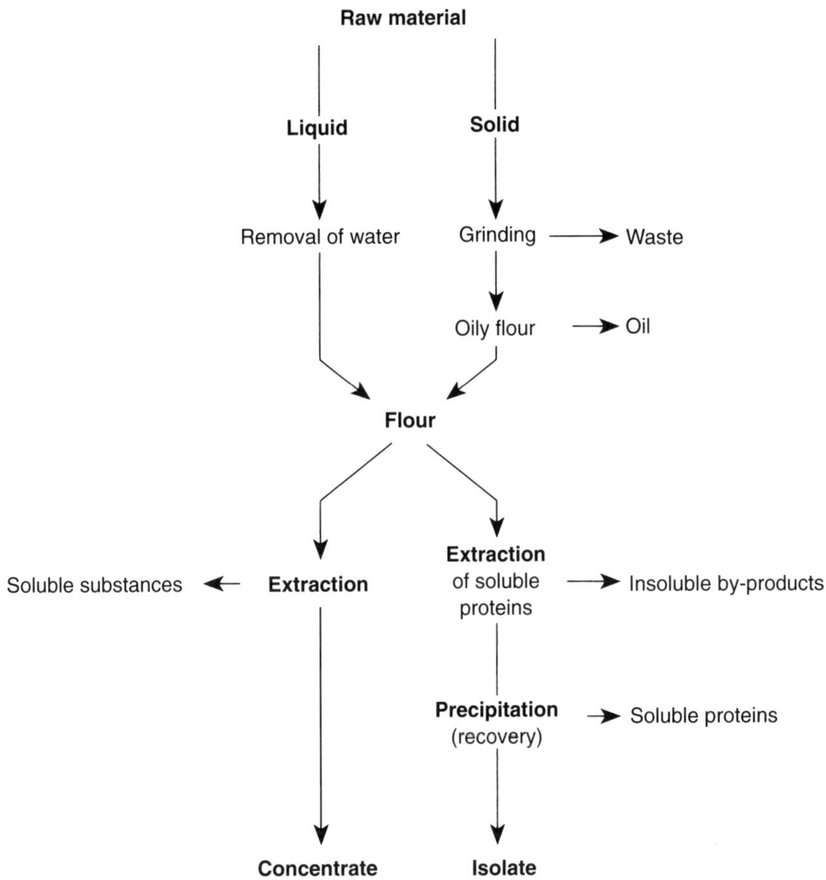

Fig. 3.2 Different forms of protein extracts.

- **Binding to charged sites of solid polymers** (resin, silica, glass, etc.). The charge depends on the pH_i of the protein and the pH of the medium (amphoteric nature). Ion exchange followed by elution with either a pH or ionic strength gradient allows proteins (whey, blood, etc.) to be extracted and selectively fractionated. Specific binding on a medium may also involve using specific affinities for enzyme substrates, or for antigens.

Types of extract obtained

Depending on its intended end-use or the extraction methods used, the protein concentration of the finished product can vary from between 15 and 20% (mechanically separated meats) up to 90% (soya isolates, caseinates).

These products can be classified according to their increasing levels of purity: flours <70%, concentrates 70–85%, isolates >85% (Fig. 3.2). Product differentiation according to protein content conceals differences of a technological kind:

- Isolates resulting from specific isolation.
- Flours and concentrates resulting from simple enrichment, by eliminating undesirable substances (starch, lipids, water, etc.); in the case of concentrates, carbohydrates, mineral salts and non-protein nitrogen are also eliminated.

Actual methods of extraction
A distinction is made between methods of protein enrichment (negative) and methods of isolation (positive).

Negative methods of enrichment
Four stages are usually employed (Fig. 3.3):

- Dispersion of the various elements in order to facilitate separation operations which use mechanical processes: breaking, stirring, crushing, chopping, grinding. The purpose of this is to increase the surface area/volume ratio.
- Thermal processing (50–140 °C for varying lengths of time). This operation prepares and facilitates separation by means of:
  - liquefying one phase (melting grease, etc.);
  - solidifying one phase (coagulation of proteins by thermal denaturation of proteins, etc.);
  - dehydration: elimination of water (concentration) facilitating the removal of lipids by extraction. This operation can also sterilise the product.
- Actual extraction of the solid phase containing the proteins and elimination of the lipid phase containing water and lipids by means of pressing or centrifuging.
- Drying in order to eliminate the residual moisture, and extraction of residual lipids by solid–liquid extraction.

Separation can be carried out directly on powders using the process of air classification, used in the protein enrichment of flours and beans. Unlike flours, concentrates require total removal of lipids because any residual oxidised lipids are responsible for protein-oxidised lipid interactions, resulting in insoluble products which are most frequently found in fish flours.

Several extraction systems are used:

- *n*-hexane + ethyl acetate + isopropanol (Moroccan process);
- 95% ethanol (Chilean process);
- *n*-hexane on its own for lipids (Peruvian) or with ethanol (odour extraction).

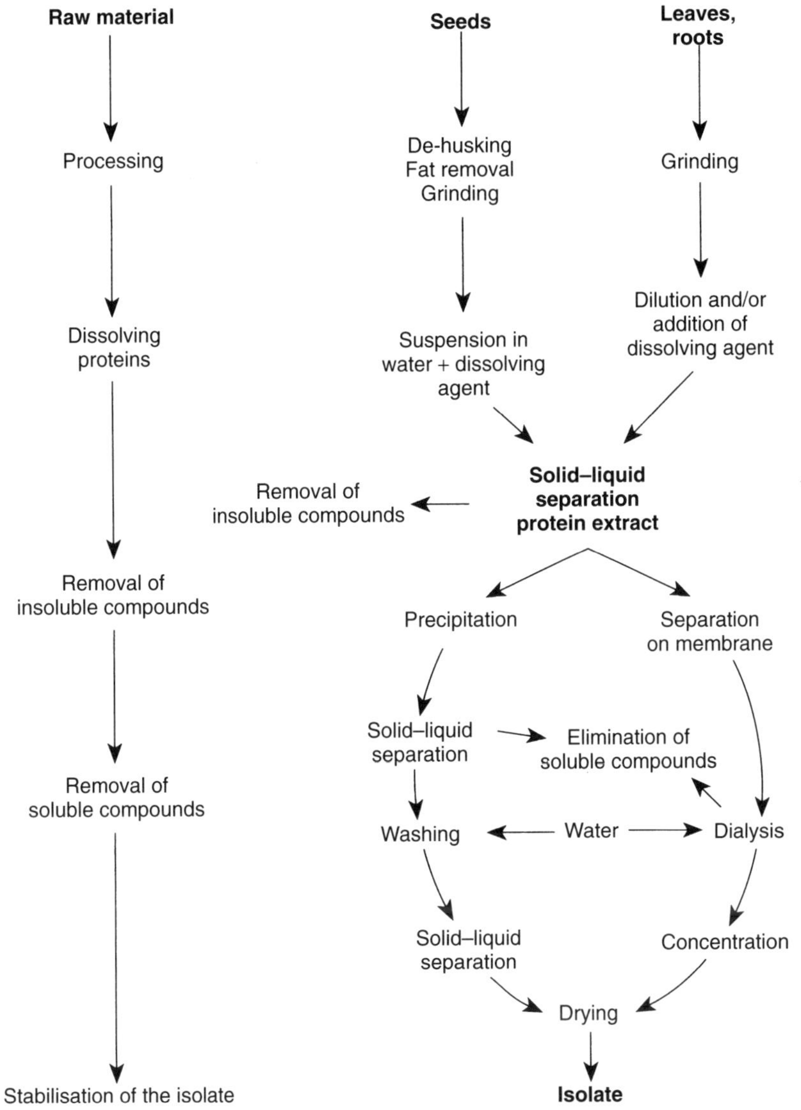

Fig. 3.3 Examples of fractionation of plant proteins.

For fish, isopropanol is often chosen because of its weak denaturing action and its reasonable cost.

Positive methods of isolation

The recovery of components from whey, blood, legumes and especially soya proteins is already being carried out on an industrial scale. Among animal proteins, only those from milk or whey have already undergone fractionation in order to produce proteins which are either very pure (caseinates, β-lactoglobulin) or associated with salts or residual lactose. Fish proteins are sometimes extracted as isolates (*surimi*) (see Chapter 8), whereas meat proteins are rarely purified.

After dissolving, the isolation process consists of extracting the protein either by precipitation and solid/liquid extraction, or by filtration depending on the molecular size, or by binding to an 'active' medium followed by elution. These methods are more selective (better return) and cause less denaturation; however, they eliminate a larger quantity of potentially useful by-products.

Three parameters must be taken into consideration:

- **Yield:**

$$y = \frac{\text{mass of proteins extracted}}{\text{mass of proteins to be extracted}}$$

This is often lower in plant proteins except for soya which contains soluble globulin.

- **Selectivity** characterised by the degree of purity and homogeneity; rarely required in the food industry.
- **Cost of extraction,** which depends on speed (mass of proteins/time) and concentration of extract.

It is often impossible to reconcile yield and selectivity. As proteins are rarely directly soluble (20% in the case of fish), yield can only be improved if solubility is increased by chemical or enzymatic means.

1 Solubilisation using chemical means. At least four factors are involved simultaneously in solubilisation: pH, ionic strength, temperature and presence of Ca^{2+}.
   (a) **pH:** solubility is minimal at pH_i, especially for denatured proteins (generally between 4.5 and 6). Alkaline pHs are preferable as the solubility maximums are higher than in an acid medium or ones closer to neutrality: however, above pH 9 they cause racemisation of amino acids, and the formation of intra- or intermolecular covalent bonds which alter the digestibility of proteins.
   (b) **Ionic strength:** this effect is marked especially at neutral pH or close to their pH_i. At extreme pHs, the increase in ionic strength has a tendency to lower solubility.
   (c) **Temperature:** low temperatures are preferred because they will reduce denaturation and are unfavourable to microbial growth. When proteins are not easily denatured, extraction at high temperatures can be profitable and selective (bone proteins are extracted at 90°C, without depositing collagen).

   Generally speaking, the effect of the first two factors can be shown in three-dimensional graphs.

   In selecting these conditions we need to take into account the type of recovery process used; if this is isoelectric precipitation, the adjustment of the pH often leads to significant formation of salts and is likely to disturb the precipitation process itself (solubility improved by salts at pH_i). If the process involves ion exchange or gel permeation, it is necessary to avoid dilution and to adjust the pH to the constraints imposed by the medium used.

2 Solubilisation using enzymes. This method, used for dissolving fish or plant proteins, is rather costly but would be an excellent method if hydrolysis could be limited. In fact, hydrolysates, which often have a somewhat bitter flavour, are obtained. They are very soluble and it is difficult to separate the other soluble ingredients from them; the residual enzymes must moreover be recycled or inactivated.

3 Purification-recovery of soluble proteins by precipitation.

(a) **Enzymatic precipitation:** the coagulation of milk using rennin is a process used in the purification of casein in the form of curds in cheese-making. The liquid phase drains spontaneously when the curd is cut.

The coagulation of blood fibrinogen using thrombin is also a spontaneous separation of the erythrocytes of the liquid serum, through the formation of a network of fibrin.

(b) **Isoelectric precipitation of denatured proteins:** a protein that has previously been denatured (by thermal or acid processing) precipitates almost completely. Recovery from this type of precipitation process will vary according to the pre-treatments that have been used (electrodialysis, ultrafiltration, etc.). This process is traditionally used for recovering whey proteins and for reincorporating them in curds for cheese-making (Centri-Whey process).

Other types of denaturation process, apart from thermal ones, can also be used: alcohol (soya), acid (casein).

Precipitation after thermal denaturation produces products whose functional properties may be mediocre unless certain pre-treatments are applied.

(c) **Co-precipitation with various adsorbents** followed by elution of the support protein: this requires: utilisation at all protein concentrations, possible recovery, no further processing of the product.

Various precipitation agents can be used (Table 3.1). **Hexametaphosphate** precipitates over 90% of the proteins from whey can be used if prior demineralisation is carried out at pH 3; the phosphates must then be eliminated by ion exchange, gel permeation or solubilisation at pH 3. **Polyethyleneglycol** can be used for fractionating blood proteins at various pHs (4.6, 6 and 7) and with molecular masses from 6000 to 20000 Da. **Polyacrylic acid** permits a protein concentrate to be obtained from whey with properties similar to those of egg white. The proteins are initially precipitated at pH 4 in the presence of polyacrylic acid, then, after solubilisation at pH 6.5, the acid is eliminated using magnesium carbonate.

Table 3.1 Comparison of composition of various whey concentrates (%)

	Recovery	Proteins	Ashes	Ashes*
Carboxy methyl cellulose	68.9	60.7	1.7	–
Iron (II) chloride	88.7	79.1	18.3	1.4
Iron (II) polyphosphate	91.7	69.5	28.4	4.7
Polyacrylic acid	63.3	82	12.9	–
Hexametaphosphate	75.5	84	12.8	1.1

* After desalination on Sephadex G25 pH 2.

(d) Other precipitation agents have been investigated on laboratory scale: bentonite, chitosane, iron chloride, lignosulphonates and numerous negatively charged hydrocolloids.

4 Purification by ultrafiltration. This technique allows macromolecules to be separated according to their size, using a semipermeable membrane (molecular sieving) (Fig. 3.4). Two streams can be obtained:
(a) the retentate or concentrate containing the macromolecules which have been retained in a more concentrated state;
(b) the permeate mainly consisting of the aqueous phase and some small molecules (simple carbohydrates, salts, etc.). It should be pointed out that these small molecules are at the same concentration in the retentate as in dialysis, the permeate and the food medium subjected to ultrafiltration.

Depending on the pore size of the membrane, the following options are possible:
(a) **Microfiltration:** very large molecules are retained which can form a polarisation layer having an ultrafiltering function (use of aluminium or ceramic membranes) (Fig. 3.5).
(b) **Ultrafiltration:** macromolecules (molecular mass (MM) from 10000 to several hundred thousand) are retained and can be modified: **organic** or **mineral** membranes in zirconium or aluminium oxide.

Various geometrical configurations for the membranes can be adopted: flat, tubular, hollow fibre, spiral. The effectiveness of ultrafiltration depends on various factors which are particularly complex since interactions between proteins and the filtering medium can influence them. The effect of protein polarisation limits the concentration of protein approximately 25–30%. In order to continue the operation, the extract must be diluted with water (operation of **diafiltration**: water output = permeate output).

The following conditions must apply for ultrafiltration to be used for the purposes of purification:
(a) precipitation methods are not suitable;
(b) the extract does not contain non-protein macromolecules, which limits its use to whey, blood and plant isolates;

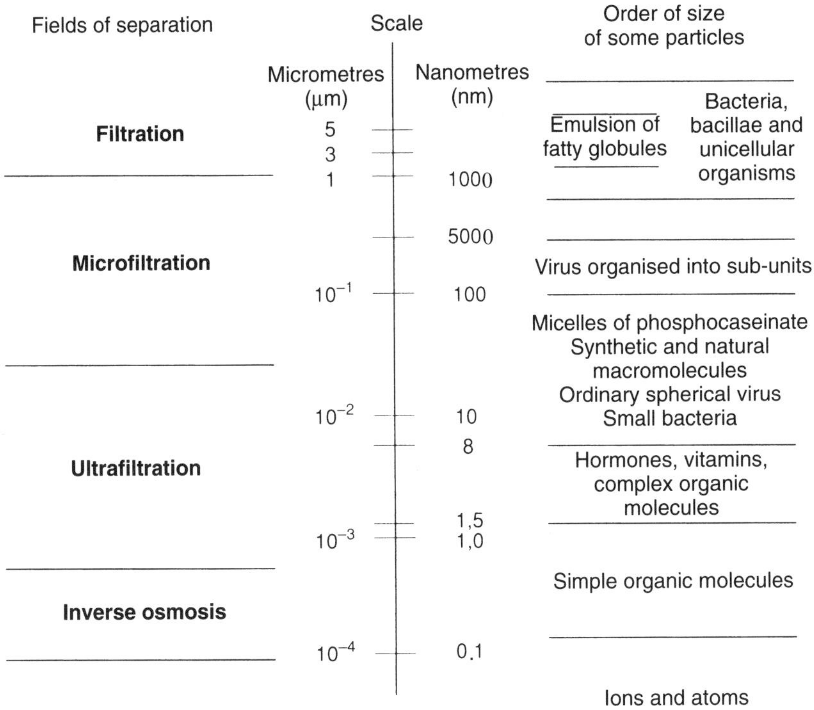

Fig. 3.4 Techniques for separating over membranes.

(c) the extract is low in peptides and amino acids, which excludes hydrolysates.

Finally, if we want to eliminate water alone, we use membranes of very low porosity; this is **reverse osmosis**, a technique that is expensive and not widely used on a large scale within the food industry. A more recent technique, **nanofiltration**, based on the same principle but using slightly more porous membranes, allows water and soluble mineral substances to be removed while retaining the small size neutral organic molecules, e.g. sugars.

5 Purification using ion exchange chromatography. Two processes can currently be used on an industrial scale:

(a) The **Vistec process** which uses types of cellulose which are weak anion exchangers; DEAE (diethylaminoethyl-cellulose);

(b) The **Spherosil process** which uses silica to which is grafted groups of strong or weak anion or cation exchangers (see Chapter 9) as a medium. This is the only technique sufficiently selective to allow clear separation of the protein fractions (Table 3.2).

The major disadvantages are that the eluate is contaminated by ions and that in the case of samples at neutral pH, 2 successive exchangers need to be used (case of sweet whey).

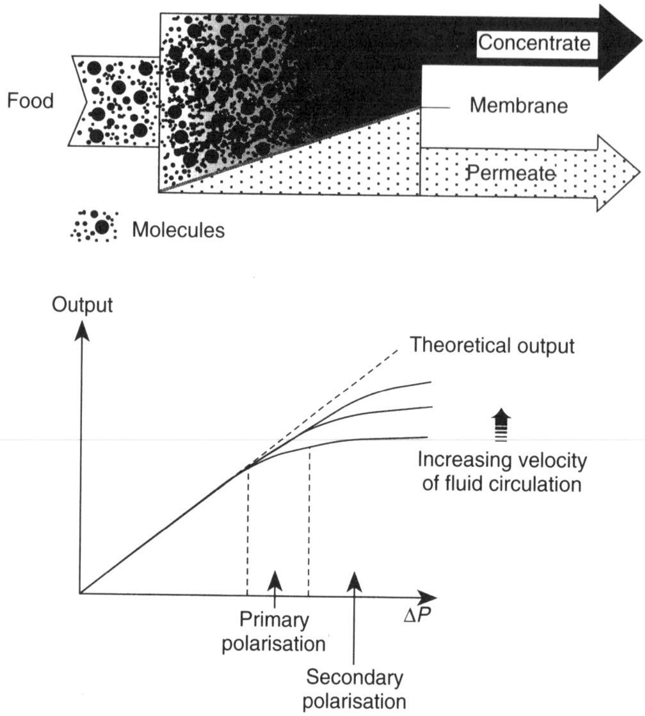

Fig. 3.5 Block diagram showing separation by ultrafiltration.

Table 3.2 Characteristics of isolates obtained using the Spherosil process

	Soya	Colza	Lucerne	Potato
Proteins (N × 6.25)	103.2	99.3	87.1	91.3
Ash	3.4	3.0	0.3	1.2
Thioglucosides	–	0.2	–	–
Colour	White	Beige	White	Creamy white
Odour	–	–	Slight, pleasant	–

3.1.1.3 Applications to the principal sources of protein

Animal protein

Myofibrillar proteins
Extraction from meat is rare except in the case of by-products; in addition it is difficult to obtain good quality concentrates because of the high oxidisation potential of the myoglobulin.

Fish protein flours are often very insoluble if effective lipid removal has not been carried out. Effective exploitation of fish by-products involves manufacturing concentrates such as *surimi*; this is a tasteless, odourless paste which is very viscous and elastic but unstable when frozen in the absence of any cryo-protectors such as sorbitol or glucose. By using a texturisation process such as extrusion, it is possible to obtain crustacean meat substitutes (see Chapter 8).

Blood and offal

Blood can be upgraded after centrifuging in the form of plasma used in animal feed and cruor; in this case it must be bleached in order to produce a white globin having emulsifying and foaming properties which can be used equally well in animal or human foodstuffs (see Chapter 9).

Milk proteins

Casein and caseinates have excellent emulsifying and binding properties which are exploited in foodstuffs (cooked meats, ready-made meals, etc.) and in non-food uses (glues, gums, etc.). They can be isolated by acidification down to pH 4.6 by mineral acids (HCl, H_2SO_4, etc.) or by an organic acid (lactic fermentation) or by coagulation by rennin (calcium caseinate). If the milk has been previously heated, part of the precipitate comes from the denatured proteins of whey (see Chapter 5).

Whey proteins from the cheese industry or from casein factories can be recovered using several techniques (ultrafiltration, ion exchange, thermal coagulation). Depending on the method chosen and the pre-treatment applied (heating, demineralisation using electrodialysis, ultrafiltration associated with heating), a whole range of protein concentrates with wide-ranging functional properties can be obtained. These can also increase the yield in terms of cheese (Centri-Whey process) or enrich cereal-based preparations deficient in certain essential amino acids (maize-based tortillas) (see Chapter 9).

Plant proteins

Proteins from grains (cereals, legumes)

We first consider *legume proteins*, starting with soya and oleaginous grains. By extracting the oil (by pressing or using hexane), oil-free cakes can be obtained which are then converted into flour.

The isolate or concentrate is obtained by means of the following:

- precipitation at pH 4.5 and solubilisation at neutral or alkaline pH;
- precipitation using alcohol;
- thermal or heat precipitation.

Often this involves extracting water-soluble or alcohol-soluble ingredients or using other precipitation agents (Table 3.3). Their uses are extremely diverse: meat substitute following texturisation, incorporation in meat products.

Table 3.3 Composition of soya protein concentrates obtained using various processes

	Method of extraction		
	Ethanol	Acid	H_2O + heat
Proteins	70.7	70.7	72.2
Fats	0.3	0.3	1.2
Fibre	3.7	3.6	4.5
Ash	6.0	5.1	3.8
Nitrogen solubility	5.0	69.0	3.0

The advantage of these isolates lies in their good viscosity, gelling and binding properties. Sometimes, certain cakes such as rape, sunflower, cotton and groundnut cakes contain anti-nutritional substances (goitrogenic substances, polyphenols, etc.) which it is advisable to eliminate (see Chapter 4).

The techniques of ultrafiltration and ion exchange are increasingly used in order to improve the purification of isolates.

Precipitation techniques are used in bean and other non-oleaginous protein-containing substances but yields are often poor. Air-classification is often preferred for obtaining 65% protein concentrates. The co-products are extremely rich in starch and difficult to improve (animal feed).

However the problem of persisting anti-nutritional substances (fermentable carbohydrates, α-galactosides) has not been resolved. The concentrates have excellent functional properties, but their bitter flavour persists.

The principal example of **protein extraction from cereals** effected on an industrial scale concerns extracting gluten from wheat flour (see Chapter 4). The aim of this operation is to reduce the gluten content of flour for dietary purposes, or to extract gluten for the purposes of improving the baking value of certain soft wheat flours (weak flours).

Extraction essentially consists of forming a paste (0.6–1 l water/kg of flour) and then washing it in hard water to eliminate the starch. In the case of maize, the gluten extracted is a co-product of starch preparation. After wet-processing has been applied directly to the grains that have been left to swell, they are crushed and the proteins are extracted by centrifuging.

Leaf proteins

Alfalfa is pressed prior to drying. Purification involves techniques of precipitation by solvent. During pressing (screw press), it is necessary to add sulphites to prevent oxidation. The juice contains 10% dry matter, 30–40% of which consists of proteins (3–4 g/l). After thermal or heat coagulation (80 °C) and centrifuging, 85% protein isolates are obtained (see Fig. 3.3).

Proteins from micro-organisms

These are difficult to extract because of the resistance of their cell walls (especially yeasts); hammer or roller mills must be used in association with plasmolysis processes (thermal, chemical or enzymatic) and the extraction yields remain mediocre.

3.1.1.4 Conclusion

Any enriching or extraction of proteins leads to the formation of a reduced product for which a use needs to be found in order to make the sector profitable. Few examples of integrated systems can be given, apart from that of whey. Therefore, before the protein concentrate is processed or used, we need to ask ourselves the following questions:

- What level of protein enrichment is a 'plus' from a nutritional and functional point of view?
- Are co-products useful on their own or do they have to be processed to generate other by-products?
- From an economic point of view is the price of the concentrate competitive with the proteins we wish to replace (animal proteins that are too expensive and too wasteful of energy)? This question is the most difficult to answer because of fluctuating world prices for cereals and soya and the policy of subsidised prices.

3.1.2 Glycans

Because glycans have a smaller number of active groups than proteins, techniques of extraction and purification are fewer and based solely on precipitation reactions which involve changes in the polarity of the medium or the molecules themselves.

In addition, as these polysaccharides do not have a well-defined molecular mass, the techniques of molecular sieving or filtration through calibrated pores are not very suitable. Finally since most glycans form very viscous solutions at low concentration, problems of fouling frequently arise if processes using membranes are employed.

3.1.2.1 Preparation of glycan concentrates

Figure 3.6 shows how glycan concentrates may be prepared. Starting from the raw materials, there are two possible scenarios:

- The polysaccharide is insoluble and no method, apart from dissociation of the microstructure, can allow it to be obtained in a dispersed form, which is both soluble and stable. This is the case with starch and cellulose. After extremely fine milling of the raw material, dry or wet centrifugal fractionation is carried out.
- The polysaccharide is soluble in hot water or made soluble by increasing the pH. After solubilisation, the polysaccharide can be precipitated

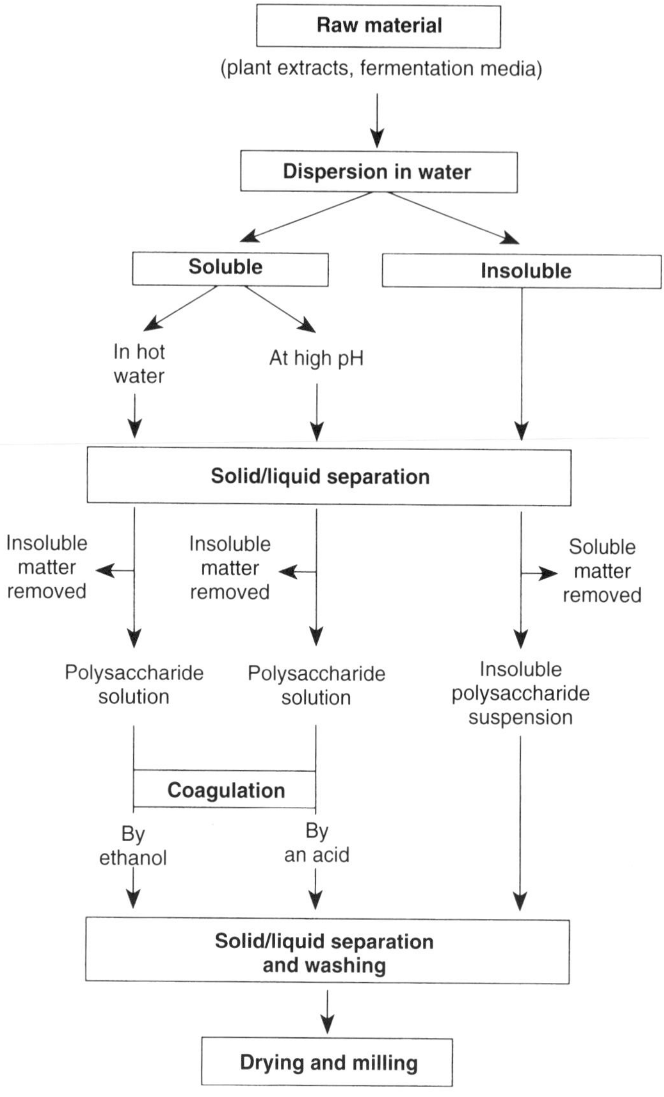

Fig. 3.6 Glycan extraction.

by reducing polarity or pH. This is the method used for extracting hydrocolloids which are gelling agents (carrageenan and xanthan using hot water, pectin and alginate using extraction in an alkaline medium).

3.1.2.2 Extraction of non-soluble polysaccharides

Starch extraction

The main sources of starch are maize, potato, wheat, manioc and rice. Maize comes mainly from countries with fairly hot climates (the USA, China) whereas potatoes are mainly grown in cool, wet climates (Europe, Russia) and wheat in countries with temperate climates; 90% of rice production is in Asia.

In the case of maize, which is the raw material predominantly used as a source of starch, the starch is extracted from the grain using a process of wet comminution. The purified starch is separated from the cellulose, lipids and proteins by means of a sequence of milling, sifting and centrifuging operations.

Operations begin by softening the grain which is immersed in a solution of dilute acid. Rough milling separates the germ, which contains lipids. Fine comminution then separates the cellulose from the endosperm; the latter is then separated by centrifuging, with the proteins being less dense than the grains of starch. Finally the starch is washed and dried.

When the raw material is wheat, the starch comes from extracting the gluten (see 'protein extraction' section). Washing the viscoelastic paste in hard water eliminates the starch in the form of milk (suspension of native starch).

Preparing cellulose

In general, pure cellulose does not need to be extracted since naturally produced cotton is practically pure cellulose. However, bearing in mind the considerable quantities of cellulose consumed by the paper pulp industry, either textile fibres or wood fibres have to be used.

After milling, the wood is subjected to a hot solution of calcium sulphate in order to eliminate the lignin which is solubilised in this way. Bleaching stages are then required to purify the cellulose.

3.1.2.3 Extraction of soluble polysaccharides

Extracting carrageenan

The raw material comes from a few types of red algae (and/or lichen): the mixture collected on the French coasts consists of *Chondrus crispus* and *Gigartina stellata*. After washing, the algae are macerated in hot water after being milled and 'digested' by an alkaline agent (sodium hydroxide, potassium hydroxide). The soluble impurities (proteins, cellulose, etc.) are eliminated by hot filtration under pressure (using a canvas filter press) in the presence of a filtering earth material. The clear filtrate is processed using pure ethanol which coagulates the polysaccharide in the form of fibres. The coagulate is washed in alcohol then dried by evaporation. The alcohol is regenerated by distillation.

Extracting xanthan

Xanthan is a secondary metabolite excreted during the aerobic fermentation of simple carbohydrates using the bacterium *Xanthomonas campestris*. Xanthan is precipitated directly from the culture medium using isopropyl alcohol. The coagulum obtained is separated, washed, pressed, dried and milled.

Extraction of pectins

Pectins are principally extracted from the rind of citrus fruits, pressed apple pulp and beetroot. Hydrolysis of the protopectin in hot water at a low pH solubilises the pectin as well as undesirable neutral soluble substances (proteins, gums, other carbohydrates). The insoluble matter is eliminated by pressing followed by filtration under pressure in the presence of a filter aid.

The clear filtrate is mixed with alcohol which precipitates a fibrous coagulum. The fibrous pectin is washed, then pressed and vacuum dried. A highly methylated pectin is generally obtained: demethylation using acids or alkalis produces low methylated pectins.

Extraction of guar and carob gums

The following products can be obtained from grains of guar and carob:

- Flours: after dehusking, the germs are separated by milling to isolate the 'splits' (albumen) which are the basis of the flour.
- Extracts: the grain is solubilised in hot water, with the insoluble matter being eliminated by filtration; the glycan is then precipitated by the isopropyl acid, then washed, pressed and vacuum dried.

Extracting alginates

Alginates are extracts of brown algae found on most rocky coasts. The main types are *Laminaria digitata* and *Fucus serratus*. The first extraction operation consists of macerating the algae in the presence of a dilute mineral acid which can convert the calcium alginate contained in the algae into alginic acid. The demineralised algae is then milled in the presence of an alkali or an alkaline salt in order to neutralise the alginic acid and solubilise it in an ionised form.

The insoluble compounds (cellulose, proteins, etc.) are removed by filtration, flotation or decantation. The alginic acid is re-precipitated by adding mineral acid; the precipitate is then washed and dried.

3.2 Structurisation/texturisation

Numerous protein sources which are intended to enrich foodstuffs with proteins to improve their nutritional and organoleptic properties are incorporated in powdered form. When these added proteins are substituted partially or totally for other proteins (or lipids: fat substitutes) in traditional foodstuffs, they must be modified or structured in order to give them an attractive texture.

3.2.1 Biochemical bases of texturisation

Texturisation processes in general involve several successive processes which correspond to different conformational states (Fig. 3.7). **First, denaturation** with partial destruction of the spatial native structure: rupture of

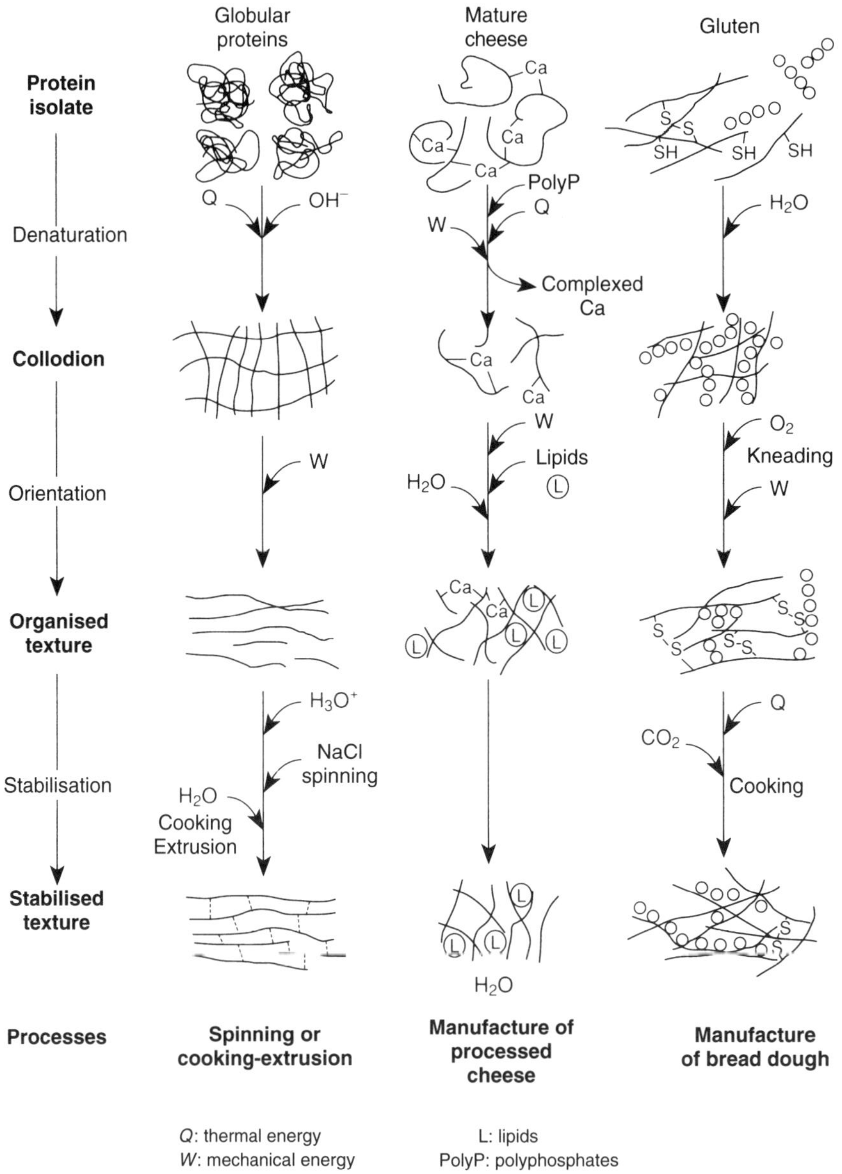

Fig. 3.7 Diagrammatical representation of the structural modifications of proteins during texturisation.

bonds with low intermolecular binding (hydrogen and hydrophobic interactions and disulphide bridges).

This rupture can be caused by **thermal** (cold or heat) and **mechanical** processes, by high pressure, **oxidising** or **reducing** processes (oxidation of disulphite bridges) or treatment using **hydrophobic agents** (organic solvents, detergents, gases under pressure).

The change in conformation can also be caused by limited proteolysis especially when this is very specific (for example, the hydrolysis of κ-casein by rennin with corresponding change in the surface hydrophobicity of the casein micelle) or when it is applied to very large molecules which are in the polymeric state: for example the case of leguminous proteins (soya) or myofibrillar fish proteins.

Although this stage does not involve any degradation of side chains, certain chemical reagents contribute to the breaking of the calcium bridges which are involved in the native micelle structure of the caseins by blocking ionised side groups (ε-amino groups of lysine residues).

Finally, modifications in pH or in salt concentration can also dissociate ionic bonds. This first stage generally requires prior solubilisation, which needs to be as complete as possible, in an alkaline (pH 8) or salt medium, for example.

Second, **organisation** and **orientation** of macromolecules which are partially or totally unfolded: gelling, fibre orientation with, where appropriate, reorganisation of molecular aggregates (peptisation of processed cheeses, for example). Generally speaking, molecule orientation is obtained either by extrusion through a die (fibre manufacture), or by using other stratagems such as rolling/spreading, surface precipitation, which are frequently used in traditional texturising techniques (manufacturing films, for example).

Third, **binding** and **stiffening** of the organised structure obtained by redistributing the intra- and intermolecular bonds which have been broken in the course of the unfolding stage.

Different types of bonds are involved in the cross-linking of the peptide chains and the formation of a network which is ordered to a greater or lesser degree: highly ordered in the case of extrusion, and disordered in thermoplastic gels.

3.2.2 Techniques of thermomechanical and thermal texturisation

3.2.2.1 Extrusion process

The use of extrusion (Fig. 3.8) in food manufacture dates back to 1954 when Boyer produced protein fibres that resembled meat fibres.

Principle of manufacturing protein fibres

The denaturation phase is accompanied by an increase in the viscosity of the protein solution called **collodion**. Three conditions are required in

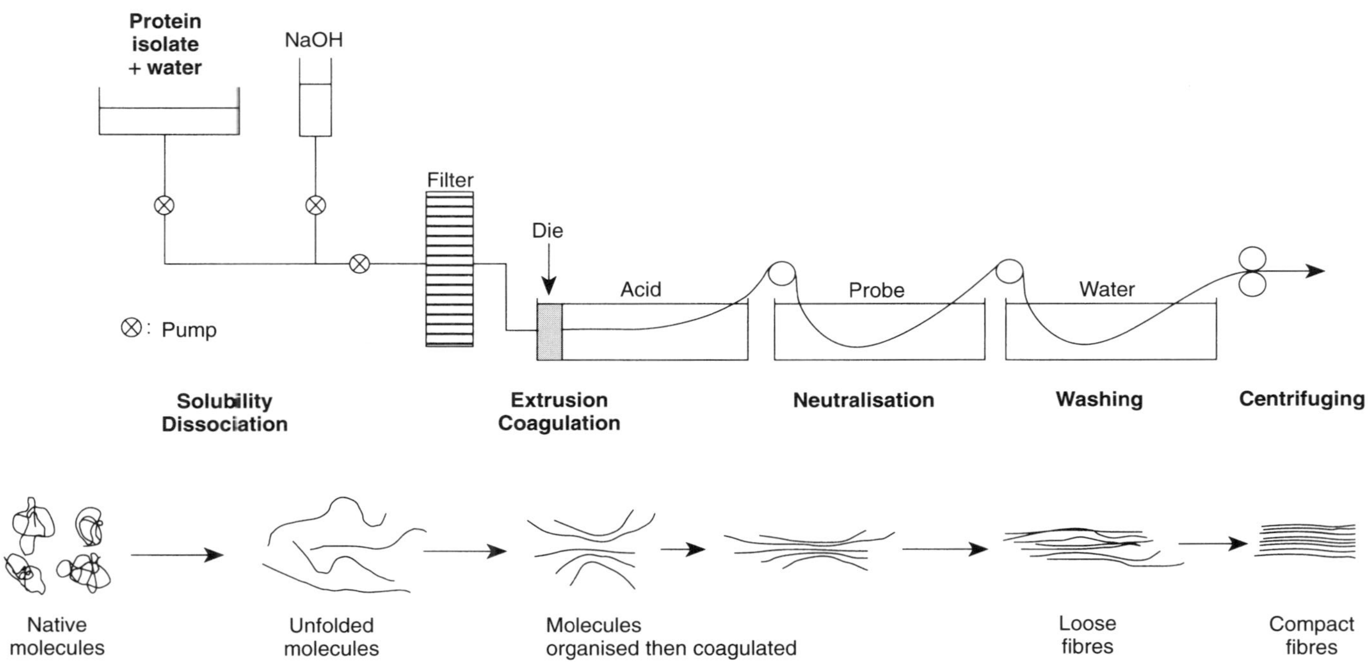

Fig. 3.8 Diagrammatical representation of the extrusion process.

order to obtain this: purified isolate with a high protein content (>85%), a high protein concentration of collodion (10–30%) and a high pH (8–10).

The extrusion of the collodion, which aligns the molecules in the direction of flow, must be effected through a die whose holes have a diameter of between 50 and 250 µm. The die is immersed in an acid and/or saline coagulation bath: lactic, acetic or phosphoric acid, having a pH of between 2 and 4, in the presence of between 5 and 20% sodium chloride. When the filaments come into contact with the bath two exchange processes take place:

- The water in the filaments migrates towards the bath.
- The ions in the bath migrate into the filaments.

The negative charges (COO^-) are thus neutralised, as is the electrostatic repulsion. This results in the insolubilisation of the protein and shrinkage of the filaments.

The filaments are then drawn out between two cylinders and transferred to washing baths having a slightly acid to neutral pH (pH 5.5–6) then brought together in parallel bundles by means of either **protein** binders (ovalbumin, gelatin, casein, gluten or soya isolate) or **polysaccharide** binders (starch, carrageenan, alginates, etc.). This operation involves soaking followed by heating, making the fibres circulate in a counter-current direction.

In order to be suitable for extrusion the proteins must combine several structural characteristics:

- Linear unfoldable molecules having a minimum length of 100 nm, without ramifications or bulky active groups.
- Molecules containing a large number of polar active groups permitting the establishment of intermolecular bonds (Ser, Thr, Asp, Glu, Lys).
- Molecular masses (MM) of between 10 000 and 50 000 Da. If the MM is high, then viscosity is high and extrusion difficult; if it is too low, the filaments have little 'stability'.

Proteins that can be extruded consist mainly of ovalbumin, zein and collagen which form fibres which are very resistant to traction. Other proteins such as casein or leguminous isolates can produce fibres and, more particularly, films.

This extrusion capacity can be improved by means of chemical additions using formol or acetic anhydride. Extrusion adjuncts are also used: the limited addition of gluten (a few per cent) improves mechanical resistance, lipids make the protein fibres succulent and tender, whereas starch increases their hydration and water-holding capacity. Finally, if the salt concentration of the collodion is high, the mechanical resistance of the fibres is high, but their diameter and water-holding capacity are low.

Extrusion techniques

In general several adjuncts are added to the collodion: ovalbumin, gluten, soya flour, protein hydrolysates, salt, glutamate, colourant, flavours. Often these are protein/polysaccharide mixtures which are submitted to extrusion (protein/pectin at neutral pH with $CaCl_2$, protein/carrageenan, etc.). Various types of coagulating agents are added.

In hydrodynamic extrusion, the coagulating bath is driven with a movement that is faster than the speed at which the fibres leave the extruder, and so they are drawn out. Fibres can also be coagulated using the hydrogen chloride gas or by drying at high temperatures (90–140 °C).

3.2.2.2 Cooking-Extrusion

Principle of the process

This is a process derived from extrusion; in a hydrated medium it is a process well known historically for manufacturing pasta. The material, to be texturised, which must be fairly viscous, is pushed along and kneaded in the screw (Fig. 3.9). Compression is provided by narrowing the thread of the screw, and by reducing the space between the screw and the wall of the cylinder and the counter-pressure due to the narrow die at the outlet. The compressed product undergoes shear and increases in heat due to passing between the screw and the serrated wall of the cylinder.

Two types of apparatus are used: single screw extruders and double screw extruders. In this type of more recent equipment, two identical co-penetrating screws turn in the same direction within the bore of a fixed barrel; the material, which is compressed all along the wall of the barrel and between the screws, advances in a figure-of-eight trajectory. The screw profile can be changed using different combinations of screws and paddles. Other variations can be used: different dies, zone temperature control,

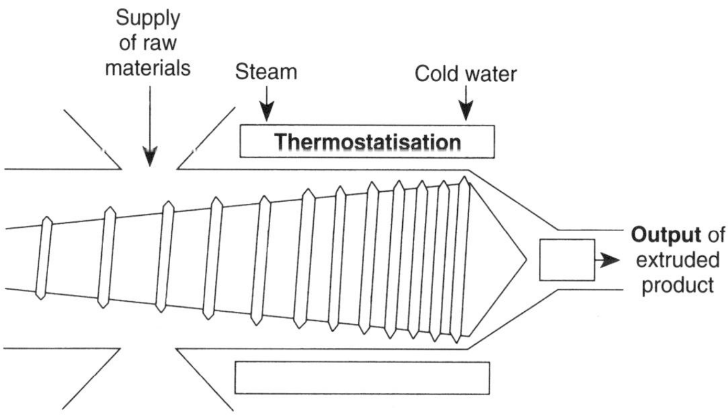

Fig. 3.9 Diagram of a cooker-extruder screw.

incorporation of other functions at a desired point (degassing, introducing a liquid or a solid, etc.).

The double screw system has the following advantages: it is suitable for a wider range of products and incorporates self-cleaning, more uniform operation and better product quality.

Process in a low-moisture medium

This process is derived from the heat shaping of plastic materials; the initial uses of cooking-extrusion within the food industry were for the production of expanded structures from starch or wheat flour. For example, protein flour is heated to 65–100 °C by means of steam injection and then brought to a 20% moisture content. The mixture is then kneaded in the extruder at 85–110 °C under pressure of 50–100 bar, then the temperature is brought to 130–180 °C for the last 10–20 seconds it remains in the equipment. The product which leaves through a die is expanded and dried under the effect of the decreasing pressure on exit and the vaporisation of superheated water.

For this type of cooking-extrusion, a defaulted product whose nitrogen or protein content represents 40–70% of the dry extract is required; solubility must be sufficiently high (50%), which excludes raw materials that have undergone intense thermal processes (as in protein cakes). The water content must be between 5 and 30% and the pH between 6 and 8.5.

The average processing time varies from 30 to 120 seconds. The various stages involved in producing the extruded product are summarised in Fig. 3.10.

The product characteristics are as follows:

- Expanded products, not particularly dense, alveolar but having low solubility: these are very hydratable solid foams.
- Some, though very limited, formation of intermolecular amide bonds, and very few covalent bridges formed.
- Products whose density can be modified by variations in die size and temperature (increased density if the temperature is reduced and the die larger).

After rehydration these structures produce spongy, stretchy products which are stable and can be used in meat balls, croquettes, certain cooked meat products or pastries. Thermoplastic cooking-extrusion is useful for cereal products (unleavened bread, textured starch, etc.).

Process within a high moisture medium

When the water content is high (45–70%), thermocoagulation leads to the formation of films or non-expanded hydrated gels. Numerous applications have appeared recently, for increasing the value of the mechanically separated proteins from whey, blood, meat and fish.

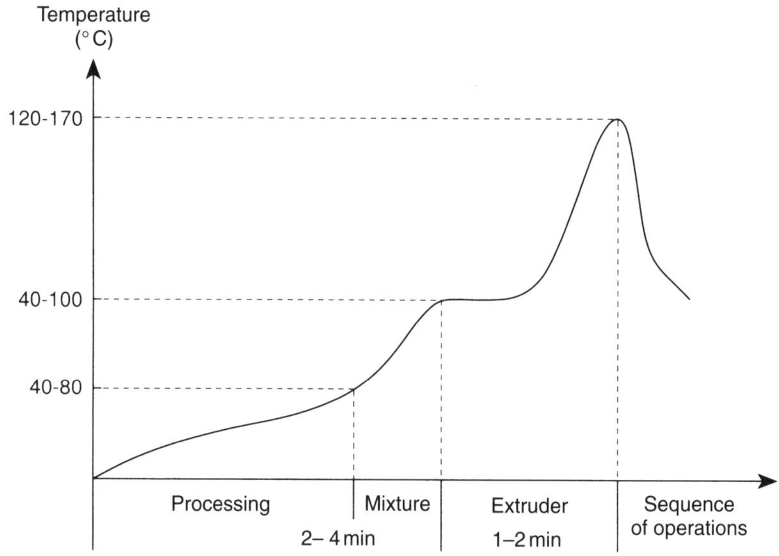

Fig. 3.10 Sequence of cooking-extrusion operations.

In order to obtain a uniform texture, a long die which is cooled is fitted to the exit of the extruder; in this way cheese substitutes can be prepared from calcium caseinate and butter oil; several additives are, however, necessary in order to obtain a uniform and stable texture.

The techniques for manufacturing meat substitutes (restructuring poultry meat separated mechanically using two thermal or heat coagulation processes at different temperatures), crabmeat substitutes (thermal or heat coagulation of rolled fish which is then cut into strips which are wound round and reassociated) are based on this same principle of cooking-extrusion in a high moisture medium. Here again the aim is to exploit fish that are not generally eaten directly (pout, types of mackerel, sardines, etc.).

Cooking-extrusion of protein mixtures is being extensively developed in order to obtain a wide range of textures: whey caseinates/proteins, egg white/plant proteins.

The association of gel-forming polysaccharides (alginates, starch) is also widely used; these glycans have a texturising effect and facilitate the transfer of proteins in the extruder, especially at high temperatures, however with the risk (or advantage) of Maillard reactions.

Finally, cooking-extrusion can be used for solubilising protein into proteinates by means of alkali addition. This is the case when caseinates are manufactured from insoluble acid casein (pH 4.6).

3.2.2.3 Other thermal texturisation processes

Thermal coagulation of soya proteins
This is the basic process for manufacturing tofu; it works in several stages:

- Manufacturing soya milk: soaking grains of soya for 10 h in water then crushing, cooking and filtration (milk with 5–6% dry matter).
- Adding a suspension of calcium sulphate: a hydrated gel is formed which traps lipids.
- Washing the curds then draining, to permit the elimination of indigestible carbohydrates. The product obtained contains 88% water, 6% proteins and 3% lipids and is used in numerous culinary preparations of Asian origin.

A variant on this procedure (Chinese *Yuba* manufacture) consists of heating the soya milk; the protein film forming on the surface is recovered, then air dried. These films which contain 10% water, 52% proteins and 24% lipids can be rehydrated and used as edible envelopes or packaging.

Thermal coagulation using steam under pressure
This is based on processing a flour which has been hydrated to 20% moisture content using steam under pressure; after bringing it down to atmospheric pressure, swelling and evaporation of flavours takes place.

'Thermal gelling' of processed cheeses (Fig. 3.11)
In this case, the thermomechanical process associated with the addition of polyphosphates is intended to do the following:

- Dissociate the calcium *para*-phosphocaseinate complex contained in the raw material (Cheddar or Emmental);
- Emulsify the aqueous phase containing the phosphocaseinate, which is dissociated and therefore less surface active, with the lipid phase.

Texturisation only occurs during cooling which takes place in a mould; the texture is closely associated with the extent of an additional thermal process (80 °C, 20 min) after the dissociation/emulsification phase (105–110 °C for a varying number of minutes). It also depends on the speed of cooling and on incorporating a small portion of the cheese that has already been melted several times (repeated use).

3.2.2.4 Low-temperature texturisation processes

One of these techniques involves freezing the curds from washed and crushed soya milk to –5 °C for several hours. After thawing, water is eliminated by pressing; a spongy texture is obtained, without any indigestible carbohydrates (Fig. 3.12).

A recent process consists of spreading a protein paste with 66% moisture content in the form of a thin layer (4–5 mm) onto a plate cooled to

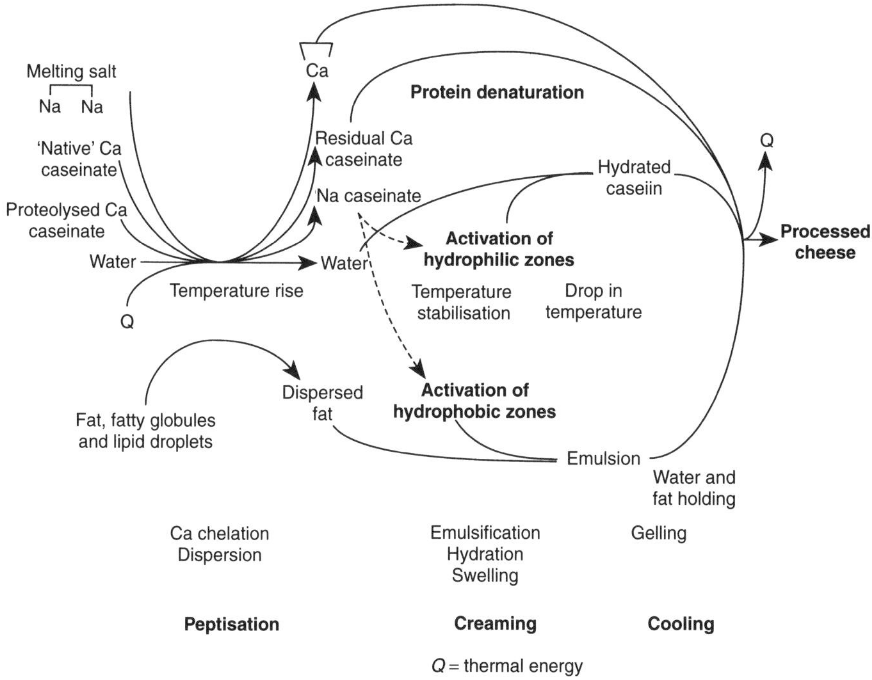

Fig. 3.11 Diagrammatical representation of cheese melting process.

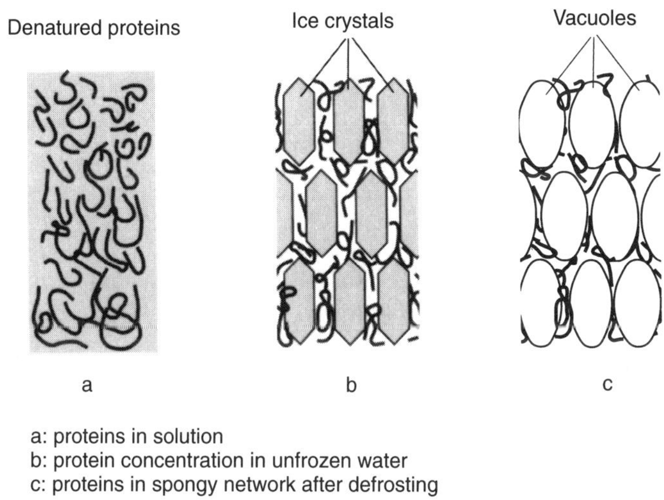

Fig. 3.12 Process of texturisation by freezing.

–20 °C. A multilayer system is obtained by the successive accumulation of previously folded layers. After applying pressure and then freezing, microwave heat treatment is applied.

3.2.3 High-pressure texturisation process

This process has been developed recently and its applications are potentially innumerable since it enables food to be cold-sterilised or pasteurised without loss of nutritional value and without changes of organoleptic qualities. The application of very high pressures (up to 5000 bar) to the microbial cells modifies the osmotic balance, especially if the shock is attained by means of rapid rises in pressure. The enzyme activity can itself be modified because of the change in protein conformation. This is the very reason why the functional properties of the polymers are modified; this is particularly true for the pectins of fruit juices or jams which combine in fibres. The denaturing effect of high pressures manifests itself through enzyme deactivation and more particularly through lowering of solubility and/or gel formation.

As far as the mechanisms involved are concerned, there are more exchanges of disulphide bridges in thermal gelling than in high-pressure gelling. In fact, the gels obtained via thermal gelling are firmer, more brittle and less viscous. The effects differ according to the nature of the proteins and the pressure applied; thus, in the case of a 30 minute treatment, gelation requires a minimum pressure of 2000–3000 bar for the myofibrillar proteins of fish and 4000–7000 bar for the proteins of white and egg yolk. Adding sorbitol or carbohydrates reduces the coagulating effect whereas sodium chloride seems to have the opposite effect.

Several types of texturisation can be envisaged: restructuring of mechanically separated fish or meat flesh and the manufacture of fibres from soya protein isolate. The advantage of this technique lies in the possibility of working at low temperature (0 °C) in flexible packaging in order to bring about a reduction in microbial load without loss of flavour or changes in colour.

Other applications are proposed: accelerating the rennin clotting of powdered reconstituted milk and increasing the sensitivity of globular proteins to proteolysis because of the partial unfolding of molecules containing few disulphide bridges (e.g. β-lactoglobulin).

3.2.4 Texturisation process using chemical means

3.2.4.1 Texturisation of proteins by oxidation

One of the texturisation technologies based on oxidation, and the oldest known, is the manufacture of bread dough. Gluten, which is the insoluble protein fraction of cereal flours used in bread-making, can, after hydration and mechanical kneading, form a dough with viscoelastic properties (see Chapter 4).

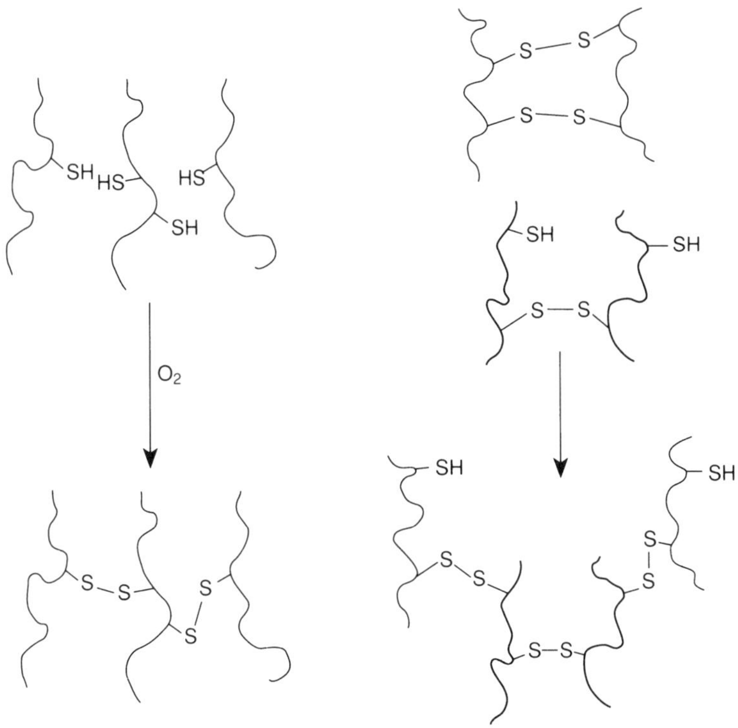

Formation of disulphide bonds by oxidation of thiol groups

Polymerisation of polypeptide chains by disulphide bond exchange

Fig. 3.13 Diagrammatical representations of protein polymerisation by oxidation of thiol groups.

This texturisation is particularly due to the oxidation of numerous thiol groups to form intermolecular disulphide bridges (Fig. 3.13), the result being a finely cross-linked network of glutenins which enclose strongly hydrated gliadins; consequently for some time it was believed that cross-linked glutenins were mainly responsible for elasticity whereas the hydrated gliadins were thickeners. However, recent knowledge of the structure of gliadins shows that their conformation, in loose spirals due to the uniform repetition of sequences, which include residues of proline, glycine and valine, is also responsible for properties of elasticity.

These oxidation reactions can be accelerated, not only by incorporating oxygen from the air in the course of kneading, but also through the presence of a superoxide group linked with the oxidation of polyunsaturated fatty acids (Fig. 3.14). Adding bean flour rich in lipoxygenase also accelerates the formation of disulphide bridges; soya flour which is much richer in lipoxygenase is preferred as its presence has no effect on taste.

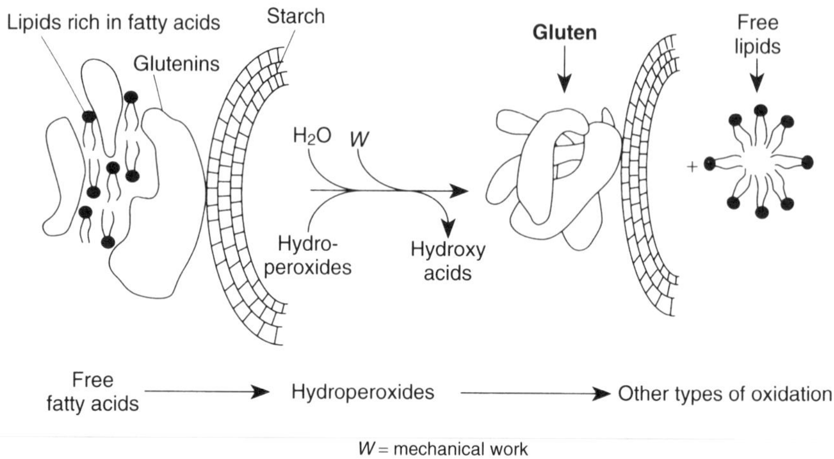

Fig. 3.14 Formation of gluten during kneading and liberation of lipids.

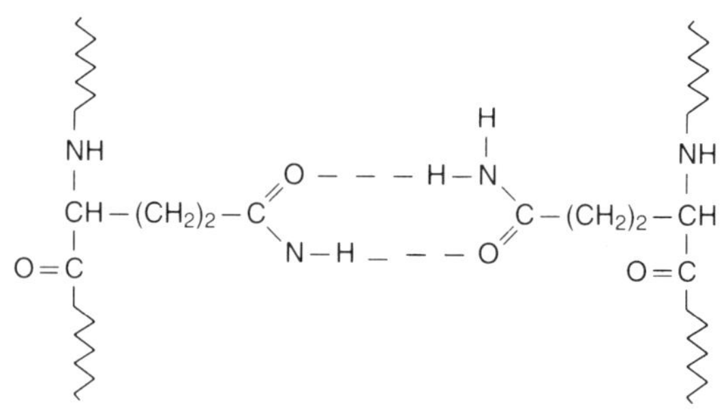

Fig. 3.15 Association of glutaminyl residues by hydrogen bonds.

Other bonds are also involved in the formation of bread dough: during hydration, hydrogen bonds are established between the amide groups of glutamine and asparagine and hydrophobic bonds result from the association of the extremely large number of non-polar side chains of the Phe, Tyr, Ile, Leu and Val residues (Fig. 3.15).

3.2.4.2 Texturisation of proteins by chemical (or enzymatic) cross-linkage
Cross-linkage can be obtained through the formation of electrostatic bonds. In this way succinylation is involved in increasing the number of negative charges (a carboxylic group is placed on the end of an ε. NH_2 lysyleside chain); the result is an increase in ionic bonds and hydrophobic interactions. The formation of covalent bridges (by glutaraldehyde, for example, or by chemical phosphorylation) remains restricted to the manufacture of gels or films for non-food use.

Enzymes such as transglutaminase can also cross-link proteins by creating isopeptide bonds between the side chains of glutamine and lysine, but the cost is not compatible with industrial use.

3.2.4.3 Texturisation using gas under pressure

Recent research has shown that gases such as carbon dioxide and nitrogen monoxide deactivate peroxidase at pressures greater than 3 bar. Textural modification can also be observed; the chemical mechanisms are still poorly understood and experimentation is at a pilot stage.

4

Intermediate food products of plant origin

Within the food industry, plant products are an essential source of proteins, lipids and starch. The properties and uses of starch and its derivatives are discussed in Chapter 11. In this chapter, therefore, we will be considering only proteins and oils of plant origin.

4.1 Plant proteins

4.1.1 General information and definitions

Because of their diversity, differences in terms of physico-chemical properties and amino acid composition, grain and seed-reserve proteins have considerable commercial potential. In addition, they are cheap in comparison with most animal proteins. Most techniques for preparing plant protein substances (PPSs) have been developed for soya, but more recently these have also included legumes (peas, beans), cakes from oil-producing plants (rape, sunflower) and cereals (wheat). In these raw materials rich in starch, the main non-protein components are insoluble in normal solvents, whereas in the case of oil-producing plants a considerable fraction of these constituents can be extracted using water. Therefore, simply removing the husks and oil from oil-producing seeds can produce flours whose protein content is often 50% higher than that of protein-producing plants, such as peas or beans, and especially cereals (Table 4.1).

Differences in biochemical composition can partly explain the different technologies involved in dealing with oil-producing plants, on the one hand, and legumes and cereals on the other. The other important parameter likely to influence the technique used is the nature and physico-chemical proper-

Table 4.1 Composition of different plant grains (%)

	Proteins	Lipids	'Soluble carbohydrates'*	Fibres	Ash	Water
Cereals						
Wheat	7–18	1.5–2	60–69	2–2.5	1.5–2	8–18
Maize	7–12	4–8	67–72	2	1.5–1.8	11
Rice	7.5–9	2	63	9	6	12
Legumes**						
Containing protein						
Beans	26	1.2	61	6.5	4	
Peas	27	1.5	60	2–6	3	
Chick peas	21	7	65	2–4	3.5	
Lentils	22–30	3	62	3	2.5	
Broad beans	28–33	2	58	2–7	3	
Containing oil						
Soya	35–50	22	15	≈10	5	
Peanut (almond)	25–30	48	12	3	3	
Sunflower (almond)	20–38	35–50	4–6	10	3–5	
Rape (non de-husked seeds)	15–30	35–40	17	4	3.5	

* By difference; fibres not included.
** Composition in comparison with the dry matter since the humidity of the raw materials is very variable.

ties of the proteins that make up the seed. Depending on their intended use and the methods used, the protein concentration of end-products can vary from 15–20% (mechanically separated cereal proteins) to 90% (isolates).

Using as a measure increasing levels of protein content, we can differentiate: flours < 70%, concentrates 70–85%, isolates > 85%. PPSs are not only characterised by their protein content but can also be differentiated by their physical form. Indeed, product particle size varies in its degree of fineness, ranging from powders of tens or even hundreds of micrometres, up to fairly coarse semolina. They can also be presented in a texturised form obtained using techniques of extrusion or cooking-extrusion (see Chapter 3).

4.1.2 Physico-chemical properties of plant proteins

Legumes (peas, beans, lupin, soya) are fairly similar in their protein composition and mainly consist of globulins (60–90%) and albumins (10–20%). This distribution of albumins and globulins can also be seen for the sun-

Table 4.2 Distribution according to criteria of solubility (%) of plant protein substances

	Albumins	Globulins	Prolamins	Glutelins
Protein producing				
Field beans	20	60	–	15
Peas	21	66	–	12
Lupin	10–20	80–90	–	0
Oil producing				
Soya	10	90	–	–
Sunflower	16–23	50–60	3–4	6–10
Rape	44–52	20–25	3–4	6–9
Cereals				
Wheat	5	10	45	40

flower (Table 4.2). On the other hand, rape and wheat have a different composition. Rape is particularly rich in albumins (44–52%) and wheat in prolamins (45%) and glutelins (40%).

4.1.2.1 Major proteins of protein- and oil-producing plants
The major fraction of rape and sunflower globulins is represented by a type 11–12S protein. In seeds of legumes we also find a type 7S protein. Type 11S globulins have a molecular mass that is generally between 300 000 and 400 000 Da except in the case of rape where it is much lower (129 000–163 000). This protein can also be differentiated from others by its high isoelectric point (7.25 instead of 4.6–4.8). These proteins have other similarities: denaturation temperature and enthalpy, average hydrophobicity, behaviour on dissociation, etc.

The quaternary structure of these proteins is complex and differs according to whether the legumes produce oils or proteins. In the case of soya, peas and beans, the type 11S protein is made up of 12 secondary units, which are acidic (α) (40 000 Da) or basic (β) (20 000 Da) characterised in $\alpha\beta$ by means of disulphide bridges, and organised into two stacked hexagonal structures. In this structure the α and β secondary units are arranged alternately. As far as sunflower and rape are concerned, studies similar to those carried out for soya have shown that the 12S globulins of these grains have a quasi-spherical shape corresponding to an oligomeric structure, made up of six secondary units, themselves arranged in a trigonal anti-prism. For globulins of protein-producing plants, different monomers are formed of acid and basic polypeptides linked by disulphide bridges.

Fewer similarities in the physico-chemical characteristics of the type 7S globulins can be seen between the various types of legumes (peas, beans, soya). These are generally glycoproteins with a trimeric structure having a molecular mass of between 140 000 and 180 000 Da and an isoelectric point close to 5.5. The complex quaternary structures of the globulins of legumes and the oil-producing seeds formed of several secondary units explain why these proteins aggregate or dissociate depending on the ionic force and the nature of the ions present.

Type 2S protein from the *Brassica* rape species differs considerably from these globulins. Its molecular mass is between 12 000 and 14 900 Da according to different sources, and its isoelectric point is over pH 9. The particularly alkaline character of this protein is the result of the very high amidation level of the aspartic and glutamic acids. It consists of two polypeptides containing 90 and 29 amino acid residues respectively, linked by two disulphide bridges.

4.1.2.2 Wheat proteins

Wheat grains contain a large number of proteins: structured proteins, reserve proteins and proteins that have a biological function. Osborne's classification separates them into four major groups according to their solubility in water: albumins and globulins are soluble, gliadins and glutenins are insoluble (Table 4.2). Albumins and globulins display the properties characteristic of their group; they are globular, with molecular masses of between 10 000 and 100 000, representing approximately 20% of the total proteins in the grain. Overall they can be compared to biologically functional proteins.

Gliadins which are soluble in high-strength alcohol solutions and glutenins which are insoluble have common characteristics. Each of these two groups represents 40–45% of the total proteins and constitutes the store of nitrogen, carbon and sulphur required for the development of the germ at the time of germination. The amino acid composition of these proteins is very rich in glutamine and proline, but low in alkaline amino acids.

However, within the four groups of gliadins the composition of amino acids differs. The ω-gliadins are even richer in glutamine and proline than the other three groups. In addition most of these can be distinguished from the other gliadins by an absence of sulphur amino acids. Moreover, the basic amino acid content decreases from the α to the ω-gliadins. All these variations have repercussions as far as their properties are concerned.

Since almost all glutamic and aspartic acids are in an amide form, and have a low basic amino acid content, gliadins are not highly charged. They also differ from the majority of other proteins because they combine low charge with strong hydrophobicity. This explains their special properties of solubility, such as in a water–alcohol medium.

In addition, their high proline content results in numerous breaks in the secondary structure because of the rigidity of this amino acid which, by combining with a high proportion of non-polar residues, gives gliadins a non-globular conformation. In their native state the ω-gliadins take the form of semi-rigid molecules with a low level of organisation.

The lysine, glycine, alanine, serine and tyrosine residue content of glutenins, the last group of grain proteins, is considerably higher than that of the gliadins, whereas their glutamic acid and cysteine acid content is lower. Traditionally three kinds of glutenins can be differentiated according to their molecular mass:

- Those whose molecular mass is higher than 200 000 Da and whose composition is similar to that of albumins and globulins.
- Those whose molecular mass is between 100 000 and 200 000 Da and which contain a high proportion of glutamic acid and glycine.
- Those whose mass is lower than 50 000 Da and whose hydrophobicity is similar to that of the gliadins.

Hydrophobic interactions are significantly involved in the formation of glutenin aggregates. An aggregate structure superimposes itself on the secondary and tertiary structure of the secondary units. However it seems certain that the large sub units are linked by covalent bonds and the small sub-units by non-covalent bonds.

The classification of wheat grain proteins has been revised in order to reflect their biochemical properties and genetic origins. The name 'prolamin' is reserved solely for the storage proteins whose polypeptides are soluble – but not necessarily fully extractable – in alcohol solutions (65–75% ethanol) with or without the use of a disulphide bridge and acetic acid reducing agent. Prolamins have been separated into three groups:

- Prolamins rich in sulphur.
- Prolamins low in sulphur.
- Prolamins with a high molecular mass.

The corresponding old names are given in Table 4.3.

In **sulphur-rich prolamins** we can differentiate α, β and γ-gliadins and glutenin secondary units with a low molecular mass. Gliadins are monomeric proteins but they can combine by means of non-covalent bonds, whereas the glutenins with low molecular mass (LMM) display very strong aggregating behaviour and form polymers with the glutenin secondary units of high molecular mass (HMM) through the formation of intercatenary disulphide bridges.

Monomers of sulphur-rich prolamins have isoelectric points of between pH 5 and pH 9. Their molecular mass is between 32 000 and 40 000 Da. In addition to the amino acid composition characteristics already referred to

Table 4.3 Old and new names of prolamins

Old name		New name
Gliadins	ω	Prolamins low in sulphur
Gliadins	α, β or γ	Prolamins rich in sulphur
Gliadins	LMM	Prolamins rich in sulphur
Gliadins	HMM	Prolamins with a high molecular mass

above, these proteins contain 2–3% cysteine residues which are generally bound in the disulphide bridges. The conformation of the monomeric prolamins (types α and γ) depends closely on the composition of the solvent medium. This can change from a state of random coil structure to that of a compact globular protein and even form aggregates of a fibrous type. These changes of conformation are accompanied by a variation in the surface hydrophobicity of the molecules.

Prolamins that are **low in sulphur** have a molecular mass of between 50 000 and 75 000 Da. As far as their amino acid composition is concerned, they are lacking in cysteine and their methionine content is lower than 0.1%; they have fewer alkaline amino acids than the other prolamins but on the other hand contain more glutamine, proline and phenylalanine.

Prolamins with a **high molecular mass** have a different amino acid composition from that of sulphur-rich prolamins because of their higher glycine and lower proline content. Their molecular mass is between 90 000 and 140 000 Da. As all the cysteine residues are located in the N and C-terminal regions, they are undoubtedly significantly involved in the properties of association of these secondary units which are essential in the visco-elastic properties of hydrated gluten.

4.1.3 Functional properties of plant protein substances

4.1.3.1 Proteins from legumes

Proteins from soya and oil-producing seeds

The extraction processes are described in Chapter 3. There are numerous different forms of protein derivatives of soya:

- Whole or oil-free flours, raw or cooked.
- Concentrates obtained from flours with a low PDI (Protein Dispersibility Index) from which the soluble carbohydrates have been eliminated, in particular stachyose and raffinose.
- Isolates obtained from flakes of oil-free soya which have the advantage of containing proteins in a virtually pure state with a neutral flavour.

The **solubility** pH curves vary considerably depending on the protein substances under consideration (Fig. 4.1). In general, legume proteins have minimum solubility at pHs of between 4 and 5 and two maxima, one at acid pH 1.5 and the other at alkaline pH 9. The fact that the solubility curve for rape has few variations is related to the presence of two protein groups, one acid and the other basic. In addition, protein solubility depends on the thermal process undergone, the pH and the ionic strength. Predictably, the more severe the thermal process used for drying, the poorer the solubility and dispersibility.

Solubility can be improved by using reducing agents capable of

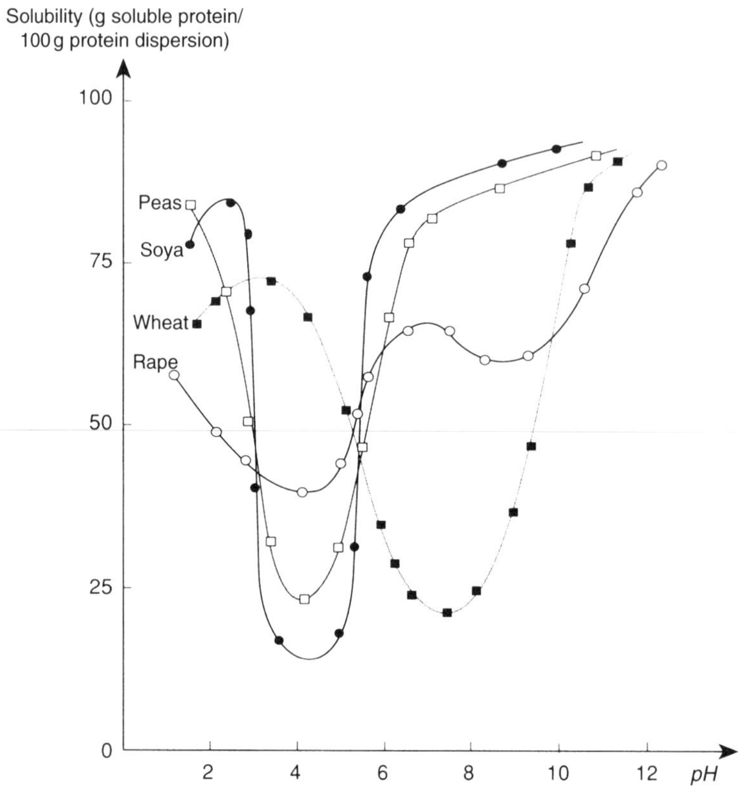

Fig. 4.1 Solubility of different plant proteins depending on pH.

breaking the disulphide bridges existing either in the native protein or formed in particular at the time of acid precipitation. The solubility of a protein can also be improved by using controlled enzymatic processes which reduce the length of the polypeptidic chains, thus encouraging hydration and solubility.

Water absorption and holding capacity is a property linked to the previous one. Soya proteins are very hydrophilic – certain concentrates absorb up to 10 times their weight in water – so they allow the water content of the products in which they are incorporated to be increased, and losses on cooking to be reduced, thus improving yield. By virtue of their affinity with water, soya proteins also compete for water in the first instance with starch, whose gelatinisation they delay, and secondly with gluten proteins, limiting the phenomena of shrinkage due to the formation of an over-rigid glutinous network.

Any parameter affecting solubility acts on the **viscosity** and **formation of gel** since these parameters (pH, temperature, salts) influence the ionisation of the polypeptide chains and therefore their capacity to interact with each other. If they are not heated to the required temperature, no pro-gel is formed but increased viscosity can be observed, mainly because of the unfolding of the polypeptidic chains and the increase in the number of interactions. These properties are very advantageous in the search for improvements in the texture of products such as sauces, ready-made meals and cooked meats.

Soya proteins are also useful in the **stabilisation of emulsions** because they accumulate at the oil–water interfaces, and therefore limit fat globule coalescence and rupture of emulsions. This stabilising property associated with the gelation properties of proteins is useful in emulsion-based products which have to undergo cooking or sterilisation (sauces, creams, stuffings, etc.). In the case of whole flours, their emulsifying power is further increased by soya lipids which are rich in natural phosphatides.

By mechanisms comparable to that of emulsification, the soluble soya proteins are concentrated at the air–water interfaces in order to encourage the **formation of foams** which are stable and can be coagulated by heat. As a result of their smaller sized polypeptide chains, proteins that have been partially hydrolysed using chemicals or enzymes are more suitable for this type of application. In addition, surface protein coagulation results in the formation of films which act as impermeable barriers which first of all limit the absorption of lipids, when deep-frying, for example, and secondly limit the loss of water and exudates.

Finally, lipoxygenase is present in some raw materials: soya flour contains four or five times more than bean flour. However, this enzyme is actively involved in bread-making where it makes dough easier to work and increases the uniformity of end-products. It intervenes in the oxidation phenomena of the non-esterified and polyunsaturated fatty acids contained in flour, thus promoting the aggregation of gluten proteins in the presence of water and the oxidation of carotenoid pigments, resulting in bleaching. The mechanism is summarised in Fig. 4.2.

To sum up, the good functional properties of these proteins from oil-producing seeds explain their extremely varied uses: introduction in meat products, meat substitutes after texturisation, etc. In addition one of the major advantages of soya bean lies in its protein amino acid composition. Apart from a slight deficiency of sulphurous amino acids, it is well balanced (Table 4.4). This composition is totally complementary with that of the cereals, particularly with regard to the lysine content.

Moreover, a great deal of work has been carried out in order to define the hypolipidemic effects of plant proteins accurately in comparison with proteins of animal origin (see Section 4.1.4.1). These studies have demonstrated the relationship between food proteins and their content of certain amino acids (in particular the Lys/Arg ratio), hormones regulating glycaemia (insulin and glucagon), apo-proteins and the metabolism of cholesterol, especially the activity of the receptors. It should be borne in mind that part of the nitrogen in the soya seed, from 3 to 7% of the total nitrogen, is

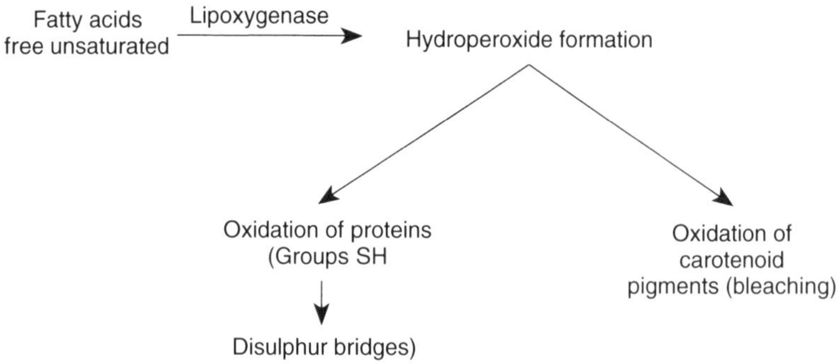

Fig. 4.2

Table 4.4 Comparison between the essential amino acid content of protein preparations of soya and two reference proteins (g/16 g nitrogen)

	Soya			Wheat flour	Whole egg
	Flour	Concentrate	Isolate		
Lys	6.2–6.9	5.1–6.8	4.9–6.6	1.9	7.2
Trp	1.2–1.4	0.9–1.5	0.7–1.4	0.8	1.5
Phe	4.9–5.4	5.1–5.6	4.9–5.5	5.5	6.3
Met	1.3–1.6	0.9–1.4	0.9–1.3	3.0	4.1
Thr	3.9–4.3	3.7–4.4	2.9–4.0	2.7	4.9
Leu	7.5–7.9	7.8–8.5	6.9–8.2	9.1	9.2
Ile	4.6–5.6	4.5–5.3	3.9–5.0	4.4	8.0
Val	4.8–5.4	4.7–5.7	4.0–5.2	5.0	7.3

in a non-protein form. This includes free amino acids, peptides, nucleic acids and phospholipid nitrogen.

Bean flours and concentrates and other non-oil-producing protein-producing plants

The influence of the variety on composition seems more marked in these types of legumes than in groundnuts and soya. In comparison with soya cakes, beans and peas contain approximately half the amount of proteins. Lupin has an intermediate content (approximately 30–35% nitrogenous material) and is fairly rich in lipids.

The **solubility** curve for pea protein concentrates shows minimum solubility for a pH value close to 4 and two maxima, one at acid pH 1.5 and the other at very alkaline pH (Fig. 4.1). This significant concentration within pH is characteristic of this family of proteins. For slight salt variations ($\mu < 0.5$), the solubility of the proteins increases exponentially with an increase in ionic strength. The extent of the phenomenon depends on the state of ionisation of the salts. So, at pH 9 and ionic strength of 0.1 obtained with $CaCl_2$, only 33–45% of the pea proteins are soluble whereas under the same conditions, but in the presence of sodium chloride, 95% are soluble.

Moreover, numerous studies stress the importance of the drying conditions on the water-holding capacity of the proteins. Thus, pea isolates dried over a drum dryer have a better water absorption capacity than spray-dried products. Likewise, in the case of the protein concentrates of lucerne, the water-holding capacity is zero when the temperature on leaving the pulveriser is 85 °C, but goes up to 42.8% when this temperature is 140 °C. This increase in water absorption capacity can be explained by an increase in the quantity of water retained by capillarity in the network formed by the proteins rendered insoluble by the treatment.

It is therefore essential when intending to use a plant protein in a given foodstuff to first evaluate its properties under environmental conditions which resemble those of the foodstuff as closely as possible.

The **viscosity** of protein solutions decreases slightly in the presence of salts but increases with temperature following denaturation. This phenomenon varies according to the origin of these plant proteins.

As the **swelling** effect of proteins is associated with the unwinding of the peptide chain, it is therefore normal for the proteins to present minimum solubility at the isoelectric point, the point at which the chains fold up to prevent water from penetrating the molecule. The curves for variation in swelling show more or less the same profiles as those for solubility. However, it should be borne in mind that protein swelling capacity can vary by significant amounts when there are only slight changes in pH. For example, the water absorption capacity of a lupin-based mixture goes from 7 g to over 18 g when the pH increases from 5.68 to 6.0.

Moreover, the temperature at the onset of **coagulation** varies according to the pH and the highest gel strengths are obtained for pH values close to 7.

4.1.3.2 Cereal proteins

Wheat is undoubtedly the earliest cereal to have been used for producing IFPs. Wheat flour is the earliest example of this. In addition, the principal example of protein extraction from cereals, practised on an industrial scale, is the extraction of gluten from wheat flour.

Composition and structure of gluten

Gluten consists of more or less equal proportions of gliadins and glutenins. Low proportions of albumins and globulins are present. The functional properties of gluten are the result of the characteristics of these components and the interactions that occur between them. Glutenin macromolecules, which develop a large surface area over which numerous non-covalent bonds can be established with gliadins or other glutenin molecules, fulfil an aggregation and elastic role in this complex. During kneading, covalent intermolecular bonds – essentially disulphide bonds – form an entanglement of molecules which are also aggregated by means of non-covalent bonds. This structure is not strictly accurate, since recent studies have shown the presence of lipid bodies inserted in the protein network. The gluten may behave as a micro-emulsion stabilised by the protein network (see Chapter 3).

Different types of gluten

Gluten is the name given to the insoluble protein fraction contained in cereals (wheat, maize, barley, etc.). However, **wheat** gluten can be differentiated by its characteristic of forming a continuous elastic network, impermeable to gases, after hydration. This characteristic is the basis of bread-making technology. Wheat gluten is a food product obtained after wet extraction of certain non-protein constituents of wheat or wheat flour (starches, other carbohydrates) so as to obtain an 80% or higher protein content, on a dry weight basis.

Vital gluten is characterised by its strong visco-elastic properties when hydrated. **Devitalised** gluten is characterised by the loss of its visco-elastic properties when hydrated, as a result of protein denaturation. Finally, the term **green** gluten is applied to gluten in a hydrated form obtained in the course of a manufacturing process.

Industrial processes for manufacturing gluten

Industrial processes for manufacturing wheat flour are of course derived from those that enable wheat starch to be separated. The processes can be classified in several ways, according to the following:

- The raw material used (wheat or flour).
- The method of separating the starch and gluten (mechanical or chemical processes).
- The quality of the gluten obtained (protein residue, vital gluten or devitalised gluten).

The standard extraction technology for wheat exploits the gluten-forming capacity of these proteins. The dough formed from flour and washed by water results in gluten (Fig. 4.3). The various processes differ mainly in terms of the water/flour ratio used (see Chapter 3).

The protein content of the gluten is generally between 72 and 80% and varies according to the variety of wheat used. Gluten is rarely marketed in this hydrated form, except in some Far Eastern countries, including Japan where it may appear in the form of frozen sheets, or in a textured form. A drying operation is therefore employed in order to obtain a powdered product which is easy to store and handle. After drying it is absolutely essential for the gluten to have retained the maximum characteristics of 'vitality', that is to say water absorption and visco-elasticity, properties that distinguish it from the other proteins of plant origin.

While it is accepted that the quality of the wheat and flours used has a

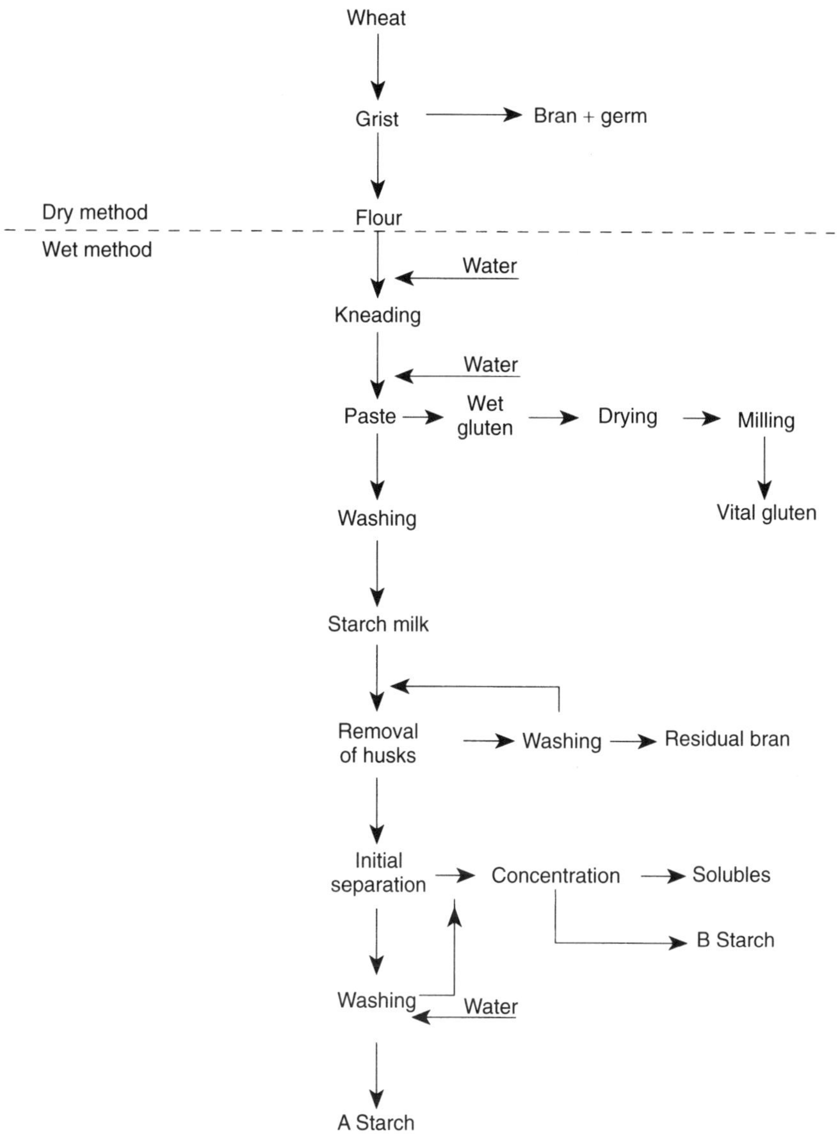

Fig. 4.3 Gluten extraction (Martin process).

considerable effect on the quality of gluten obtained, it is also acknowledged that the drying conditions constitute the most essential factor influencing its properties. In fact denaturation, that is to say the loss water absorption capacity or visco-elasticity, can intervene very rapidly once the product is subjected to temperatures over 60 °C.

Although the technique of freeze drying is acknowledged as being the one that least alters these properties, it is expensive to implement on an industrial scale. Other current drying techniques allow very satisfactory results to be obtained, provided that they are properly implemented and controlled. The dryers currently used the most are of a 'flash' type, using hot air, and internal or spray recycling.

Composition and properties

Wheat glutens produced on an industrial scale world-wide take the form of powders whose **colour** varies from cream to dark brown. The qualities of the wheat or flour, and also the drying conditions, mainly affect the colour of the gluten. **Moisture** content varies from 6% to 10%. The qualities of the gluten alter rapidly during storage if its moisture content is over 10%.

The **protein content** is normally determined by multiplying the Kjeldahl nitrogen by the factor 5.7 and depends essentially on the extraction process and the washing conditions applied. This protein content is not necessarily a criterion that can adequately define the quality of the gluten. Glutens with a higher protein content (82%) are preferred for incorporation in flours sold with a guaranteed protein content, or for manufacturing gluten biscuits, for example. However, at the same rate of incorporation in a basic flour, for bread-making the improvement caused by gluten with 75% or even 65% protein can be better than that of glutens with 80% or more proteins.

The **particle size range** of commercial gluten is very variable and can reach a size of 3 mm and over. In order to be incorporated in flours its granularity must be at least equal to that of the flour. Finer glutens hydrate more rapidly. A good quality vital gluten must be able to absorb approximately twice its weight in water when it is fully re-hydrated. The **water absorption** of the glutens remains the criterion that best allows the vitality of a gluten to be evaluated: a devitalised gluten can no longer absorb water. We also know that the best wheats, the best-extracted flours and the most sophisticated manufacturing processes will never produce a good vital gluten if ideal drying conditions have not been used.

Gluten conversion

Chemical and enzymatic hydrolysis

This operation firstly allows the solubility of the gluten to be improved by reducing the size of the polypeptide chains by increasing the number of C and N terminal hydrophilic extremities and then by hydrolysis, in an acid medium, which causes partial deamidation of the glutamine residues. These deamidation reactions lead to the suppression of the hydrogen bonds that are established between the NH^- and CO^- groups of the residue. They reduce surface tension and significantly increase surface hydrophobicity which correlates in a linear fashion with emulsifying capacity, giving the gluten which has been treated in this way, emulsifying and foaming properties. In fact it is solubility which is mostly increased.

For optimum swelling capacity, the rate of deamidation must not exceed 40%. Hydrolysis in an alkaline medium is not carried out, as, unlike hydrolysis in an acid medium, it gives the product a bitter taste. The rate of deamidation which produces optimum dispersibility is low: 5–20%. No more than 5% of 'starch-removed gluten' must be incorporated in fruit juices, as beyond this results in an unpleasant taste. Gluten, whether alone or mixed with other plant proteins (soya, maize) subjected to acid hydrolysis, becomes a flavour enhancer.

Chemical modifications

A chemical hydrolysate of gluten, whose small peptides are removed prior to being mixed with ethylene oxide or ethylene-imine, produces a clear liquid, soluble in water and in various organic solvents. **Epoxidised gluten** gives acrylic resins adhesive and plastic properties; it permits the production of films which can be used as special packaging (Fig. 4.4). In an alkaline medium at ordinary temperatures, gluten and acrylonitrile can be made to react. The **cyanoethyl** gluten obtained is used to lubricate threads in the textile industry. Among the chemical modifications of gluten used in foodstuffs are the following:

- **Phosphorylation** of gluten which increases its water-holding capacity.
- **Acylation** which increases the solubility of gluten in water between pH 7 and pH 9 but reduces it below pH 4.
- **Complexing** of gluten by chelating agents such as sodium or calcium phytates, phosphates and sodium or potassium citrates, results in a gluten which hydrates better and more rapidly.
- Gluten **reduction** allows it to be texturised and made into a crabmeat substitute! The reducing agent can be a sulphite or disulphite of sodium or potassium. A foaming agent is incorporated as well as other ingredients (starch, casein, soya flour).

Gluten emulsification

In order to eliminate the taste and smell of gluten, and to be able to incorporate it more easily in the mix, different emulsification techniques have been tried. Some of these are as follows:

- Coating gluten with certain non-ionic hydrophilic lipids (monoglycerides of lactic esters of fatty acids, polyoxyethylene stearate, monoglyceridyl citrate).
- Dispersion of the fine gluten particles in an inert diluent and complex formation by a complexing agent such as palmitoyl chloride, phosphatidylethanolamine or sodium stearoyl-2-lactylate (see Chapter 13).
- Mixing gluten with lecithin.
- Enzymatic transformation of diglycerophospholipids naturally contained in gluten in acylated monoglycerophospholipids.

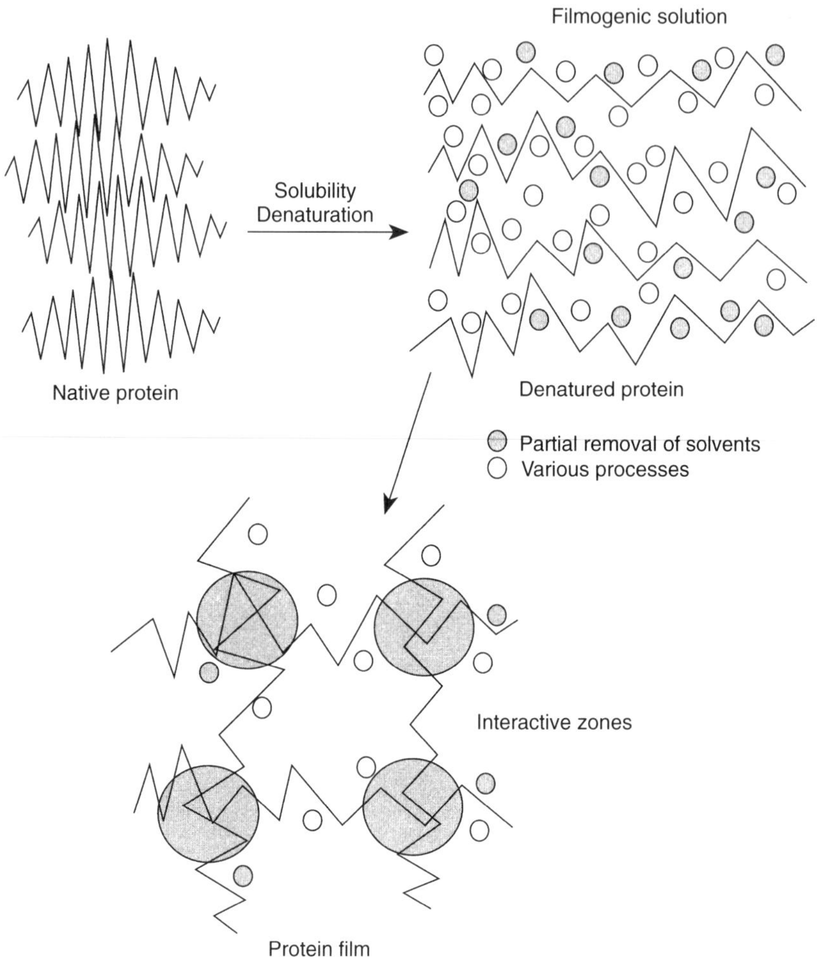

Fig. 4.4 The formation of a protein film.

4.1.3.3 Leaf proteins

The abundance and quality of leaf proteins across plant species is remarkable. The content in total nitrogenous substances matter (TNS) varies between 10 and 60% dry matter (DM) (coriander or amaranth). Lucerne, which is the only plant that has been used for industrial extraction, can contain between 12 and 25% TNS and produce 1500 kg to 2500 kg per hectare per year, in three or four cuts in a European climate.

The extractability of these proteins varies considerably according to species, promoted by the effects of favourable constituents such as tannins, and slowed down – or even sometimes prevented – in the presence of a high content of pectic substances. Moreover, unlike seed proteins, the proteins contained in the green parts of plants are closely associated with a number of non-nutritional or even anti-nutritional elements.

In this respect, lucerne proteins are neither the most abundant per hectare nor the most easily recovered, and they contain anti-nutritional elements. However, the complementary nature of the extraction techniques with dehydration and the agronomic qualities of lucerne explain why it has been the first source to be exploited and probably the best in temperate regions. In an irrigated tropical zone other plants have not been ruled out as possibly being more suitable for the production of leaf proteins. The use of certain aquatic plants (Jacinthe, Hypomea) with a high protein content is even being considered. Before looking at the technological aspect, with particular reference to lucerne, we will provide a brief description of leaf proteins in general, and the other compounds that accompany them.

Composition of the protein fraction of the leaves

The proteins are essentially enzymes associated with physiological activity, in particular photosynthesis. These proteins are frequently divided into chloroplastic and cytoplasmic proteins, but obviously this difference does not coincide with the criteria of ease of solubility or precipitation which govern recovery processes. Chloroplast proteins represent approximately two-thirds of the mass of leaf proteins and are themselves split half and half between insoluble or poorly soluble membrane proteins and soluble chloroplast proteins.

Other proteins, finally, are located in the cytosol, the mitochondria and other cellular elements.

The **membrane proteins** which are very numerous, are essentially involved in photosynthesis and are located in the thylakoides. The most important is light harvesting chlorophyl (LHCP). This antenna which collects photons contains approximately 50% of membrane proteins and 50% chlorophyll as well as carotene and lutein.

The **soluble chloroplast proteins** contain the most widely distributed protein in the world: ribulosediphosphate carboxylase (Fraction I). Fraction II which exists in smaller quantities, and is also soluble, precipitated at low pH (<4.5) cannot be crystallised like Fraction I. It is a mixture of chloroplast and cytoplasm proteins having a molecular mass of between 10000 and 300000 Da, with two major constituents of a large molecular size and a large number of smaller molecules.

Soluble non-chloroplast proteins do not represent a significant quantity and essentially consist of numerous proteins having an enzymatic function. Alongside these abundant proteins whose balance of essential amino acids is excellent, leaves contain a very diverse range of anti-nutritional elements, which vary in their toxicity, their nature, their function and their binding capacity with the proteins, and these also vary from one plant to another (see Section 4.1.4.2).

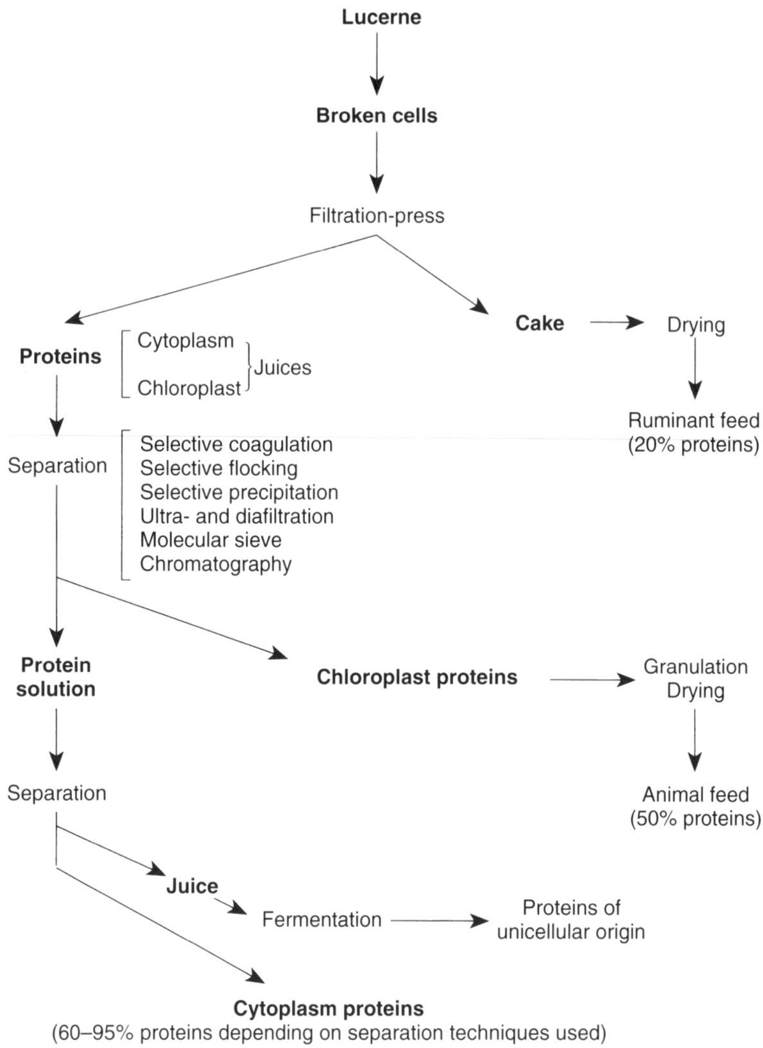

Fig. 4.5 Extraction of leaf proteins.

Extraction and fractionation

Figure 4.5 represents the general flow chart for the extraction of leaf proteins. After filter-pressing, acid precipitation or thermal coagulation, the green concentrate is divided into two fractions: chloroplast proteins and cytoplast proteins. These preparations are used in animal feed in particular, as they are significant sources of xanthophyllian pigments for eating and laying hens and sources of β-carotene for cows or sheep deficient in vitamin A during gestation periods.

Using leaf proteins for humans would have even more nutritional and economic value and with this in mind, work has branched out in two directions:

- Studying the production of purified isolates and their functional properties.
- Direct introduction of leaf proteins into food, a preferred method of use for many developing countries.

This family of proteins has very good nutritional and functional properties (solubility for certain fractions, emulsifying power, foaming capacity). However, unlike other food proteins of animal or plant origin which have a storage function, the very function of leaf proteins associates them with a number of other organic elements which are either non-functional or undesirable and from which they must be separated.

4.1.4 Biological properties of plant protein substances

4.1.4.1 Nutritional value

PPSs provide essential amino acids which complement those from other ingredients making up food products. Knowledge of the composition in amino acids – briefly touched on in previous paragraphs – must be accompanied by knowledge of protein quality. This is based on the coefficient of digestibility (COD), the biological value (BV), and the protein efficiency coefficient (PEC). So, the CODs of plant protein substances lie between 80 (rape) and 90 (heated soya), whereas the PECs vary from 0.7 (lupin) to 2.6 (rape). The techniques for producing concentrates or isolates can result in proteins that have a different content of amino acids from that of the raw material. It can also result in the elimination or concentration of enzyme inhibitors or toxic factors in the end-product.

In the case of concentrates, carbohydrates can be dissolved in water in a slightly acid medium (pH 4.5) or in a water–alcohol solution. The alcohol also allows pigments and the substances responsible for bitterness to be dissolved. As far as the isolates are concerned, they contain a very low level of carbohydrates, including fermentable carbohydrates. In the case of soya, for example, the carbohydrate concentration is 0.2% compared with 16–18% in concentrates and 30% in the initial flour. The lipid content is also negligible (0.5% compared with 1.5% in concentrates, 2.5% in defatted flour and 18% in the seed).

Lipid extraction influences the nutritional value of the product insofar as their presence, when associated with that of lipoxygenases, can result in phenomena of oxidation and formation of hydroperoxides of fatty acids, responsible for undesirable flavours. Inactivation, by subjecting these enzymes to thermal processing, also improves the organoleptic qualities of the products.

The effects of alkaline treatments on the nutritional value of proteins depend on their severity: pH, duration, temperature (Table 4.5). Modifications in nutritional value can be attributed either to the formation of covalent bonds which reduces the action of the digestive enzymes, bringing about a reduction in digestibility and unavailability of certain amino acids (such as alkaline amino acids), which are blocked by reacting with a carbohydrate during the Maillard reaction, or to the disappearance and isomerisation of essential amino acids which are primarily responsible for the chemical index and biological value.

The thermal treatments involved in the course of the processes for obtaining concentrates and isolates are used to reduce the moisture content of initial raw materials, or to remove the solvents for extracting lipids or residual alcohol from concentrates, or to dry the products obtained, or finally to sterilise them. These treatments, depending on the operating conditions used, also result in reduced digestibility and availability of amino acids through the formation of inter- or intramolecular bonds, which are difficult for proteolytic enzymes to attack. In fact the intermolecular bonds can result from aldehyde or ketone formation on a basic amino acid causing specific inhibition of the trypsic attack owing to the secondary amino functions being blocked and a drop in nitrogenous digestibility, and a reduction in the nutritional efficiency of these unavailable amino acids. When thermal processing is applied using dry heat in the presence of oxidisable lipids and especially in the presence of reducing carbohydrates, more damage is caused. On the other hand, when it is applied in the absence of reducing carbohydrates and in a wet environment, there is an improvement in organoleptic qualities and digestibility.

The **hypolipaemiating and hypocholesterolaemiating character** of the plant protein substances has mainly been studied using soya ingested in forms varying from flour made from whole grain (40% protein) to protein isolate (90% minimum protein) under varied experimental conditions. Studies in humans have revealed the existence of an effect which becomes more perceptible as the cholesterolaemia rises.

In broad outline it is possible to distinguish three levels of hypocholesterolaemic effects: effects due to the protein, those due to the non-protein part of the PPSs (compounds accompanying the protein) and those due to the diets into which the PPSs are introduced.

Numerous protein accompanying compounds can be involved. We can list, without going into greater detail, soluble fibres, lecithin, the ratio of saturated fatty acids to polyunsaturated fatty acids, saponin, the zinc/copper ratio and phytosterols. The factors linked to diet composition are the percentage of proteins and lipids, whether or not food cholesterol and choline are present, and their amount.

Two broad types of mechanism are involved in the way proteins affect cholesterolaemia. Their effect can be pre-absorbent or post-absorbent depending on whether it occurs at an intestinal stage or at a metabolic stage. The effect of proteins on cholesterolaemia seems, at the very least, complex:

Table 4.5 Effects of industrial processes on the nutritional quality of plant protein substances

Industrial processes	Effect	
Heat treatment		
≤100 °C	Denaturation, coagulation Inactivation of enzymes (lipases, proteases, etc.)	Digestion ↗ Organoleptic quality ↗
100–140 °C	Maillard reaction in the presence of reducing carbohydrates, drop in lysine availability Formation of intra- or intermolecular covalent bonds Destruction of amino acids: cystine, cysteine	Nutritional value ↘ Digestibility ↘
>140 °C	Destruction of amino acids: tryptophan, glutamic acid Isomerisation of amino acid residues (L/D) Formation of isopeptide covalent bonds (lysinoalanine)	Toxic by-products ↗ Nutritional value ↘ Digestibility ↘
Alkaline process		
	Increased protein solubility Destruction and isomerisation of amino acid residues (Arg/ornithine, urea/NH_3, cysteine/dehydroalanine) Formation of covalent bonds (lysinoalanine)	Nutritional value ↘ Digestibility ↘
Interaction proteins/other components		
	Reaction with lipids and their by-products of oxidation Reaction with reducing carbohydrates: Maillard reaction Reaction with plant polyphenols	Nutritional value ↘

↗ parameter improved; ↘ parameter drops.

the use of a mixture of amino acids or supplements in various kinds of amino acid does not always produce reliable results, the sequence of ingesting nutrients in a diet can play a part, and not all plant proteins have the same capacity for reducing cholesterolaemia. Certain ones, such as those from rice or lucerne, have no effect in comparison with casein.

4.1.4.2 Anti-nutritional factors of raw materials of the PPSs

Choosing species that are nutritionally useful from the choices offered by a given ecosystem forms part of human feeding behaviour. Alongside the nutrients that are essential for us, the protein raw materials of potential plant foods contain variable amounts of substances without any nutritional value, which in some cases even reduce this value. Some of these substances are even considered to be dangerous to health. These substances cause many health problems: reduction of food efficiency, inhibited growth, pancreatic hypertrophy, impairment of the intestinal mucous membrane, hypoglycaemia, goitrogenic effects and allergic reactions. These substances can be classified into two groups according to the way they function (Table 4.6).

The first group brings together compounds that, if not compensated by an additional contribution of the nutriment or nutriments concerned, have a toxic effect; we call this the group of **anti-nutritional substances**. The second group brings together compounds whose harmful effects cannot be compensated for. These compounds exercise a purely toxic effect on the organism. They react either as stimulants or inhibitors of enzymes, hormones or amino acids.

Some of these anti-nutritional factors have a **direct action** on the palatability of the food:

- Certain saponins consisting of a sapogenin such as aglycone have a bitter taste; they act by forming non-covalent complexes, by means of hydrophobic interactions, and they can bind on the taste receptor sites.
- Various tannins with an astringent taste which cause a sensation of dryness in the mucous membrane of the mouth. This action is a result of denaturation of the glycoproteins of salivary secretions or from hyposialorrhoea (reduction of salivary secretion) associated with constriction of the salivary ducts. At low doses tannins improve the palatability of foods, but at high doses they reduce nutrient uptake.
- Gossypol, a toxic polyphenolic compound mainly found in cotton, which reduces nutrient uptake by means of a process that has not yet been precisely identified. However, we know that it causes liver damage (oedema, haemorrhages). Its toxicity is significantly lowered after the addition of iron sulphate.

Other factors have an **indirect action** associated with nutrient uptake. Some examples of this are as follows:

- Lipoxygenases – these enzymes are present in soya and are believed to be responsible for undesirable smells (rancid smells, bean smells) by

Table 4.6 The various anti-nutritional factors present in raw materials from plant protein substances

Group/Name	Plant source	Principal action	Detoxification
I Anti-nutritional substances			
Phytates	Legumes	Chelation of mineral elements	Solubilisation (water, acid)
Lipoxygenases	Legumes	Destruction of vitamin A	Thermal processing
Ascorbic acid oxidase	Peas	Oxidation of vitamin C	Thermal processing
α-Galactosides	Legumes	Agents of flatulence	Solubilisation (water, water–alcohol)
II Toxic substances			
Trypsic inhibitors	Beans, peas, lucerne, soya	Inhibition of trypsin activity	Thermal processing
Alkaloids (quinolizidin)	Lupin	Neurological problems, teratogenic action	Solubility (water, apolar solvent, acid)
3N oxalyl-1,2,3-diaminopropionic acid	*Lathyrus sativus* (vetch)	Neurological problems (lathyrism), problems with bone metabolism	
Cyanogenetic glycoside	Lima bean	Neurological problems (goitrogenic)	Soaking, prolonged cooking
Tannins (flavonoid)	Bean	Change in taste, protein complexation, enzyme inhibition, anti-vitamin K action, action on muscle mobility	
Gossypol	Cotton	Oedema, haemorrhages	Addition of ferrous sulphate
Miscellaneous polyphenols	Rape, field bean, lucerne, soya, sunflower		Solubility (water, water–alcohol)
Glucosinolates	Rape	Altered taste, goitrogenic and carcinogenic action	Solubilisation (water, water–alcohol)
Vicine and convicine	Beans	Haemolytic anaemia	Thermal processing
Haemagglutinins or lectins	Beans, peas, lucerne, soya	Problems with absorption, agglutinating agents in red cells	Thermal processing
Saponin (glycosides)	Lucerne, peas, soya	Bitterness, haemolysis of red cells, complexation of food proteins	Solubilisation (water) Thermal processing

forming hydroperoxides of fatty acid. These enzymes can be partially deactivated by cooking or blanching (see Section 4.1.3.1).

- Glucosinolates – almost 80 substances belonging to this group have been identified. They are all characterised by the presence of a molecule of glucose, a sulphate group, nitrogen and sulphur atoms. The term 'glucosinolate' proposed for this group of substances emphasises the presence of glucose, whereas the suffix '-ate' identifies a compound of an anionic nature. In rape, the molecule has an isothiocyanate group and can result in the formation of various thioglucosides. When the seeds are crushed, the glucosinolates are brought into contact with the myrosinase and converted into glucose, sulphate and isothiocyanate. A reduction in the acceptability of the foodstuffs is observed, in addition to stimulation of the secretion of the hormone TSH (thyroid-stimulating hormone). The result of this is a disturbance in the secretion of thyroxin (thyroidal hypertrophy) and, as a consequence, of iodine metabolism.
- The generic term alkaloid brings together extremely varied compounds. In lupins there is a family of alkaloids from the alizarin bordeaux group. They are responsible for the bitterness and unpalatability of this legume.
- Molecules such as phytic acid are considered to be complexing agents of mineral salts. They are chelating agents which recombine with the di- or trivalent cations (Ca, Mg, Fe, Zn, Mn) and can cause a deficiency in these elements.
- The α-galactosides: flatus factors which produce uncomfortable symptoms. They are also known as fermentable carbohydrates, characterised by an α-1–6-galactoside bond which cannot be hydrolysed because of the absence of α-galactosidases in the human intestine. These carbohydrates based on galactose, glucose and fructose are either triholoside, raffinose, or a tetraholoside, stachyose, or a pentaholoside, verbascose. These holosides are, on the other hand, metabolised by the flora of the intestine which releases methane and carbon gas, the source of intestinal gases, as it ferments them. The pH is also reduced and diarrhoea can occur. They are not very sensitive to heat treatment. However, pre-soaking at a slightly acid pH, between 45 °C and 65 °C can reduce the risks of flatulence.
- Other anti-nutritional factors are considered to be responsible for slowing down growth or for metabolic disturbances in animals. These are as follows: inhibitors of digestive proteases (trypsin, chymotrypsin) and haemagglutinins or lectins. They are found in practically all legumes.

Protease inhibitors are proteins of MM between 5 and 20 kDa. They bind irreversibly to their substrates and deactivate them, which then results in organ hypertrophy. By deactivating trypsin the Kunitz inhibitor produces an increase in the secretion of cholecystokinin (CCK) which is one of the main anorexigenic hormones. The exocrine pancreas, made hyperactive by the CCK, increases in size, while nutrient uptake reduces. Moreover the proteases complexed with these molecules are not digested; the loss of endogenous nitrogen is then considerable and results overall in a very significant methionine deficiency.

Haemagglutinins or lectins are proteins that bind specifically to erythrocytes. After ingestion they bind on the membranes of enterocytes and alter some of their functions. If consumed on a regular basis they can result in delayed growth, colitis or anaemia. Their MM varies between 60 and 110 kDa. They can easily be inactivated by heat.

Vicine which is a pyrimidine glycoside and its derivatives (divicine and convicine) are substances present in beans, which are responsible for favism. Favism is a haemolytic anaemia of a hereditary type which can take acute forms. It only occurs in individuals suffering from congenital enzymopathy: a deficiency in glucose-6-phosphate dehydrogenase, an enzyme that controls the oxidation of carbohydrates especially in erythrocytes by means of pentose phosphates.

There is a goitrogenic factor in soya, different from the thioglycosides in plants from the cabbage family (brassicae). This soya anti-thyroidal has a peptide structure. It increases requirements for iodine, causing hypothyroidism (goitre). However, thermal processing destroys this anti-thyroidal.

Cyanogenetic glycosides which release cyanhydric acid through hydrolysis can also be found in certain legumes, in particular certain varieties of lima bean. Soaking and prolonged cooking hydrolyse these glycosides and eliminate the cyanhydric acid.

The amino acid present in legumes of the *Lathyrus sativus* species causes neurolathyrism which manifests itself in neurological problems, resulting in paralysis of the extremities. *Lathyrus* seeds are normally wild and are only eaten during famine periods in countries in the Far East, India and Pakistan.

Suitable treatments are necessary to deal with some of these antinutritional factors and can result in their elimination. Soaking and/or thermal processing of seeds is thought to be generally effective. In other cases, they can also be eliminated by germination or fermentation. With certain treatments there is the problem of losing some of the soluble proteins. This occurs, for example, when phytic acid is eliminated from various seeds or when lupin alkaloids are eliminated by soaking, resulting in the leaching of soluble proteins.

4.2 Plant oils and fats

4.2.1 Composition

Oils, which are widely used in foods, can be classified according to their composition:

- Oils rich in saturated fatty acids and in oleic acid: groundnut, olive.
- Oils rich in polyunsaturated acids: safflower, sunflower, soya, maize.
- Intermediate oils: rape.

Table 4.7 shows the average values (as percentages) of the main fatty acids contained in the fats of plants. This composition is important both from a nutritional and a functional point of view.

4.2.2 General principles of processing

The treatment of seeds that are rich in lipids will involve the following stages: cleaning, de-husking, crushing, rolling, cooking, pressing, extraction; that of seeds that are low in lipids, such as soya, will consist of the following operations: cleaning, drying, maturing, de-husking, rolling, pelletisation if necessary, extraction and drying.

In oil mills, generally speaking, two products are produced: first of all crude oil and, secondly, the solid part which is known as a 'cake'. The cake always contains a certain quantity of oil which will sometimes be deliberately increased to improve the value of these by-products. It also contains a certain amount of water which is fixed by law at 12%. The value of a cake depends upon its protein content which needs to be as high as possible, and by its cellulose content which needs to be as low as possible.

We will now briefly comment on the operations that have important repercussions on the nutritional qualities and functional properties of the end-products.

4.2.2.1 Drying

A moisture content of 8% is generally the maximum value allowed for most seeds. It is also essential for most grains. It is also essential for the moisture content of the defatted portion not to reach 15%.

Soya, for example, is dried to 10% and the dried seed is returned to a silo where it remains for one or two days. Without this time for maturing which allows the moisture content to equilibrate, the shells are difficult to separate on de-husking. With sunflowers and rape, the seed is dried to 5–6% moisture content before de-husking.

4.2.2.2 De-husking

This operation has two advantages:

1 To eliminate the substances of no value for animal feed.
2 To facilitate further processing.

There are numerous variants on this operation depending on the raw material involved. So, although all groundnuts are de-husked prior to pressing, de-husking of soya grains is effected according to the amounts of protein and oil contained in the seed in order to obtain a cake with 45–50% profat (protein + lipids).

Table 4.7 Average fatty acid composition and solidification temperature of the principal vegetable fats and oils

	Proportion of fatty acids (%)							Solidification temperature (°C)
	Lauric C12:0	Myristic C14:0	Palmitic C16:0	Stearic C18:0	Oleic C18:1	Linoleic C18:2	Linolenic C18:3	
Cocoa butter			24	35	39	2		22
Copra oil	44	18	10	4	6	2		14 to 22
Palm kernel oil	50	15	8	3	15	3		20 to 25
Palm oil		1–10	45	4	40	10		35 to 42
Olive oil			10–17		50–80	10		–6
Rape seed oil			3		15	15	9	–10
Canbra* oil					60	25	9	
Groundnut oil			10	3	60	22		3
Soya oil			11	3	25	55	6–9	–10 to –16
Sunflower oil			8	5	20	65	1	–17
Cotton oil		1	25	2	18	53		+12 to –13
Maize oil			13	2	30	55	1.5	–10 to –20

* Genetic modification of rape.

4.2.2.3 Crushing and rolling

Crushing and rolling treatments differ according to the seed involved. Crushing soya is carried out in grooved crushers, followed by rolling on smooth cylinders. This rolling operation is very important, as it will influence the success of the direct extraction of hexane. A temperature of 65 °C is necessary in order to reach the thermoplastic state essential for producing flakes that do not crumble. This temperature will also act as a heat source for the extraction which must be carried out at over 52 °C.

Rape is treated on roller crushers similar to those used for soya. As rape is a seed rich in oil (approximately 45%), rolling is effected at a maximum temperature of 40 °C in order to prevent the cylinders becoming clogged and choked up. The same equipment is used for rolling sunflowers by slackening the cylinders to produce flakes of 0.35 mm thickness.

4.2.2.4 Cooking

Not all seeds require cooking (e.g. copra) but when this is necessary, the moisture level is the critical point in the cooking cycle. It is determined according to the dry oil-free extract. The higher this is, the more moisture is required to cook the grain.

Heat treatment serves many purposes:

- Adjusting the moisture content of the seeds (between 3 and 5%).
- Increasing the plasticity of the seeds.
- Increasing the fluidity of the oil.
- Coagulating the protein fractions.
- Deactivating the enzymes (myrosinase).
- Destroying pathogenic micro-organisms (salmonella) and toxic substances.

4.2.2.5 Pressing

Hydraulic presses that supply pressure of 400–500 bar can be used to carry out this operation. The two important parameters for this operation are the texture of the grain and its oil content. We can take as the two extremes copra (very fibrous) and groundnut (very soft seed).

4.2.2.6 Extraction

Bearing in mind the energy used to increase pressure and the resulting wear and tear, it is preferable to increase extraction yield by dissolving the remaining oil in a solvent.

The moisture content of the extract is very important. Problems with sticking begin when the extract has a moisture content of over 9% of the oil-free base. The problem of the direct extraction of rape or sunflower grains with 9% moisture content for 45% oil is explained by this level of water content. It is therefore essential to dry the seed to 5% before processing.

The quantity of solvent to be used must also be optimised. Current practice uses a ratio of 1–1.25 l/kg of product.

The temperature of the solvent is also very important. It is advantageous to work at a temperature of just under 60 °C to improve product diffusion. Distillation is used to separate the dissolved oil from the hexane that has been used to extract it. This operation involves an evaporation unit which converts the liquid hexane into gaseous hexane and a condensation unit which makes the opposite phase transition.

4.2.2.7 Refining

Refining is applied both to pressed oils (virgin oil) and to extracted oils. It aims to produce a product that has a neutral flavour, is resistant to oxidation, suitable for its desired end-use and free from toxic or harmful substances.

Free fatty acids (FFAs) are naturally present in the seed but can also be produced by reactions of enzymatic hydrolysis. Enzymatic hydrolysis can only occur in the presence of water; therefore care must be taken to ensure that the crude oils are as anhydrous as possible. The water content of these oils must not exceed 0.2%.

Given that the resistance of fatty acids to oxygen in the air is low, even in very favourable conditions, the presence of FFAs in a fat can be compared with that of an oxidation catalyst. In addition the FFAs are also catalysts of hydrolysis, a reaction that produces new FFAs and also partial glycerides which are equally undesirable.

The FFAs are eliminated either chemically (neutralisation using soda) or physically if they are volatile (steam distillation). This is the most delicate and important operation involved in refining.

Phospholipids are polar compounds (pK = 9.5 for cephaline and 13.9 for lecithin) and their emulsifying and surface-active properties hinder refining operations so that they must be eliminated as the first stage of all the processes (see Chapter 13). The phospholipid content of crude oils depend on the method of extraction used and above all on the oil variety: although almost absent in copra palm oils they can represent 2% of soya oil.

Among the colourants found in vegetable oils, we can list the following:

- β-Carotene present in all vegetable oils (over 0.1% in palm oil).
- Chlorophyll present in appreciable quantities in olive oil and rape seed oil and to a lesser degree in soya and sunflower oil.
- Colourants resulting from oxidation, responsible for the brown colour of certain oils.

Crude vegetable oils also contain the following:

- Traces of paraffins with 11 to 36 atoms of C.
- Carbohydrates and glycolipids.
- Esters of fatty acids (20–28 C) and fatty alcohols with long chains (22–30 C) present in sunflower, safflower, maize and cotton seed oil.

Table 4.8 Refining operations and their effects on the minor components of vegetable oils

Operation	Conditions	Components removed
Demucilagination or boiling off	Processing with water at 70–80 °C + H_3PO_4 Centrifuging	'Mucilages', phosphatides, glycolipids, protein fractions
Neutralisation	Chemical method: addition of soda Physical method: steam distillation	Free fatty acids, residual phosphatides, degradation compounds of oxidative origin, gossypol, aflatoxin, organophosphorated pesticides
Bleaching	Adsorption of pigments on bleaching earth, of active carbon, special silicas or combinations of these substances	Carotenoid and chlorophyllian pigments, soaps, hydrocarbons
Dewaxing/ wintering	Dewaxing by filtration or centrifuging, fractional crystallisation	Waxes and substances insoluble at low temperatures
Deodorisation	Injection of dry vapour in oil maintained under vacuum at high temperature (200–230 °C for 90–180 min, or 240–250 °C for 30 min)	Free fatty acids, volatile compounds responsible for smell and taste, peroxides and products of degradation, sterols and tocopherols, organophosphorated pesticides

- Compounds responsible for smell and flavour, such as the sulphur compounds at the origins of the flavour and smell characteristic of the Cruciferae (rape) or ingredients supplying aldehyde and ketone functions resulting from oxidation. In this respect, the sulphur in rape oils due to the presence of thioglucosides in seeds must be eliminated prior to hydrogenation. As the chemical treatment of refining is ineffective, the only method remaining is steam processing.

Operations of chemical refining and their effects on the minor constituents of vegetable oils are summarised in Table 4.8. The different stages of this refining strategy are not always applied in the same way. Some operations are sometimes omitted, depending on the raw material and also on its quality.

4.2.3 Properties and applications

We will now review the main properties of fatty plant substances by looking at two cases:

- The manufacture of margarines.
- Use as heat-exchange fluids.

4.2.3.1 Manufacturing margarines

Margarine is a water/oil (W/O) type emulsion which has two phases: the continuous or lipid phase and the dispersed or aqueous phase. It also contains additives (emulsifying agents, salt, colourant, antioxidants, preservatives, vitamins, etc.) which are distributed between the fatty stage and the aqueous phase. The role of the emulsifiers is to reduce the amount of work required to form a homogeneous mixture from two non-miscible phases, and the final stability of the product will be obtained through crystallisation of the fat within the emulsion. Margarine is therefore a polydispersed system of fats (in the solid and liquid state), water and/or milk and various other ingredients.

Formulation of the fat phase

The fat phase represents the most important part of the emulsion, 82–84% in traditional margarines which must have a maximum water content of 16% but only 6% in the new so-called 'light' margarines.

As margarine is stabilised by cold, it goes without saying that the **solid factor**, present at varying temperatures in the course of crystallisation and also inversely during melting, is doubtless the most important parameter, the one that will specify the fatty phase. It is established that:

- solid factors from 0 °C to 10 °C control the spreading behaviour of the product;
- solid factors from 15 °C to 20 °C are important factors for hardness and oily exudation;
- solid factors from 20 °C to 25 °C control the stability of the product under heat;
- solid factors from 30 °C to 35 °C are involved in the increased organoleptic value of the product.

The melting points of the various natural raw materials are distributed in accordance with Fig. 4.6. In the main, only liquid oils or vegetable fats with a relatively low melting point (e.g. copra) are suitable for preparing margarines whose melting points are between +28 and 42 °C. Only palm oil and animal fats lie within this temperature range. Such oils cannot be used in their natural state. We need to modify these oils and natural fats (see Chapter 13).

Hydrogenation, cracking and interesterification can be used to obtain a suitable raw material. A suitable mixture might contain inter-esterified oil, hydrogenated oil and palm olein (obtained by fractionation). It is possible using a computer program to determine the part played by each of the constituents in achieving the solid factor of the mixture at a given temperature (°C). If E_{nT} represents the value of the statistical equivalents of solid at T°C and if X_n is the concentration of the various ingredients, we can perform a calculation to determine the solid factor (the percentage of solid fat present):

$$\text{Percentage of solid fat in the mixture at } T\,^\circ\text{C} = S_T = \sum_{0}^{n} (X_n E_{nT})$$

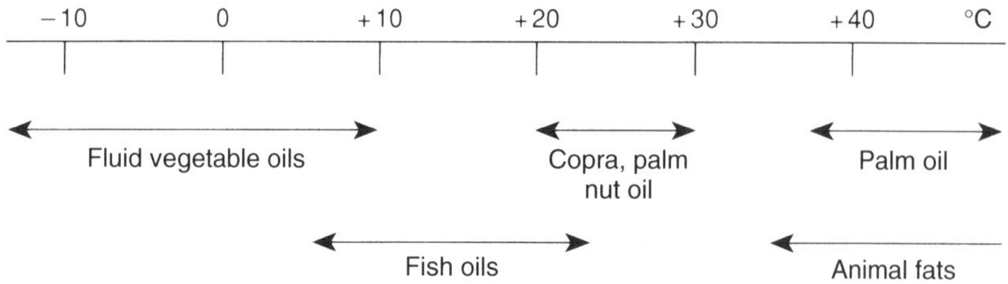

Fig. 4.6 Melting points of various natural raw materials.

Into such a program for formulating margarines, it is then necessary to introduce the **rate of crystallisation** which will in particular determine the hardness and plasticity of the product.

The polymorphism of triglyceride crystals is discussed in Chapter 13.

In the mixtures of oils and solid fatty substances that constitute margarines, transitions generally take place in a liquid form:

$$\alpha \rightarrow \text{melting} \rightarrow \text{crystallisation} \rightarrow \beta' \rightarrow \text{melting}$$

The unstable α form changes rapidly into the β' form by releasing transitional heat. As a result of rapid cooling, the glycerides do not have the time to crystallise individually so a random crystallisation process takes place, resulting in the formation of mixed crystals. These crystals have melting points lower than those of the crystals of the ingredients with the highest melting points. Properties such as spreading capacity, elasticity, oil exudation, and organoleptic qualities such as freshness and 'melt-in-the-mouth' qualities depend on the percentage (by weight) of crystals, their form and melting point, as well as the types of bond combining them.

For example, in a traditional margarine, the average surface area of the crystals is in the order of 60–70 m^2 per solid cm^3 and the size of 15–20% of crystals is less than 0.3 µm (the remaining 80% have a diameter lower than 1 µm).

There is a good ratio between the solid factor S and the hardness D_T:

$$D_T / S_T^2 \approx 0.70 - 0.75$$

For example, with a solid content in the order of 25%, hardness will be in the order of 440–500 g/cm^2. For spreading margarines, this value is lower than 300 g/cm^2, and it is between 300 and 1000 for a margarine intended for manufacturing puff pastry. In margarines the crystals are associated with each other by two types of bonds:

- The secondary reversible bonds due to the van der Waals interactions which are easily broken by shearing and stirring and fairly quickly re-established as soon as this action ceases.

Table 4.9 Formulation (%) of traditional margarines

Oils	Melting point (°C)	Margarine	
		Table	Baking
Soya		20	20
hydrogenated	28	–	10
hydrogenated	36	30	20
Palm	44	20	–
hydrogenated		10	35
Palm kernel	38	20	15
hydrogenated			

- The primary irreversible bonds due to the precipitation of the crystals during formation on the weak bonds, requiring considerable energy to break them.

Therefore if a strong standard shearing action is exercised (shearing–kneading) we can go from an initial hardness D_i to a new hardness D_f. The ratio $[(D_i - D_f)/D_i] \times 100 = Ws$ is an expression of **plasticity**. For a Ws value of between 30 and 40% a sticky product is produced. For values of over 80% a brittle product is produced. To sum up, in order to obtain margarines, the fat phase to be used should be defined by a solid percentage/temperature curve and a solid percentage/time curve.

As there is a strong relationship between hardness and the amount of solid fat present, an effective manufacturing process will aim to crystallise the emulsion (depending on the solid curve and the speed of crystallisation) and obtain the desired hardness and plasticity.

In addition to these characteristics, the lipid phase must be as pale as possible and stable under oxidation. So, as we have already emphasised, the process of refining and deodorising oils and fats intended for the manufacture of margarine must be carefully monitored.

For the purposes of illustration we can propose two types of traditional margarine formulations (Table 4.9). Nowadays these traditional margarines are being replaced by products that are richer in essential fatty acids, that is to say margarines containing very large quantities of liquid oils (for example, sunflower and maize products) whose formulations are completely different. They are mixtures of fluid oils/inter-esterified fatty oils. The quantity of the inter-esterified phase which is incorporated in the final composition of the fat phase is between 20 and 40%.

4.2.3.2 *Uses as a heat exchange fluid (fats for frying)*

Using fats for frying, cooking or roasting certain foods is unavoidably accompanied by thermal oxidation reactions which raise the content of adulterating products and modify physico-chemical and nutritional characteristics. The main changes fatty substances undergo when heated are of a hydrolytic, oxidising and thermal nature. So, water contributed by the food can lead to hydrolysis of the fatty substance with liberation of fatty acids, appearance of partial glycerides and glycerol, which can generate acrylaldehyde. The high temperature encourages oxidation and polymerisation reactions with the formation of hydrocarbons, ketones, alcohols, polymers, cyclic monomers. These numerous modifications of a chemical nature inevitably involve modifications to their physical properties and characteristics (viscosity, density, refractive index, etc.).

All these modifications and their intensity will depend on the way in which the fat is used. When the fat is used for cooking, it is only used once for a very short time at a temperature that is generally lower than 100 °C. Under conditions described as being 'favourable', few or no modifications of the fatty substance are observed. At higher temperatures a distinction must be made between shallow frying and deep frying.

Shallow frying

Shallow frying, which consists of preparing certain foods by cooking them in a frying pan or in a dish can be carried out using numerous fats: butter, margarine, lard, olive oil, groundnut oil, etc. This type of cooking is characterised by the fact that the oil is used only once, and by the very high value of the ratio of the surface area presented to the oil and the weight of the oil. This ratio is often between 70 and 200 depending on the food under consideration. The parameters that can be varied are therefore numerous and it is difficult to establish any general rules regarding the amount of damage suffered by the fat. One of the methods currently preferred for estimating the effect of frying on the chemical characteristics of various fats is to analyse the quantity of new chemical entities (NCEs = oxidised acids and oxypolymers).

The difference in behaviour during shallow frying is associated with the degree of insaturation of the fat. For example, if under identical conditions, cooking with lard results in an NCE content almost two times lower than shallow frying with groundnut or sunflower oil. In addition, cooking with margarine rich in linoleic acid can – for relatively short cooking times (6–10 min) – result in relative losses of this essential fatty acid in the order of 8%. To sum up, bearing in mind the large contact surface area between the air and the oil, the changes are mostly of an oxidised type.

Deep frying

In deep frying, the oil is brought to a temperature which is generally between 160 and 200 °C. As the relationship between the surface presented to the oil and the weight of the oil is between 1 and 2, the products formed will result from oxidising and thermal reactions. In addition, a significant difference in comparison with shallow frying lies in the fact that the oil is used several times.

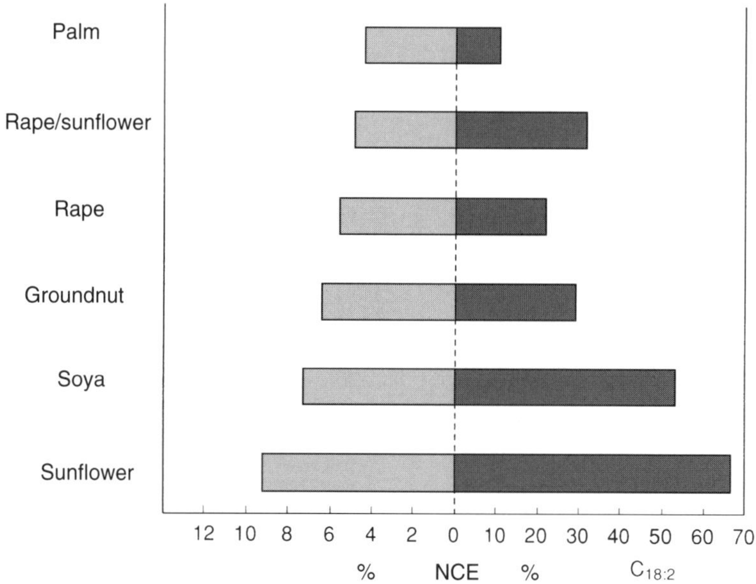

Fig. 4.7 New chemical entities (NCEs) formed during deep frying depending on the initial content of linoleic acid (15 × 40 min frying processes at 200 °C).

The behaviour of various fats that have undergone 15 fryings at 200 °C is illustrated in Fig. 4.7. These results show that the rate of NCEs formed during these tests increases with the content in polyunsaturated fatty acids (linoleic and linolenic acid). From the different oils examined, it is sunflower oil that undergoes most chemical changes during heating. Also the higher the temperature of the oil, the greater the chemical changes. However, temperatures in the order of 220–240 °C must be reached before significant quantities of fatty acids and cyclic monomers can be detected.

Among the oils used most in deep frying are groundnut oil, palm vegetable fat, sunflower and soya oils. Finally, undesirable lipolysis considerably reduces the decomposition temperature (smoke point) and therefore makes the oil unsuitable for use in frying.

5

The dairy industry

The dairy industry has traditionally taken milk and converted it into such products as milk for drinking, cheese, butter, etc. By-products were rejected or used directly in animal feed. Recently scientific and technological progress, and the search for more profitable outlets, have encouraged manufacturers within the dairy industry to develop the potential of the biochemical constituents of milk. The dairy industry has acted as a pioneer for other sectors of IFPs, both in the extremely diverse range of technological methods employed and in the range of products manufactured. This chapter will deal only with cows' milk, which has for some time been the subject of significant industrial development.

5.1 Introduction

5.1.1 Characteristics of the raw material

As this involves obtaining very well-defined uniform IFPs, one obvious fact should be borne in mind: the dairy industry does not have a source of storable raw materials, to be used when required. It has to use its raw material and process it as quickly as possible. In fact milk is a complex raw material which is heterogeneous, microbiologically active and difficult to store.

Three phases are present in milk: an aqueous phase, whose predominant ingredients are lactose (50 g/l) and a number of globular proteins (7 g/l); and two other phases, consisting of fat globules (98% triacylglycerol) and casein micelles (92% of whose dry matter consists of proteins and the rest of mainly minerals) respectively. The concentrations of the casein and lipid fraction are 24–28 g/l and 35–45 g/l, respectively.

In raw milk, these phases are naturally differentiated, but most technological processes modify this distribution. This can be broadly summarised as follows:

- Refrigeration results in some of the caseins and minerals moving into the aqueous phase.
- Heat treatments have virtually the opposite effect with – depending on the heating conditions – interactions between lactose and the protein fractions.
- Homogenisation disperses the fat globules which are covered by proteins.

5.1.2 General properties of the ingredients

5.1.2.1 Proteins

Figure 5.1 shows that out of the 32–35 g of total nitrogenous substances in cows' milk there is only approximately 5% of non-protein nitrogen (NPN) substances. The remaining proportion consists of many proteins. There is no appreciable quantity of peptides in milk. The NPN substances remain in solution when the proteins have been precipitated, for example by adding 12% trichloroacetic acid. These comprise numerous substances having a low molecular mass (less than 500 Da); the most abundant is urea (0.25 g/l); alongside this there are metabolic intermediate products (orotic acid) and free amino acids. This fraction is significantly involved in the growth of bacteria.

α_{s1}-, α_{s2}-, β- and **κ-caseins** are the molecules that form the casein micelle. With the γ-caseins they form over 75% of the total nitrogen. They are phosphorylated by an ester bond with serine or threonine, which accentuates their acid behaviour. They have the essential characteristic of precipitating at pH 4.65 at ambient temperature and not being made insoluble by heating to 100 °C. These are small proteins, with a molecular mass in the region of 20 000–25 000 Da; however, in their native state they are always found strongly combined with each other in relatively large micelles which also contain minerals.

Whey proteins are much less abundant than the caseins. They are mostly not associated and do not participate in enzymatic coagulation; however, they have a higher nutritional value, mostly because of their sulphuramino acid and lysine content. The proportions given in Fig. 5.1 show the predominance of β-lactoglobulin (approximately 3 g/l) which despite its name ought really to be ranked alongside the albumins because of its low molecular mass of 18 360 Da, its high solubility, its electrophoretic mobility and holoprotein nature. α-lactalbumin (MM 14 200), in addition to its role as a regulator of the activities of lactose-synthetase, is characterised by its high tryptophan content.

These two proteins are dealt with in more detail in Chapter 9 (Section 9.1.3). **Immunoglobulins** (Ig) are present in all milks. In cows' milk they

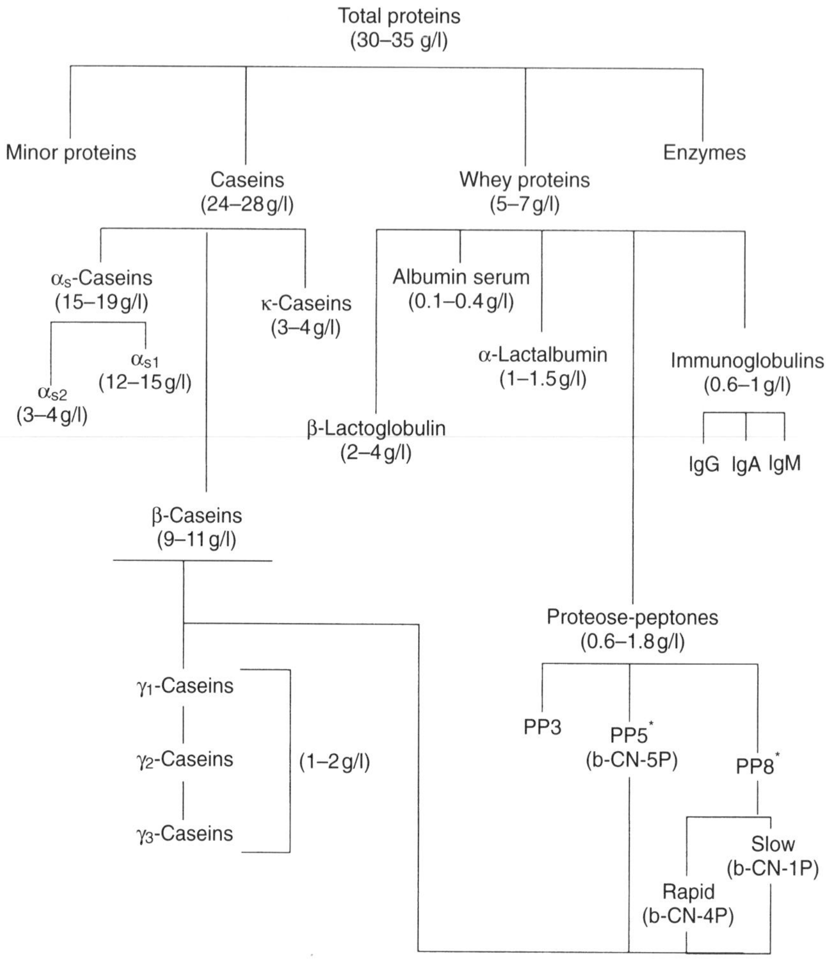

Fig. 5.1 Composition of the protein fraction of cows' milk.*
The PP5 and PP8 components from the plasminic hydrolysis of β-casein have been given the names shown in brackets.

form only a tenth of the whey proteins (0.5–0.7 g/l); however, their proportion increases significantly in colostrum (12 g/l at the end of the first day and 80 g/l in the first hour's colostrum). This increase is the result of the movement of the blood Igs, because the biosynthesis of the other components is not reduced. However, not all the milk Igs come directly from the blood; some are synthesised in the mammary gland.

Proteose-peptones constitute the minor protein fraction of whey. They remain soluble after heating to 95 °C and acidification at pH 4.6. Two principal classes can be differentiated:

- Constituents resulting from the enzymatic proteolysis of major proteins.
- A fraction of thermostable glycoproteins having a marked hydrophobic character.

The spatial structure of caseins is very different from that of whey proteins. It is very open because of the high content of uniformly distributed proline residues which limits the α helix and β sheet formation, and because of the absence of disulphide bridges. This results in a conformational state which is not particularly sensitive to thermal denaturation but can easily be accessed by enzymes. The role played by the nature and distribution of amino acids, as well as by the number and position of certain groups with regard to the functional properties of caseins, is also of interest. The distribution of seven phosphate groups located between the amino acid residues 42 and 80 as well as the distribution of the amino acid residues provides the casein α_{s1} with a very hydrophilic region and three hydrophobic regions. β-Casein has an N hydrophilic terminal region and a hydrophobic C terminal region. The amphiphilic nature of these proteins therefore gives them the ability to orientate and diffuse towards the oil/water or air/water interfaces, thus producing good emulsifying and foaming properties.

On the other hand, the structure and the properties of whey proteins differ from those of caseins because of the uniform distribution of hydrophilic and hydrophobic amino acid residues all along the polypeptide chain. In addition, a low concentration of proline residues and the presence of disulphide bonds give whey proteins a very compact globular conformation. In comparison with caseins, whey proteins are found in a soluble state, they do not precipitate at their isoelectric point but are quite unstable to heat and precipitate during thermal processing in the following order: immunoglobulins, β-lactoglobulin, serumalbumin and α-lactalbumin. At around the denaturation temperature, the protein structure opens up and unfolds. This reaction can be reversible, but more often the irreversibly denatured proteins aggregate and precipitate. Under certain conditions, denatured proteins can result in a gel.

The amphiphilic character of dairy proteins allows them to move towards the oil/water or air/water interface and to lower surface tension. For example, in an emulsion or a foam, the proteins will migrate towards the interface where they will form a film and lose some of their conformation and hydration energy in reducing surface tension. The stability of the system will be strongly dependent on the nature and properties of the film (steric and electrostatic factors, thickness, elasticity and viscosity of the film) (see Section 5.2.1).

5.1.2.2 Lipids

The fat in milk, constituted by predominantly glycerides, is solid at ambient temperature (Fig. 5.2). It is almost completely free and is finely dispersed in the fat globules. Polar lipids which are mostly phospholipids form part

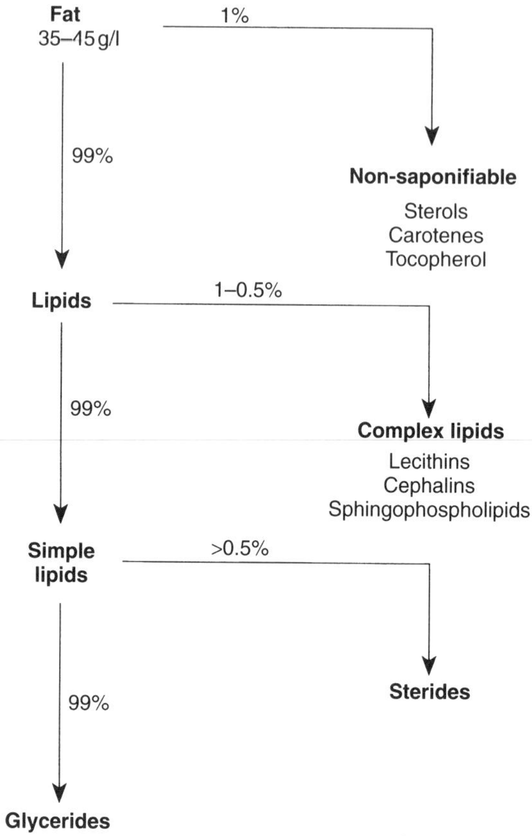

Fig. 5.2 Composition of the fat in milk.

of the lipid fraction; they are mainly found in a combined form in the globular membrane. Liposoluble, non-saponifiable substances, principally carotenes and vitamins A and D, form the rest. An analysis of the composition of the fatty acids in milk reveals that two-thirds of fatty acids are saturated and the remaining third remains unsaturated. Although palmitic acid is predominant, short chain (C_4–C_{10}) fatty acids are found in a significant proportion compared with most everyday fats and oils apart from coconut and palm-kernel oil (see Chapter 4). Note that lactose constitutes almost the whole of the carbohydrate content. Its physical and chemical properties as well as its separation by crystallisation are dealt with in Section 10.3.

5.1.3 Technological aims

As this involves controlling and conserving heterogeneous biological substances in stable forms, the procedure usually consists of the following:

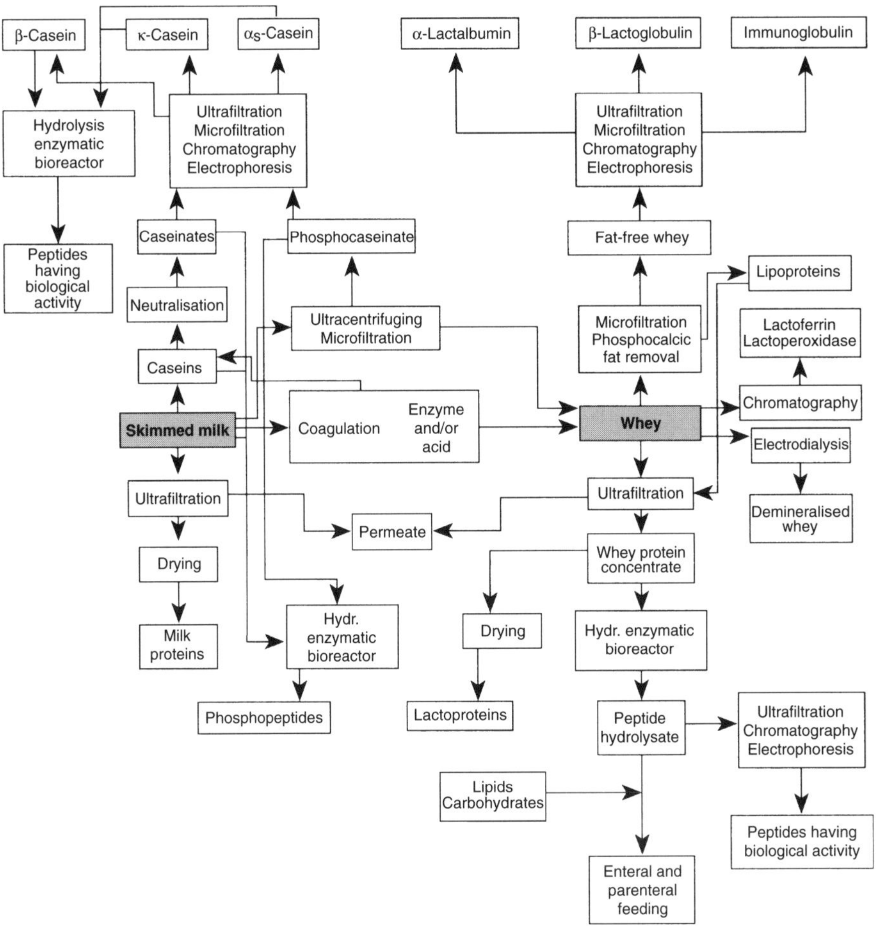

Fig. 5.3 Separation and uses of the protein fraction of milk.

- Isolating the constituents of milk.
- Re-formulating them either for nutritional purposes, or for technological purposes.
- Recombining them with other ingredients according to requirements and the properties sought after.

Figure 5.3 summarises the many possibilities for separating the protein fraction and the equally numerous possibilities for economic development. The functional properties and the nutritional qualities of the protein fraction of milk, in particular, have made them choice ingredients in a certain number of secondary food processors and in other industrial sectors (pharmacy, cosmetology, etc.). Alongside the development of technologies of extraction and purification, technologies that aim to modify existing products with a view to improving their functional and nutritional properties have been developed. These various aspects will be dealt with below.

5.2 IFPs based on dairy proteins

5.2.1 Functional properties of dairy proteins

The fundamental physico-chemical differences between caseins and whey proteins (distribution of hydrophilic and hydrophobic regions, tertiary structure, presence of free sulphydryl groups) are reflected in terms of functional properties (Table 5.1).

The 'random coil' structure of the caseins as well as the different localisation of the hydrophobic and hydrophilic groups give them a more amphiphilic character than whey proteins. Caseins can easily diffuse to and unfold at interfaces. Therefore, these proteins have a high capacity for reducing surface tension correlated with their foaming properties. The emulsifying capacity of skimmed milk powder is higher than that of the caseinates because of the respective molecular state of the casein in the two

Table 5.1 Main functional characteristics of milk proteins

Characteristics	Caseins	Whey proteins
Solubility	Insoluble at pH 4.6	Very soluble at all pH, but insoluble at pH 5 if thermally denatured
Viscosity	Very viscous solutions at neutral and alkaline pH Minimal viscosity at pH_i (4.6)	Solutions not very viscous unless thermally denatured
Hydration	High water retention with formation of glue at high concentrations Minimum at pH_i	Water retention increases with denaturation
Gelling	No thermal gelling except in presence of calcium Gelling of micelle by chymosin	Thermal gelling from 70 °C: influence of pH and salts
Emulsifying characteristics	Excellent emulsifying characteristics especially at neutral and alkaline pH	Good emulsifying characteristics except at pH 4–5 after thermal denaturation
Foaming characteristics	Good swelling but poor foam stability $\kappa > \beta > \alpha_{s1}$	Good swelling and excellent foam stability $\beta Lg > \alpha La$
Flavour holding	Good holding	Holding very variable depending on state of denaturation

types of product and the presence of whey proteins which improve the strength of the interfacial films. β-Casein is more effective in reducing surface tension, followed by α_1-casein, then κ-casein.

The most remarkable functional property of whey proteins is their solubility over the whole pH scale and their water-holding capacity. They can bind twice as much water as the caseinates or plant proteins. This capacity to bind and structure water results in properties that are essential for the food industry both from a technological point of view (good development of enzymatic processes) and from an economic point of view (water holding all along the production and marketing chain). The differences in composition of individual whey proteins result in differences in their functional properties. Because of its high molecular mass, serumalbumin has mediocre functional properties whereas α-lactalbumin gives the emulsion a stability that is three to four times higher than β-lactoglobulin. β-Lactoglobulin has better surface tension properties, which can be further improved by heat denaturation or by limited proteolysis. The proteose-peptone fraction of whey proteins has excellent foaming properties. In order to possess good functional properties, protein must therefore be soluble, amphiphilic and flexible, and must be able to change position and conformation easily while retaining minimum tertiary structure.

5.2.2 Preparation and applications

Dairy ingredients integrated in the composition of a foodstuff are mainly introduced in powder form. Although the trend for using ingredients that come from concentrating or fractionating skimmed milk or whey is constantly growing, we should say a few words about milk powders or whey that has been dried without prior processing.

5.2.2.1 Powder from raw milk products

Because of its high cost and the development of undesirable flavours released by the oxidation of fatty substances during storage, **whole milk powder** is not widely used as an ingredient. The addition of lecithin which acts as an anti-oxidant, however, allows whole milk powder to retain an acceptable flavour. Skimmed milk powder is incorporated in a variety of products because of its solubility, and emulsifying and water-holding capacity. The product is available in three forms: 'low, medium and high heat'.

'Low-heat' powders contain a small quantity of denatured proteins and are used in products in which the properties of solubility, gelation, emulsion and low protein denaturation are needed (fortified yoghurts, cheese made from reconstituted milk, creams and frozen desserts, powdered and frozen soups, powdered sauces and salad dressings). Solubility is characterised by the WPNI (whey protein nitrogen index), which represents the soluble protein nitrogen at pH 4.6, in mg of nitrogen per gram of powder. For 'low-heat' powders, the WPNI is equal to, or higher than, 6.

'Medium-heat' powders have good water-binding capacity and surface activity and are used in ice-cream, frozen desserts, frozen soups, acidified cream, etc. (WPNI of between 1.5 and 6).

'High-heat' powders are highly denatured and not so soluble (WPNI of less than 1.5); their water-binding capacity is increased and their bread anti-swelling factor reduced. Because of their poor solubility and the reduction of their action as a depressing agent on bread dough, this type of powder is used in structured products (baking, biscuit-making, confectionery and seasoning) contributing to the cohesion, texture, flavour, colour and nutritional value of the product. There are even powders with reduced calcium content that improve flavour, texture and water binding in meat and spreadable products.

The composition of **buttermilk powder** is more or less equivalent to that of skimmed milk powder. Incorporated in chocolate, it contributes to the product's characteristic flavour as well as to the stability of the emulsion. In baking it is incorporated in cakes because of its flavour and its ability to turn gluten yellow.

Chapter 9 deals with the economic development of whey. We will briefly mention the properties of whey powder here. Progress made within the field of spray-drying has resulted in the production of non-hygroscopic **whey powder** which has a proportion of non-denatured proteins which varies according to the type of powder ('high heat', 'medium heat' or 'low heat'). Proteins from 'high-heat' whey powders are highly denatured and relatively insoluble and are used in foods in which high solubility is undesirable or not necessary (baking products, pasta, breakfast cereals, etc.). 'Low-heat' powders contain very soluble proteins and are used in food products in which flavour, colour, physical stability and solubility are required.

The properties of whey powders also depend on the type of whey (sweet or acid) and the processes it can undergo before drying (demineralisation, crystallisation of lactose, ultrafiltration, etc.). Sweet whey offers some advantages for cooked cereal products. The amino acid residue composition of whey proteins is complementary to that of plant proteins. In addition, sweet whey improves the nutritional and organoleptic value of the end-products. In some cases, the softening effect of bread dough has more of a positive than a negative influence on the manufacturing process, by firming up the soft part of the bread and improving resilience when whey powder is added to rolls (3–6%), biscuits (2–20%) and pie pastry (2–10%).

Section 9.1.2 deals with the economic development of demineralised wheys and/or those from which the lactose has been removed.

5.2.2.2 Caseins and caseinates

Whole casein, often simply called casein, corresponds to all the micelles of caseins, after the minerals and citrate that are bound to them have been removed. In the laboratory simple high-speed centrifuging of skimmed milk allows the micelles to form a sediment and the micellar casein to be prepared.

Acid casein is obtained through the acidification of skimmed milk at pH 4.6 at 43–45 °C, using mineral acid (hydrochloric or sulphuric). Lactic casein results from the precipitation of caseins after acidifying the medium using mixed acidifying strains of bacteria. After approximately 16 h fermentation at 26–27 °C, the temperature is raised to 55–65 °C in order to encourage the agglomeration of casein particles. The advantage of lactic casein is that it contains two to three times less chloride than hydrochloric casein prepared using mineral acid, although this is still the most common practice. Ionic pre-acidification reduces this disadvantage. Passing skimmed milk over a cationic resin can obtain both demineralisation and pre-acidification up to pH 4.9–5.0, because the Ca, K and Na ions in milk are exchanged with the H^+ ions in the resin. Final acidification is then obtained by adding mineral acid up to pH 4.5–4.6. Unlike milk powder and rennin casein, the micellar structure of acid and lactic caseins is lost by acidifying and dissolving the colloidal calcium.

Rennin casein results from the enzymatic coagulation of caseins by pepsin and chymosin. After 20–30 minutes coagulation, the temperature is raised to 55–65 °C, as in the case of acid casein. A quantity of micellar casein is still present in rennin caseins and makes this type of casein insoluble and not very useful from a functional point of view, except for complexing calcium.

Caseins are used in confectionery because of their water-holding capacity (over twice their own weight) and thus they inhibit the formation of sugar crystals. They also help to glaze sweets and aid in the development of flavours attributable to the aldehydes formed by the Maillard reaction. Apart from this caseins are not particularly useful in the food industry because of the difficulty in dissolving and dispersing them.

By neutralising the precipitates of acid and lactic caseins, by sodium, potassium, ammonium or calcium hydroxide, a soluble product more commonly known as **caseinate** can be obtained. In the form of either monomeric secondary units or aggregates, caseinates can migrate rapidly to the oil/water interface of emulsions or foams and can stabilise them to prevent coalescence or melting. Caseinates also display good emulsifying properties, whatever their pH and ionic strength, with a maximum at pH 10.4.

The properties of caseinates are greatly influenced by manufacturing processes. Na, K and NH_4 caseinates are fully soluble from below pH 5.5 to higher values and form viscous solutions that possess excellent functional properties (emulsifying, foaming, thickening, water absorption), higher than those of calcium caseinate, which remains in the form of a colloidal suspension and absorbs much less water. Consequently sodium caseinate is incorporated in food products with high or intermediate water content whereas calcium caseinate is incorporated in products with low water content. However, by replacing calcium with various quantities of sodium, and sometimes by incorporating whey proteins, we can clearly increase the stability of calcium caseinates at acid pH and incorporate them as bulking agents, emulsifiers and stabilisers, additionally because of their water-holding and aerating capacity in foods whose pH is between 5.0 and 5.5. Moreover, the properties of the caseinates and whey protein concentrates are extremely complementary. Caseinates have good emulsifying power and are stable to heat, and whey protein concentrates form a gel during cooking, which means that these ingredients can be used in combination in cooked foods.

Sodium, potassium or calcium caseinates are currently used in food products because of their solubility, water absorption, stability under heat, surface tension properties (emulsification, foaming), owing to their random coil conformation, as well as for their viscosity. Low-viscosity sodium caseinate can be added during yoghurt manufacture in order to standardise protein content and viscosity. By mixing sodium and calcium caseinate at varying concentrations, it is possible to obtain a whole range of caseinates of decreasing viscosity.

Because of their nutritional value and the absence of NaCl (especially calcium caseinate) caseinates are used in dietary and nutritional products such as 'light' butters and spreads. As far as meat products are concerned, adding dairy proteins to poultry reduces cooking losses as well. The stage of mechanical de-boning of chicken tends to produce a dark product, which calcium caseinate, in emulsion form, can also bleach. Caseinates are also used for their bleaching, stabilising, emulsifying and viscosity-regulating effects, and for their water-binding capacity in the following products: coffee whiteners, cake mixes, desserts, whipped fillings, creams, breakfast cereals, confectionery, powdered sauces and salad dressings, drinks, cheese substitutes and cream-liqueurs. When combined with other food proteins (whey, soya and cereal proteins) they are used in baking, cake-making and in frozen desserts.

5.2.2.3 Co-precipitates

Co-precipitates are obtained when severe heat treatments (90 °C for 1–20 min) are applied to skimmed milk, followed by the addition of acid or calcium chloride. A complex is then formed between the denatured whey proteins and the caseins through the disulphide bonds. Acidifying to pH 4.6, or adding calcium chloride, allows the protein complex to precipitate so that approximately 96% of the proteins can be recovered. The protein content is in the order of 80–85%. By altering the quantity of calcium added, the pH and the duration of the thermal processing, it is possible to obtain different qualities of co-precipitate. Co-precipitates with a high level of calcium have the best organoleptic qualities when stored, whereas undesirable flavours can develop in other types of co-precipitates.

Co-precipitates are formed to the detriment of solubility (especially at pH 6.7–7.2), since between 4 and 15% of whey proteins are denatured.

However, solubility can be improved by the complexing agents of calcium, which increase dispersibility, or by a series of stages involving pH adjustment, heating and cooling. The result is a 95% soluble product. The emulsifying capacity of co-precipitates with a high calcium content is higher than that of other co-precipitates and that of sodium caseinate. The whipping capacity of the co-precipitates is lower than that of egg albumin but higher than that of sodium caseinate.

Soluble co-precipitates are used in products with a high or intermediate water content. In chocolate production they contribute to flavour stability and to anti-oxidising properties. The good stability to heat of co-precipitates allows them to be incorporated in ultra heat treated (UHT) products, yoghurts, whipped creams and spreadable products with a low fat content. Co-precipitates with low solubility can also contribute to the texture of breakfast cereals, baking products and pasta, and increase their nutritional value. They are used in cooked meat-salting processes because of their capacity for binding water and fat. Because of their content in whey proteins rich in lysine, tryptophan and sulphur amino acids, co-precipitates have a nutritional value that is higher than that of caseins, and they are unquestionably advantageous in infant nutrition.

There is a new type of ingredient whose composition is very similar to that of the co-precipitates: whole milk proteins prepared by adjusting their pH to 10, heating to approximately 70 °C in order to dissolve the casein micelles, acidification to pH 3.5 in order to complex the caseins and whey proteins, adjustment of the pH to 4.6 in order to precipitate the proteins which have been complexed, then washing and drying; preparations of whole proteins have very good solubility and the non-denatured whey proteins retain their functional properties intact.

5.2.3 Modification and improvement of functional properties

Applying physical, chemical or enzymatic treatments to dairy protein-based IFPs allows some of their functional properties to be improved and their use to be extended.

5.2.3.1 Modifications made by changing the composition of the medium

In general, eliminating mineral substances and a number of other dialysable substances (electrodialysis, gel filtration) improves most functional properties. Ultrafiltration treatment applied to whey eliminates salts and lactose and can be combined with thermal processes in order to produce better functional properties. Overall, all the processes for purifying proteins tend to improve their interfacial properties: as in the case of β-lactoglobulin purified by the Spherosil process. These concentrates are excellent gelling agents at a protein concentration of over 5%, at neutral pH and a temperature of 70–85 °C. These firm and elastic gels have a good water-holding capacity. At pH_i (5–5.2), aggregation reactions result in a granular coagulum which can also be obtained at a pH higher than 5.5 in the presence of the Ca^{2+} and Mg^{2+} ions. In a partially denatured form at neutral pH, these concentrates are also excellent bulking and emulsifying agents.

Among the most recent applications of the technique of ultrafiltration, its use as a process for extracting β-casein from skimmed milk can be quoted. This process is based on the partial dissociation at low temperature of the casein micelle and the 'leaking' of β-casein, as well as on the characteristics of the surface properties and permeability of new mineral membranes. In this way, concentrates enriched with β-casein which possess improved surface active properties can be obtained.

5.2.3.2 Physical and chemical modifications

Partial denaturation resulting from moderate **thermal processing** makes the protein unwind and exposes the hydrophobic groups so that they can then move to the air/water or lipid/water interface. The amphiphilic character of the molecule is reinforced, and so consequently are its properties as a surfactant and water-retention agent. The effects are strongly dependent on the pH at which the process is carried out.

By means of **acetylation** and **succinylation**, the α and ε-amino groups, whose charge is strongly positive, are converted into anionic groups with negative or zero charge. Modifying electrostatic repulsion by creating negative charges (succinylation), or by reducing positive charges (acetylation), can draw out or unfold proteins, thus reducing the surface tension of the protein solutions. Acetylation increases the solubility of casein at acid pH and its stability to heat. Essential amino acids such as lysine can have their ε-NH_2 side chain involved in the covalent complex, but these products are not approved for food use at the moment.

As in the case of succinylation, **phosphorylation** creates new electrostatic repulsion within the polypeptide chain by increasing the number of anionic hydrophilic groups. Chemical phosphorylation is obtained by the covalent attachment of phosphorous atoms ($POCl_3$) to the protein. Internal bonds can also be created through phosphate bridges. Phosphorylation of casein improves its viscosity and water holding but slightly reduces solubility and emulsifying capacity.

5.2.3.3 Enzymatic and chemical hydrolysis

Apart from the use of rennin for producing casein (see Section 5.2.2.2), enzymatic hydrolysis can modify the functional properties of caseins. As an example, hydrolysis of casein using pancreatin increases its bulking capacity to the detriment of its emulsifying capacity; limited hydrolysis by trypsin improves the emulsifying capacity of caseins at the isoelectric point. Finally, the manufacture of casein hydrolysates produced by heat in the presence of hydrochloric acid has been carried out for some time. The products obtained after neutralisation and drying in fact essentially consist of free amino acids and sodium chloride. They are used to manufacture soups and micro-organism culture media. Moreover, the production of peptides exhibiting biological activity by means of the hydrolysis of dairy proteins is currently being fully developed. This matter is dealt with in Chapter 14.

5.2.3.4 Improvements due to association with other ingredients (mixes)
Milk proteins can be made into mixes by being combined with glycans (see Section 12.5.1.4) or with other proteins. Combining skimmed milk proteins with those of egg white, part of which is in a micro-particulate form, can result in a product used as a fat substitute. The manufacturing process starts by mixing liquid egg white, skimmed milk, lecithin, vegetable gums and some organic acids, for example citric, lactic or malic. After water absorption and adjustment to pH 6–7, the mixture is brought to approximately 60 °C. Equipment with a high shearing capacity is then used at a temperature of 85 °C to make the proteins coagulate in the form of microparticles. The product obtained is then cooled and stored at a temperature of between 2 and 4 °C. It has a total protein content of approximately 15%, over 40% of which is in the form of microparticles, whose diameter is between 0.1 and 3 µm.

This type of mix can replace the fats in a large number of food products such as ice-creams, yoghurts, mayonnaise, dressings, desserts. Its caloric value is 5.5 kJ/g; in addition the process for preparing these mixes does not have any harmful effects on the nutritional quality of the proteins they contain.

The protein microparticles actually have the capacity to reproduce the taste of fat substances, in particular their creamy flavour and texture.

5.3 Proteins exhibiting biological activity: lactoferrin and lactoperoxidase

Improvements in purification techniques have allowed dairy proteins to be prepared in sufficient quantities to envisage using these macromolecules for high added-value purposes. This is currently the case with lactoferrin and lactoperoxidase.

5.3.1 Lactoferrin

Milk contains between 20 and 200 µg/ml lactoferrin whereas human milk contains over 2 mg/ml; bovine colostrum also contains appreciable quantities: between 2 and 5 mg/ml.

5.3.1.1 Molecular properties
Lactoferrin is a glycoprotein whose molecular mass is equal to 77 000 Da. It contains two sites, each capable of binding a ferrous ion. The lactoferrin–iron complex is very resistant to proteolysis and stable up to

pH 2. Native lactoferrin has an iron saturation rate of between 10 and 30%. The presence of citrate can displace the bicarbonate from the lactoferrin, thereby salting out the iron. This binding of iron depends on the bicarbonate/citrate ratio, thus influencing the bacteriostatic activity of the lactoferrin.

There is a close homology of structure between lactoferrin, blood or plasma transferrin, conalbumin (ovotransferrin) and the melanotransferrin of human melanoma cells. All these proteins have similar properties, in particular with regard to binding iron. Moreover, a curious similarity of structure has been observed between these proteins and those of bacterial proteins that bind arabinose, galactose, leucine, isoleucine, valine and sulphate. The 'sulphate-binding protein' of *Salmonella typhimurium* has the closest homology.

5.3.1.2 Biological role

Lactoferrin has an anti-toxic activity. It is involved in conveying the biologically available iron. Thus, it has been possible to show that it is involved in conveying iron to the intestine, and that it interacts with the peritoneal macrophages and the hepatocytes. It has also been identified as being 'colony inhibitory', acting on the cells of the spinal cord during myelopoiesis. Moreover, as lactoferrin is capable of delaying bacterial growth, it is thought that, together with lysozyme and lactoperoxidase, it forms part of the primary defence system against bacterial infection.

The bacteriostatic activity of lactoferrin has been demonstrated in human mucus, in bovine colostrum and in udder secretions during lactation. In addition, treating cells with lactoferrin definitively halts all their functions. This includes stopping the metabolic activity of the DNA and RNA precursors. Lactoferrin has an irreversible inhibiting effect on a large number of micro-organisms. This action is at its most powerful during the exponential phase. The bacteriostatic activity of lactoferrin can be compared to that of conalbumin. Lactoferrin has a strong inhibitory effect on bacteria that have a metabolic requirement for iron, as a consequence of its powerful competitive affinity for the iron (III) ion.

On the other hand, lactoferrin has no bacteriocidal effect on microorganisms that have very low iron requirements. So, for example, *Escherichia coli* has a high iron requirement, whereas *Streptococcus lactis* requires very little. It is therefore not surprising that *E. coli* is inhibited by lactoferrin, whereas *St. lactis* is not. However, we should add that if the antibacterial activity of lactoferrin has been proved on many occasions *in vitro*, it has not yet been proved irrefutably *in vivo*. In addition, certain specialists consider that the explanation that cell growth is inhibited solely by binding iron is not satisfactory.

In addition, the role of lactoferrin is not limited to its action on microorganisms. It increasingly appears to be an essential ingredient in maternal milk, as far as iron nutrition is concerned. A specific membrane receptor of lactoferrins has, in fact, been found in various species, on the surface of duodenal enterocytes.

This membrane receptor has a molecular mass of 100000 Da. It binds the various lactoferrins in a specific and saturatable way. This binding process does not depend on the degree of iron saturation of the protein, but on the pH, with a maximum at pH 5.5. It is also dependent on the calcium concentration. The glycan part of the protein is not involved in the recognition mechanism. Finally, several teams have shown that lactoferrin has a mitogenous activity, particularly on intestinal cells.

5.3.1.3 Industrial uses

These observations have resulted in various applications. The higher resistance of breast-fed infants to intestinal infections in comparison with bottle-fed babies has been attributed to anti-microbial substances such as lactoferrin present in large quantities in human milk. As a consequence, this glycoprotein was offered as a supplement in milks for babies and young children.

As this natural defence mechanism does not belong to the class of antibodies, it has great potential for protecting all the mucous membranes against an enormous variety of pathogenic and non-pathogenic micro-organisms. So, the remarkable anti-bacterial activity of lactoferrin and its powerful action on *St. mutans*, lie at the origins of its use, either alone or combined with lactoperoxidase, in the preventive treatment of caries, dental plaque and mouth infections (gingivitis), and antisepsis in dental surgery. Likewise, lactoferrin is recommended in the preparation of products for treating infections of the skin and mucous membranes caused by yeasts (*Candida*) or bacteria and in cases of mixed yeast–bacteria infections.

5.3.2 The peroxidase system

5.3.2.1 General properties

Lactoperoxidase (LP), which was the first enzyme to be discovered in milk, where it is found in appreciable quantities, has an anti-bacterial effect. It is generally found in a soluble form in the serum and has been identified with L2 lactenine. It is a metal-enzyme with a ferrohaem core of molecular mass 82000 Da and optimum pH of 6.8. In the presence of an acceptor (A) it catalyses the following reaction:

$$AH_2 + H_2O_2 \rightarrow A + 2H_2O$$

It causes indirect oxidation; it releases the atomic oxygen from peroxides such as hydrogen peroxide and this oxygen is accepted by a substance present in the medium. Milk peroxidase has good thermoresistance; it denatures at 70 °C for 15 min, or 80 °C for 30 s. The measurement of its activity is used as a control test for industrial pasteurisation processes.

5.3.2.2 Bacteriostatic and bacteriocidal action

Lactoperoxidase, on its own, has no bacteriostatic or bacteriocidal effect. It catalyses the oxidation of thiocyanate (SCN^-) using hydrogen peroxide (H_2O_2). The products obtained at the end of the reaction (sulphate, cyanate, etc.) have no effect but the reaction intermediates have a powerful action. The hypothiocyanite ion ($OSCN^-$) is the active principal in accordance with the following mechanism:

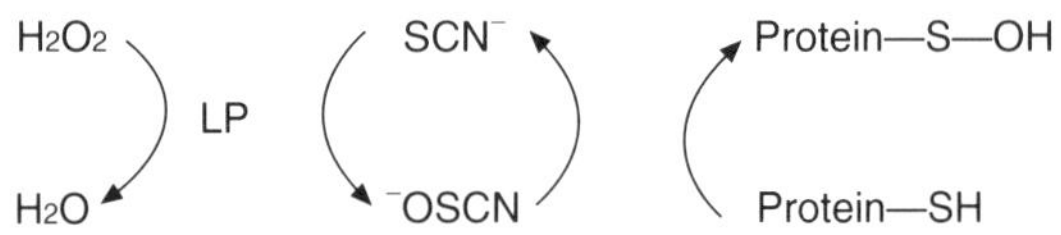

The anti-bacterial effect of lactoperoxidase is therefore reversible in the presence of reducing agents such as cysteine, and oxidation of sulphydryl groups in corresponding sulphenic acids (—S—OH) has been observed. However, the LP/SCN/H_2O_2 system remains effective in milk in the presence of small quantities of free thiol groups.

As enzyme concentration is not a limiting factor in the system, and the level of thiocyanate varies from 0.02 mM to 0.25 mM according to the animal's diet, only hydrogen peroxide is lacking; the reaction can then only occur if there is an external contribution of this.

Certain catalase-negative lactic streptococci produce enough hydrogen peroxide to activate the system, but are then inhibited.

Ideal system activation can be obtained by adjusting the level of SCN to between 0.20 mM and 0.25 mM (12–15 p.p.m.) and by generating an equimolecular quantity of H_2O_2 by means of an enzymatic system. Implementing this protective system on farm milk comprises two stages. Initially, approximately 10% of the lactose is hydrolysed by adding a lactase to produce glucose *in situ*. Then the glucose is oxidised by a glucose oxidase which is added at the first milking stage, as soon as possible. The simplified equation of these reactions is as follows:

$$\text{Lactose} \xrightarrow[\beta\text{–Galactosidase}]{\text{Lactase}} \text{Galactose} + \text{Glucose}$$
$$\text{Glucose} + O_2 + H_2O \xrightarrow[\text{Oxidase}]{\text{Glucose}} \text{Gluconic acid} + H_2O_2$$

At the end of the process, adding catalase allows the hydrogen peroxide to decompose. This process can also be applied using an enzymatic reactor where β-galactosidase and glucose oxidase are co-immobilised. Hypothiocyanite has a destructive effect on bacterial membranes since, after a few minutes' contact, the potassium and amino acids are liberated into the medium. The cell constituents oxidised most often are the sulphydryl groups and the nicotinamide nucleotide (NADH, NADPH).

The peroxidase system has a bacteriocidal effect on numerous pathogenic micro-organisms and a bacteriostatic effect on certain Gram-positive bacteria such as lactic streptococci and thermophilic lactobacilli. This difference in sensitivity seems to be dependent on the structure and the composition of the cellular membrane.

Observing the resistance of certain streptococci demonstrates the importance of this system in terms of regulating the ecological system of oral and intestinal flora.

5.3.2.3 Applications

One of the industrial advantages of this anti-bacterial system lies in the storage of raw milk. So, this treatment, combined with storage at low temperatures, can, for example, prevent the proliferation of psychotropic flora over a long period. There are active peroxidase systems present on different sites. The salivary, lachrymal and uterine peroxidases in fact have strong similarities with lactoperoxidase from a biochemical and immunological point of view. Therefore, in addition to stabilising milk and various food products, other applications of the peroxidase system have been developed:

- protection of young animals as an intestinal anti-infection and anti-diarrhoea agent;
- agent in the fight against caries and dental plaque;
- protection of mucous membranes (dermatology).

5.4 Lipid IFPs

Much research has currently been devoted to possible ways of increasing the value of dairy fat. This can be classified into two main groups of operations:

1 Extraction of selected fractions.
2 Modification in composition by means of physical or enzymatic processes.

In the first type of treatment, **fractionate crystallisation** of the **dried milk fat** (DMF) allows a range of products which possess functional properties suited to clearly defined manufacturing processes (Fig. 5.4) to be obtained.

Stearins with a high melting point are mainly used in pastry-making. They are commercially known under the name of pastry quality 'concentrated butters'. They have special organoleptic qualities in comparison with vegetable fats because of the presence of fatty acid of short chain length, very perceptible to the taste. If they are used, then the manufacturer is entitled to utilise the term 'butter', greatly valued by the consumer.

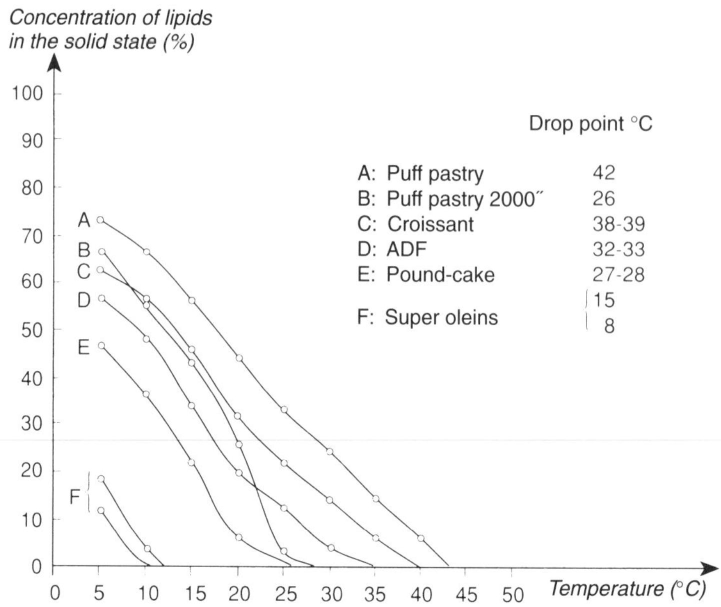

Fig. 5.4 Typical commercial fractions of anhydrous dairy fat (ADF).

Among the cake-making products available on the market, we can single out the '*mille-feuilles*' or puff pastry quality. This has excellent physical qualities of consistency, and its melting point is 42 °C. Puff pastry is used as the basis of a wide variety of cakes which are greatly appreciated for their lightness and crispness: these include the whole range of puff pastry items such as vol-au-vents, turnovers, palmiers and iced pastries.

Compared with 'puff pastry quality', 'puff pastry 2000' quality has a melting point of 26 °C which is considerably lower, even in comparison with butter. It is recommended for all puff pastry applications, except for raised puff pastry dough. It is very easy to mould and stretch at between 15 and 22 °C. In addition, it does not leave any fatty sensation behind when eaten, since it melts completely at a temperature of 26 °C. It is a second stearin obtained by re-fractionating an olein from the first fractionation process.

'Croissant' quality, whose consistency is specially adapted for manufacturing croissants, 'Danish pastries' and sweet flan pastry, is a stearin from the first fractionation process. Its melting point is between 38 and 39 °C. Hard stearins are also used in the production of ghee. On the other hand, oleins from the first fractionation process are used in biscuit-making, in the cheese industry or incorporated in DMFs. So, the 'pound-cake' quality recommended for manufacturing pound-cakes, choux pastry and madeleines is a DMF whose melting point has been lowered to between 27 and 28 °C by adding olein. The latter supplies a certain amount of 'melt-in-the-mouth' quality.

Within the range of demi-butters with 41% dairy fat instead of 82%, replacing the DMFs with olein having a melting point of between 26 and 28 °C gives low-fat butter good spreading qualities. Oleins from double fractionation whose melting points are on a scale between 15 and 8 °C are used in all products where fluidity is required at low temperatures. Among future applications, we can envisage the use of these liquid fractions with low melting points in the preparation of butters which can be spread straight from the refrigerator.

Polar lipids which constitute between 0.2 and 1% of total lipids, are essentially phospholipids (90–99%) with small quantities of ceramides (1–8%). The principal phospholipids in milk are phosphatidylcholine (PC), phosphatidylethanolamine (PE) and sphingomyeline (SPH). They each represent between 19 and 35% of total phospholipids. Phosphatidylserine (PS) and phosphatidylinositol (PI) are present in small quantities (3–5%). It should be borne in mind that the fatty acid composition of the phospholipids is characterised by the absence of short and medium chain fatty acids. Ceramides from milk are ceramide-monohexosides (CMH) and ceramide-dihexosides (CDH). They are located in the globule membrane. These are glucosyl-ceramides (β-glucosyl-1-1-*N*-acyl-sphingosine) and lactosyl-ceramides (β-galactosyl-1-4-β-glucosyl-1-1-*N*-acyl-sphingosine). The fatty acid composition of the ceramides, which contain from 3 to 5% CMH and 3% CDH, is close to that of sphingomyeline, characterised by a very high proportion of saturated fatty acids.

Lipoprotein concentrates, obtained by clarifying wheys by means of **thermal-calcic aggregation**, have a higher lipid content than the whey from which they came (2–7 g/l), namely 5–28% of the dry matter. The considerable differences in the lipid content of the concentrates can be explained by the fact that some of them have been subjected to further processing in order to demineralise them and/or remove their lactose. The fatty acid composition of the total lipid extract is typical of the dairy product with a high level of saturated fatty acids (70%) and a low level of polyunsaturated fatty acids (5%). The proportion of phospholipids has been estimated at 20–30% of the total lipids. The principal phospholipids are sphingomyelin, PC, PE, PS and PI.

The high proportion of sphingomyelin and the low proportion of polyunsaturated fatty acids differentiates milk phospholipids from other sources of phospholipids available on the market. Although this low rate of polyunsaturated fatty acids can damage the nutritional image of these lipids, it is an advantage within the technological field, in that these phospholipids are less sensitive to oxidation. Moreover, as the phospholipid composition of milk lipids is close to that of the cells in the human epidermis, cosmetology would seem to be the normal outlet for this lipid fraction.

Numerous research teams are interested in the **bio-forming of lipids**, in particular those from milk. This conversion of dairy fats, using two enzymes of high specificity, should result, by means of lipolysis, trans- and inter-esterification and desaturation reactions, in products whose physical, nutritional or aromatic properties are fully controlled. In conclusion, we can mention that dairy fat is one of the most important elements involved in the production of milk chocolate. Because of its compatibility with cocoa butter it is capable of forming a continuous phase. In addition, it contributes to the smooth taste and texture of chocolate and is less costly than cocoa butter (see Chapter 10).

6

Egg products

For centuries the egg has been regarded as a high-quality food, although in the West, when we consider its nutritional value, we are basing our assessment almost exclusively on the properties of the hen's egg. The name 'egg products' includes eggs presented in ways other than in shells. After briefly summarising the biochemical characteristics of the egg, we will consider its nutritional value, functional properties and the current economic developments of this excellent raw material.

6.1 Structure and composition of the egg

6.1.1 Whole egg

The principal parts of the egg are, from the inside outwards: the yolk or vitellus, the white or albumen, the shell membranes (internal and external) and the shell. The relative proportions of each of these constituents can vary considerably. The average values applicable to a hen's egg are: shell 9.5%, white 61.5%, yolk 29.0%. A whole egg contains approximately 66% water, 11% mineral substances and 23% organic substances (12% proteins; 11% lipids).

6.1.2 Composition of the white

The white consists almost wholly of water and proteins, with a few minerals, which is very unusual for a product of animal origin (90% of the dry matter consists of proteins). It also contains free glucose (at double the concentration of blood plasma) which constitutes a primary source of energy

Table 6.1 Proteins in white of chicken eggs

Proteins	Dry matter (%)	Molecular mass (Da)	Isoelectric point	Temperature of denaturisation (°C)	Characteristics
Ovalbumin	54	46 000	4.6	84	Gelling agent
Conalbumin (ovotransferrin)	12	76 000	6.5	61	Combines with metals, bacteria inhibitor
Ovomucoid	11	28 000	4.0	70	Inhibits trypsin
Ovoglobulins	4 + 4	36 000–45 000	5.6	92	Foaming properties
Lysozyme	3.5	14 300	10.5	75	Lysis of bacteria
Ovomucin	3	$5.5–8.3 \times 10^6$	4.5–5.0		Viscosity factor
Ovoinhibitor	1.5	49 000	5.1		Inhibits serin proteases
Ovoglycoprotein	1.0	24 400	3.9		Viscosity factor
Flavoprotein	0,8	34 000	4.0		Binds vitamin B_2
Ovomacroglobulin	0.5	7.7×10^5	4.5		High antigen capacity
Ficin inhibitor	0.05	12 700	5.1		Inhibits SH proteases
Avidin	0.05	68 300	10		Complexes biotin

available to the embryo. Each protein is known for its specific properties, both functional and nutritional (Table 6.1):

- **Ovalbumin**, the most abundant protein in albumen, is a phosphoglycoprotein. It contains 3.5% carbohydrates with the number of moles of phosphate bonded to the serine residues varying from 2 to 0. The molecule contains four free sulphydryl groups and two disulphide bridges, but the number of the latter increases during storage and an '*S*-ovalbumin' is formed which is more thermostable than the native protein. The proportion of *S*-ovalbumin, which is 5% at the time of laying, can reach 80% after six months' cold storage.
- **Conalbumin** (or ovotransferrin) is a glycoprotein consisting of two sub-units. It has the capacity to bind bi- and trivalent metal cations into a complex. At its pHi one molecule can bind two cations and take on a red (Fe^{3+}) or yellow (Cu^{2+}) colour. These metal complexes are more thermostable than protein in the native state.
- **Ovomucoid** is a glycoprotein which consists of three sub-units. It is heat resistant – except in an alkaline medium – and has an anti-trypsin activity.
- **Lysozyme** is a holoprotein with a very high pH_i. It has a β-glucosaminidase enzymatic activity which allows it to lyse the wall of certain Gram-positive bacteria.
- **Ovomucin** is a glycoprotein whose carbohydrate content is almost 30%. The stretched structure of the molecule is a result of the electrostatic

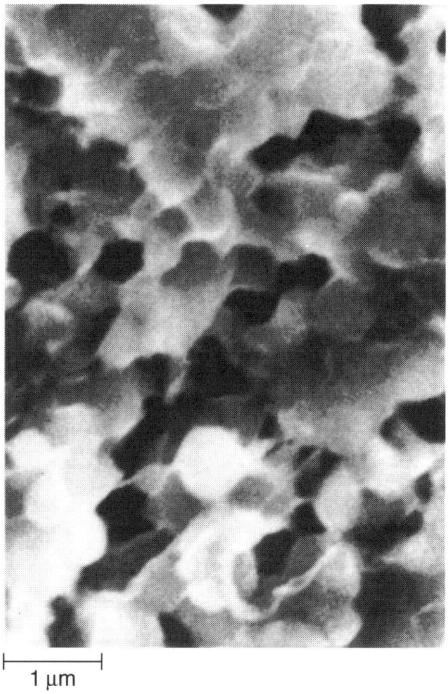

Fig. 6.1 Egg yolk granules in the native state. Observation under scanning electron microscope after freeze etching (–160 °C under vacuum) of a centrifuge residue (10 000 *g*, 45 min) of egg yolk diluted to ½. NaCl concentration: 0.085 M, 10 °C.

repulsion due to the negative charges of the residues of sialic acid which are responsible for the viscosity of the gel layer of the albumen. This protein is insoluble in water and soluble in salt solutions whose pH is higher than, or equal to, 7.

6.1.3 Composition of the yolk

The yolk is a dispersion of particles in a continuous aqueous phase or plasma (Fig. 6.1). The proteins and lipids in the yolk must be considered together, both from a chemical and a functional point of view. The yolk is actually a source of lipids which are easily dispersed in water, thus permitting emulsification of other substances. These properties are due to their high content in phospholipids and to the fact that all the lipids (including the triglycerides) are associated with at least two proteins, vitellin and vitellenin (Table 6.2). The constituents of the yolk can be separated by centrifuging into three fractions:

- A **low-density lipoprotein** (LDL) fraction (lipovitellenin) containing 90% lipids, almost all of which are triglycerides. This fraction represents approximately two-thirds of the dry matter of the yolk.

Table 6.2 Composition of chicken egg yolk

	Dry matter (%)	Proportion of yolk proteins (%)	Molecular mass	Lipid content	Phosphate content of proteins (%)	Location
Phosvitin	4	10	36 000	0	10	Granules
HDL lipovitellin	16	36	400 000	20	$\alpha = 0.5$ $\beta = 0{,}25$	Granules
LDL lipovitellenin	68	24	3 et $- 10 \times 10^6$	88	0.1	
Livetins	10	30	α: 80 000 β: 45 000 γ: 150 000	0	–	Continuous phase
Yolk riboflavin binding protein (YRBP)	1.5	0,4	36 000	0	0.2	Continuous phase

- A **high-density lipoprotein** (HDL) fraction which forms a granular sediment. It represents 23% of the total dry matter, contains phosvitin as well as lipovitellins (lipoproteins). These contain 18% lipids, divided more or less equally between triglycerides and phospholipids.
- A **soluble protein** fraction containing livetins and a few traces of other seric proteins.

The lipids in the yolk are represented by triglycerides (65–70%) and by phospholipids (25–30%), three-quarters of which have a phosphatidylcholine base. Phospholipids are richer in unsaturated fatty acids than the triglycerides but the fatty acid composition of these lipids can vary according to the food eaten by the hen.

The particles therefore contain three types of proteins. **Lipovitellins**, which constitute the HDL fraction, can be separated into α and β-lipovitellins. At a pH of under 7, they appear in the form of dimers. In **phosvitin**, serine represents 31% of the total number of the amino acid residues and over 90% are esterified by the phosphoric acid. This protein is capable of binding Fe^{3+} ions. The complexes formed are soluble and constitute an iron reserve.

The continuous phase contains two types of proteins. **Livetins** are globular proteins derived from proteins in the blood plasma of the hen and can be separated into three groups of different molecular mass (α, β, γ). **Lipovitellenins** can be separated into two fractions L_1 and L_2 with molecular masses of 10 and 3×10^6 Da respectively. In these lipoproteins, the proteins and the phospholipids are situated at the surface of a spherical structure whose core consists essentially of triglycerides and cholesterol.

Vitellenin has a balanced amino acid composition which is close to that

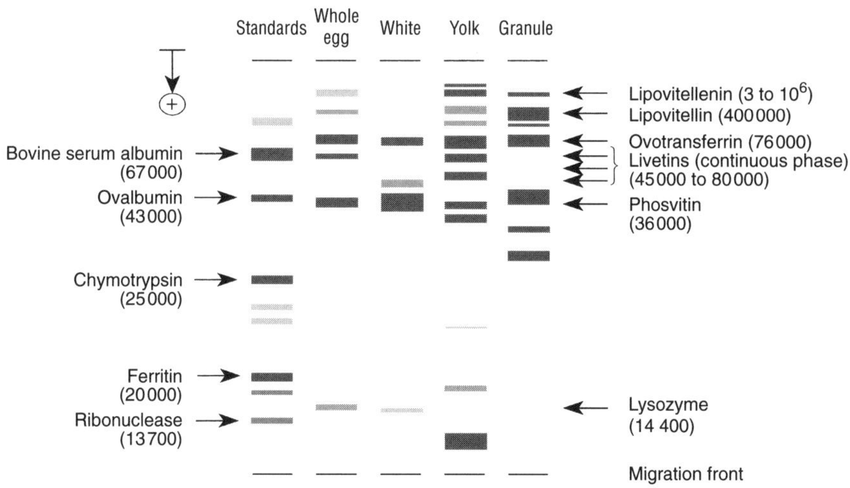

Fig. 6.2 Electrophoregram of egg proteins. Migration carried out in 7.5% polyacrylamide gel in the presence of SDS. Tris-Gly 0.04 M, pH 8.3 buffer.

of vitellin, but it contains very little cysteine. The YRBP (yolk riboflavin binding protein) is a flavoprotein which binds one mole of riboflavin (vitamin B_2) per mole of apoprotein. This glycoprotein whose pH_i is equal to 4.1 is immunologically identical to the flavoprotein in egg white. All the proteins in the white and the yolk are easily identifiable by electrophoresis in a dissociating sodium dodecyl sulphate (SDS) medium (Fig. 6.2).

6.2 Nutritional value of the egg

The egg is a fairly low-energy source of perfectly balanced proteins and easily digestible lipids. It is also an important source of phosphorus, iron and vitamins. On the other hand it is a food lacking in carbohydrates, calcium and vitamin C.

6.2.1 Biological value of the proteins

Egg proteins are well known for their very high biological value which results from the complementarity between the proteins in the yolk and those in the white, and from the balance between the amino acid residues in these proteins. The net protein utilisation (coefficient of digestive utilisation × biological value) of the proteins in the whole cooked egg reaches 94%, which is an indication of their excellent efficiency. In the raw state, proteins from the white eaten alone are only approximately 50% digestible because of the presence of anti-trypsin factors (ovomucoid) and in particular because the raw egg white is a poor stimulator of secretions of gastric and pancreatic juices. Cooking, by coagulating the proteins, facilitates the work of the digestive enzymes, and allows a 92% digestive utilisation to be reached, especially if the egg is accompanied by other foods. In addition, cooking destroys the biotin–avidin bond.

On the other hand, egg yolk is very easily digested in the raw state and any excessive cooking tends to reduce its digestive utilisation. In practice this effect is relatively weak and we can accept that all normal culinary preparations improve the digestibility of the white without adversely affecting that of the yolk. What is more, albumin proteins possess direct or indirect anti-bacterial biological properties (anti-protease activities, complexing vitamins or metals) which contribute to satisfactory egg storage. In this respect, we should stress that washing eggs encourages the penetration of bacteria, because it may eliminate the protective protein cuticle. Water penetration can then allow bacteria to enter. Eggs must therefore be dried quickly after washing.

6.2.2 Lipid digestibility

Egg yolk lipids are highly digestible by humans (between 94 and 96%) thanks to their emulsified state. This digestibility is higher in relation to the triglycerides (98%), the fraction that has the highest content in saturated fatty acids; but still reaches 90% for phospholipids. The egg yolk content in unsaturated fatty acids (approximately two-thirds that of total fatty acids), and particularly in linoleic acid, is moreover an important nutritional element for humans.

As far as the fairly high content of cholesterol (250–300 mg/egg) is concerned, several points should be made. When the daily ingestion level of cholesterol is reasonable, there is no direct relationship between this level and that of blood cholesterol. Eating an egg seems to fall below this threshold. In addition, the level of blood cholesterol depends on the other sterines ingested (plant sterine in particular) and on other dietary characteristics (caloric rate, fibre content, etc.).

6.2.3 Minerals and vitamins

Together with milk, the egg is the food richest in assimilable phosphorus whereas it supplies only a small amount of calcium in relation to human requirements. Another interesting nutritional characteristic of egg yolk is its high iron content, as an egg can meet 30% of a person's daily requirements in this element. If hens have received a balanced diet, the egg is an important source of vitamins. An egg meets between 10 and 15% of a person's daily requirement in vitamins A and D. This food can also supply approximately 5–10% of vitamin B_1 requirements, approximately 20% of vitamins B_2 and B_5 requirements and almost all the requirements in biotin (B_8). Cooking eggs for longer than 5 min can result in certain losses

– of vitamin A (up to 30%), vitamin B_1 and above all folic acid (B_9) (up to 50%).

6.3 Functional properties

Eggs and egg products are obviously used in the food industry for their nutritional value, but also for their functional properties which make them indispensable in numerous manufacturing processes. These properties, and their uses, are summarised in Table 6.3.

6.3.1 Aromatic and colourant capacity

The whole egg, and more particularly the yolk, has a characteristic and very highly valued flavour. The flavours are bound on the lipids of the yolk which contains over a hundred volatile compounds. The colour of the yolk determines the attraction and acceptability of the egg for the consumer. The colour of the vitellus depends on how rich it is in xanthophyll and carotenoid pigments, which come from the hen's diet.

6.3.2 Coagulation and gelling

Egg proteins are responsible for the coagulation which takes place as a result of the action of physical or chemical agents. The egg moves from a fluid state to a solid state called the coagulum. Thermal coagulation takes place from 62 °C upwards in the case of the white and from 65 °C upwards in the case of the yolk. The principal proteins of the white (ovalbumin and conalbumin) have good gelling properties. Ovomucoid alone does not coagulate. The proteins in the yolk are also subjected to thermal coagulation with the exception of livetins and phosvitin.

Salt and sucrose protect against heat denaturation and allow the temperature of pasteurisation to increase by 6 and 3 °C respectively. However, they also increase the resistance of micro-organisms. This protective effect can be explained by a reduction in the quantity of free water available in the soluble phase. Modifying the structure of the protein bound water improves the heat stability of the mixture and delays denaturation.

On the other hand, at the pH of egg yolk, sodium chloride reduces the protein load, affects the hydrogen bonds and increases the role of the hydrophobic bonds. Proteins can therefore aggregate with each other if the temperature is sufficiently high for denaturation to occur. The gelling properties of the yolk proteins are associated with the lipoproteins. LDLs are denatured from 60 °C upwards, lose their fluidity at 65 °C and form a gel at 85 °C. The gel obtained is more stable than bovine ovalbumin or serumalbumin gel prepared under the same conditions. Unlike these two proteins, the lipovitellenins (LDL) produce gels that are stable between pH 4 and 9.

Table 6.3 Functional characteristics of eggs and ovoproducts and their uses within the food industry

Capacities	Agents responsible	Variation factors	Products of substitution	Industrial applications
Aromatic (entire)	Numerous volatile compounds	Chicken diet Storage conditions Technological processes		All food industries
Colourant (yellow)	Xanthophylls Carotenoids	Chicken diet Light Presence of salt Drying	Colourants	Biscuit industry Baking industry Desserts Pasta
Coagulant (entire)	Coagulatable proteins	Time/ temperature pH, ionic strength Presence of sugars Dilution Technological processes	Carrageenan Alginates Modified starches	Biscuit industry Baking industry Processed meats
Binder (entire)	Proteins	Additives increasing viscosity Technological processes	Polysaccharides Pectins Gelatins Gums Proteins	Ice-cream Pasta Processed meat
Anti-crystalliser (white)	Proteins	Presence of yolk Presence of cations Technological processes	Polysaccharides	Confectionery
Foamer (white)	Globulins Lysozyme Ovomucin Ovalbumin	Age of egg Homogenisation Beating conditions pH Dilution Presence of salt or sugars Presence of yolk Technological processes	Caseins and caseinates Whey proteins	Biscuit industry Baking industry Confectionery Ready-made meals
Emulsifier (yellow)	Lecithins Lipoproteins Cholesterol	Beating conditions pH	Soya lecithins, dairy proteins	Biscuit industry Baking industry Processed

Table 6.3 *Continued*

Capacities	Agents responsible	Variation factors	Products of substitution	Industrial applications
		Dilution Presence of salt or sugars Presence of white Technological processes		meats (croquettes) Emulsified sauces

Modifications to functional properties after freezing–thawing are small and essentially associated with an increase in viscosity. The first constituents of the yolk to be affected by freezing are the LDLs and the gelling of the yolk in cold conditions is due to the protein–protein interactions following the rupture of the lipoproteins. In addition, the concentration of salts in the non-frozen phase will also be responsible for particle degradation.

6.3.3 Emulsifying properties

The high emulsifying properties of egg yolk are attributed to the phospholipids and in particular to the lecithins present in the form of lipoprotein complexes. Livetins and lipovitellins help to reduce surface tension and facilitate the formation of the emulsion, but do not influence stability. The LDLs contribute most to emulsion stability.

The hydrophobicity of the LDLs is higher than that of bovine serumalbumin or that of β-lactoglobulin. The lipid constituents that surround the apoprotein on the surface of the micelle produce a hydrophobic environment which facilitates the adsorption of the apoprotein at the interface while the emulsion is being formed. Denaturing the LDLs by heat processing reduces emulsifying activity and capacity, as well as the emulsion stability.

The high viscosity of egg yolk provides the emulsions with good stability. There is a linear relationship between the stability of the emulsion and the square root of the viscosity. Adding egg white to the yolk reduces the stability of the emulsions formed and this effect is essentially linked to a drop in viscosity. This observation is important, since industrial egg yolk can sometimes contain up to 20% egg white.

The viscosity of the yolk increases when sodium chloride is added as this improves the stability of the emulsions but results in a significant reduction in the emulsifying capacity of the constituents of the yolk. Salt causes the protein and lipoprotein complexes of the yolk to dehydrate, with the sodium chloride using some of the water to dissolve it. Dehydrated proteins would tend to group together, thus resulting in increased viscosity, but this would make migration and adsorption at the interface more difficult. In addition, salting allows functional properties to be retained more effectively following heat processing. Finally pasteurisation, freezing and concentration make little difference to the emulsifying properties.

6.3.4 Foaming capacity

Foaming capacity is a very valuable property of the white, and involves the ovomucin, the globulins and the ovalbumin. The proteins in the egg white show maximum foaming performances both at their 'native' pH (pH 8–9) and in the region of their isoelectric pH (4–5).

Sodium chloride increases bulking and reduces foam stability. This is probably the result of a drop in viscosity of the protein solution. The Ca^{2+} ions can improve stability by forming bridges between the carboxylic groups of the protein.

Carbohydrates depress the foam expansion but improve its stability. So, when manufacturing meringues or other products that need to swell, it is better to add the sucrose towards the end of the operation when the foam has already expanded. In this way, the foam-stabilising role played by the glycoproteins of the egg white (ovomucoid, ovalbumin) is linked to their capacity to retain water in the lamellae.

It is well known that low concentrations of contaminating lipids (less than 0.1%) seriously damage the foaming properties of proteins by placing themselves at the air/water interface, thus preventing, through competitive adsorption, the most favourable conformation of protein films.

Egg white is particularly sensitive to excessive beating. Beating egg white or ovalbumin for more than 6–8 min causes a partial aggregation–coagulation of proteins at the air/water interface. These proteins, which cannot be dissolved, are not properly adsorbed at the interface and do not form a coherent interfacial film, hence the viscosity of the liquid lamellae is not sufficient to create good foam stability.

Moderate heat processing used prior to foam formation improves the foaming properties of numerous proteins, one of which is egg white. These foams also have the property of retaining their structure during heating. However, severe drying processes seriously damage their swelling properties, because they reduce solubility.

6.3.5 Other functional properties

Egg white and, to a certain extent, the yolk possess excellent **binding properties**, with the latter involving the water-holding, lipid retention and adhesion properties. In addition, egg white possesses an **anti-crystallisation**

capacity; for example it delays the crystallisation of sucrose in a saturated solution and improves the homogeneity and the texture of confectionery products. Whereas the coagulating and emulsifying properties of the yolk are functional properties which are extensively involved in the physical behaviour of foods and their characteristics of taste, the binding capacity of the mineral elements is a property which is rarely brought to the fore.

However, the remarkable ability of the constituents of egg yolk to bind minerals could be used for nutritional purposes. The proteins responsible for this would actually allow the essential mineral elements to be conveyed in a form the organism can assimilate. Phosvitin has moreover been described as the protein which conveys iron for the embryo. It is possible to isolate two phosvitin–Fe^{3+} complexes of different colours, with their stoichiometric ratio equal to 2. In one case the iron is linked by a tetrahedric bond, and in the second case by an octahedric bond.

In the case of egg yolk, all the proteins, with the exception of the livetins, are phosphorylated and capable of binding minerals, but the phosvitin possesses by far the greatest chelating capacity. The Scatchard diagram shows 140 sites for binding magnesium and 160 for calcium with similar affinity constants. These values can be similar to the number of phosphate groups of the phosvitin.

6.3.6 Modifications to functional properties

When eggs are stored, the conversion of ovalbumin into *S*-ovalbumin, and the dissociation of the ovomucin–lysozyme complex, with destruction of the ovomucin gel, are important reactions from a technological point of view, since they result in at least partial loss of gelling and foaming properties and liquefaction of the egg white. These reactions are essentially due to a rise in pH. In fact, eggs permanently lose CO_2 which migrates through the membranes and the shell. This phenomenon, which can be accelerated by a rise in temperature, results in an increase of the pH of the albumen from 7.6 to a maximum value of 9.7. Several processes allow the rise in pH and its detrimental consequences to be reduced. It is thus possible to maintain an egg's quality for approximately 6 months at –1 °C (temperature slightly higher than its freezing point) and 90% relative humidity, in order to reduce loss of water by evaporation. Another solution consists of storing the eggs in an atmosphere that contains 2.5% CO_2. The reduction in the porosity of eggs, either by soaking in oil, or by briefly heating in water in order to coagulate a thin layer of proteins under the shell, or by using impermeable packaging, has also been tested. In all cases refrigeration is favourable. Later we will see that the whole liquid egg, the yolk or the white can also be stored by adding sucrose and/or salt, with or without prior concentration by ultrafiltration or even after dehydration.

6.4 Current economic developments

6.4.1 Technologies implemented

The principal operations involved in the technology of producing egg products are summarised in Fig. 6.3.

6.4.1.1 Breaking

This operation consists of breaking the eggs individually, as bulk-breaking is prohibited. The egg, placed automatically on a type of egg-cup, is struck by two blades which thus separate the egg into two half shells, with the white being separated from the yolk when it reaches a receiving spatula. Certain breaking machines are now equipped with a scanner to detect the presence of yolk in the whites.

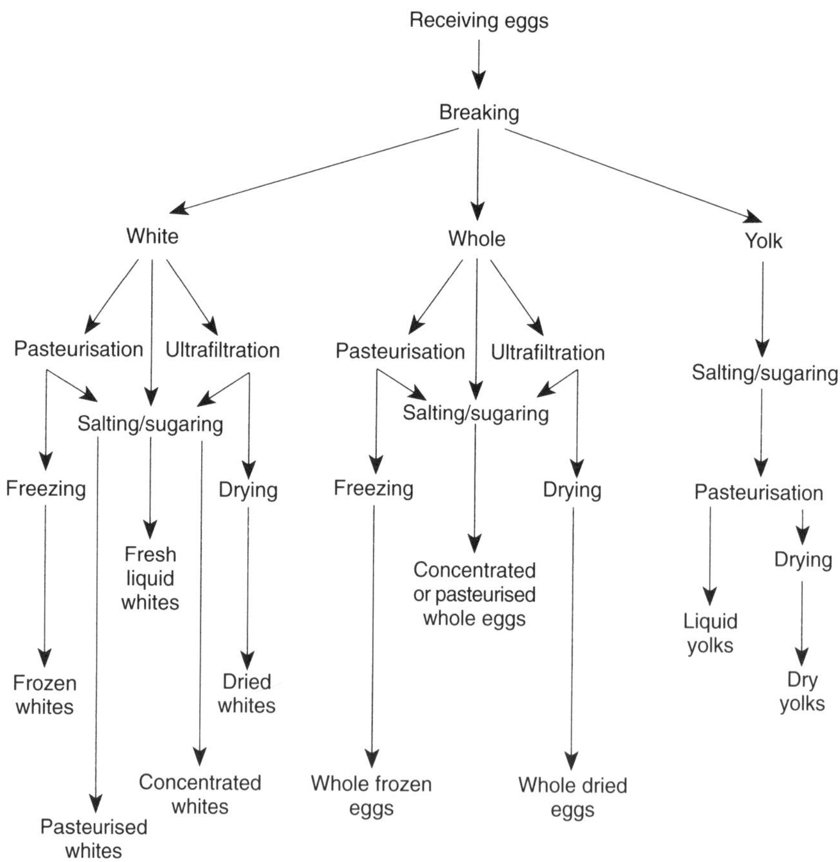

Fig. 6.3 Principal operations in the technology of egg products.

6.4.1.2 *Separation and fractionation operations*

Separation

The quality of white–yolk separation greatly depends on the state of freshness, and the storage conditions of the eggs, and therefore influences the subsequent functional properties of the egg products obtained. For example, migration of yolk into the white impairs the foaming capacity of the egg white. Further to this operation, it is possible to obtain liquid egg products in the form of yolks, whites or whole eggs, which are then strained in order to eliminate shell debris and to ensure the homogeneity of the products.

Fractionation techniques

The main techniques used to extract egg proteins for commercial use are as follows:

- Chromatography techniques:
  - by ion exchange in order to extract avidin, flavoproteins, ovoglobulins and lysozyme;
  - affinity chromatography used to extract the proteins which exhibit biological activity, such as avidin, flavoprotein, conalbumin;
  - gel-filtration used as a preparatory stage in the separation of ovomucin and as a method of effecting quantitative analysis of the lysozyme.
- Techniques of precipitation by means of:
  - reduction or increase in ionic strength, for example in order to prepare ovomucin, or to precipitate lysozyme using NaCl;
  - ammonium sulphate, in order to separate the proteins in the mixture: separation of ovalbumin and ovomucoid.

6.4.1.3 *Pasteurisation*

The purpose of this process is to eliminate pathogenic micro-organisms such as the salmonella present in liquid egg products, by applying time scales of 2.5 min at 58 or 64.4 °C, depending on whether this involves whole eggs, yolks or whites.

The treatments employed require plate exchangers, with a corrugated surface, tubular ones or those with hot incubators in order to pasteurise dehydrated whites (six days at 52 °C). Generally speaking, high pasteurisation reduces the foaming capacity of egg whites and has no effect on the emulsifying capacity of the yolks, if these are salted in advance.

6.4.1.4 *Salting and sugaring*

These operations are used to prepare the egg products for subsequent treatments, so as to retain their functional properties and improve storage. Salting is an operation employed prior to extracting lysozyme from the egg white, as a means of increasing the coagulation temperature with a view to using more severe heat treatments applied to the whole eggs and yolks. Sugaring is used for the same reasons.

6.4.1.5 De-sugaring

This is applied to egg whites in order to eliminate glucose and avoid the phenomena associated with the Maillard reaction during heat processing. It operates either by means of fermentation, by incorporating bacteria or yeasts, or by means of an enzyme (prohibited in France) using glucose-oxidase and catalase. Generally speaking, foaming capacity is improved in egg whites that have had their sugar removed.

6.4.1.6 Concentration

Ultrafiltration is the technique most frequently used for concentrating egg products which have between 11 and 33% dry matter for the white, from 24 to 48% for the whole egg and 46% for the yolk. It is used either to obtain products to be marketed in a concentrated form, or as a preliminary stage before dehydration. The advantage of this procedure is that it does not involve heat and is therefore practically non-denaturing for egg products, apart from the whites whose foaming properties reduce slightly. In addition, these egg products, which are liquid at intermediate moisture contents, can be kept from between six months and one year at ambient temperature.

6.4.1.7 Freezing

This is applied to liquid egg products which need to be stored. They need to be pasteurised no later than 12 hours after breaking. This is carried out in cells or in tunnels at –45 °C or on shelling cylinders which permit products in the form of straws to be obtained. These are easy to measure out and quickly defrost. From the point of view of functional properties, the viscosity of the yolks and the whole eggs increases after rapid thawing, whereas it is almost stable for the whites. Retaining these qualities is directly associated with the rate of freezing.

6.4.1.8 Drying

This allows the water content of the various egg products to be lowered by means of various processes:

- **Spray drying:** this is applied to previously concentrated egg products, with sucrose removed from the whites, salted or sugared to limit denaturation. Centrifugal spraying is generally preferred to spraying by pressure (nozzle) which is not as easy to use.
- **Freeze drying:** this permits products of excellent quality to be obtained from the frozen egg product, but it is still very costly for industrial use. Emulsifying and foaming properties are affected since after rehydration the yolks are more viscous and protein solubility reduces during storage.

6.4.1.9 Irradiation

To date this technique has not received authorisation, but trials carried out reveal its advantages in reducing pathogenic flora and in improving egg product storage. Doses applied using electron accelerators or X-rays vary from 2 to 4 kGy. Problems with taste and smell are encountered in connection with egg products which have been frozen in the absence of oxygen. The foaming capacity of the whites tends to increase.

6.4.2 Industrial uses

Egg products are widely used within the food industry for their first class functional properties (Table 6.4). Although their cost remains high in comparison with other protein binders (milk, blood, soya proteins, for example), the excellent foaming or coagulating properties of the white, and the emulsifying properties of the yolk, give them unquestionable advantages.

6.4.2.1 Functional qualities of some egg products

A comparative study has allowed us to understand certain functional properties of some spray-dried egg powders, by comparing the results between

Table 6.4 Advantages and limitations of using the principal types of egg products

Type	Advantages	Limiting factors
Liquid egg products	Functional properties similar to eggs in shells Great flexibility of use Preparation of à la carte products varying the dry extract, salt, sugar Depending on salt or sugar concentrations → sale limit at 3 °C varies from 5 days to to 1 month	Must be used immediately before knowing the results of the microbiological checks
Frozen egg products	Increased viscosity and return to normal value possible by adding salt and/or sugar Bacteriological quality identical to the fresh product if used immediately after thawing	Storage at –20 °C and mastery of thawing Not very flexible to use Modified functional properties and sample heterogeneous after thawing
Powdered egg	Cheap transport and storage costs Storage over 1 year at 20 °C Stable bacteriological quality Increased viscosity of rehydrated eggs	Functional properties reduced (in particular swelling capacity, coloration) Removal of sugar from whites for successful storage Rehydration operation not yet mastered

these and other egg products (fresh, frozen, etc.). As far as the results obtained for the whole egg are concerned, this study showed that the temperature and gelling time were very close, no matter what product was tested, and that in particular there were very few differences at this level between powders and fresh eggs; the gelling strength depended more on the pH than on the form of the whole egg (powder, fresh, etc.).

The emulsifying capacity of the whole eggs did, however, depend on the form of the product; it was significantly higher in the case of the fresh egg than in the case of a powder. With regard to the results obtained with egg whites, the gelling times of the white are comparable, whether these are fresh, frozen or powdered. The gelling temperatures of these different products are also very close, but slightly lower in the case of the fresh egg. However, gels of egg whites obtained from powders are firmer than those obtained using other forms of white.

The emulsifying properties of egg white powders are closely linked with the pH of the product. The emulsifying properties of fresh egg whites are consistently higher than those of powdered ones. The foaming capacity of powdered egg whites are, on the other hand, consistently higher than those of fresh or frozen egg whites. This might seem surprising because, generally speaking, drying damages the foaming properties of proteins; however, powdered egg whites undergo pre-treatment (de-sugaring) and above all post-treatment (oven-drying) which increases the foaming capacity and the gelling capacity. The various stabilities of foams from powdered whites seem, on the other hand, to be very close, at a given pH, to those of fresh and frozen egg whites. We should also point out that the emulsifying capacity of powdered yolks is clearly lower than that of fresh egg yolks. On the other hand, the stability of egg yolk emulsions obtained from powders is higher than that obtained with fresh eggs.

Measurements of viscosity carried out on egg yolks have shown that egg yolk from powder possesses a viscosity which is approximately ten times higher than the fresh yolk between 25 and 60 °C. On the other hand, at a temperature of 70 °C, the viscosities are more or less the same.

6.4.2.2 Using egg products as food ingredients

Because of their functional and nutritional properties, and in their various forms (liquids, dehydrated, frozen), egg ingredients allow the manufacture of the following types of products:

- Shelled hard-boiled eggs.
- Frozen or dehydrated omelettes.
- Hard-boiled egg cubes for egg salads and aperitifs.
- Egg sausage.
- Egg ketchup – flavoured sauce based on egg yolk.

- Drinks – orange juice mixed with whole liquid eggs, egg liqueurs, eggnog.
- Scotch eggs – hard-boiled eggs covered with sausage meat.
- Egg white yoghurts.
- Ready-to-use mixes.

Work has been carried out using dehydrated egg products, with a view to modifying them for ready-to-use mixes: powdered pancake mix, powdered flavoured omelettes, cooked meat mixes, vegetable pâtés, poundcakes, etc.

6.4.2.3 Molecules of technological and pharmaceutical use

As a result of using and developing fractionation techniques, certain egg white and yolk proteins with useful biological properties can be used and purified, such as the following.

Extracts of the white

- Lysozyme: this is well known for its anti-trypsic and anti-bacterial qualities especially in relation to the vegetative cells of *Clostridium butyricum*, hence its potential use in the dairy and pharmaceutical industries.
- Conalbumin: its chelating properties allow it to transport mineral substances within the organism.
- Ovomucoid, the ovoinhibitor: these proteins are essentially used for their anti-trypsic properties.
- Avidin and flavoprotein: these have nutritional advantages because they transport biotin and riboflavin respectively.

Extracts of the yolk

- Lecithin is used in cosmetic and food products, but for economic reasons it is extracted from soya.
- Phosvitin: on the one hand this protein provides a higher and more easily assimilated source of phosphorus than casein, and on the other it possesses antioxidant properties.

Extracts of the shell

These are hardly ever exploited commercially, firstly because of problems of collection, and secondly because of the relatively low tonnages which limits such uses as:

- incorporation in metals;
- use as fertiliser;
- use of the keratin present in the cuticle with a view to extracting cysteine, which is already carried out using poultry feathers.

6.4.3 Future prospects

The development of egg products and processed products based on eggs on the market shows how dynamic the industry is in finding solutions for the need to improve added value and increase consumption, which is an essential condition for improved commercial use.

However, although numerous investigations carried out to date have made an effective contribution to this development, efforts in terms of research must be intensified. Among the priority routes to take, we could cite the following:

- Developing products that have been adapted to changes in consumer habits (semi-processed, processed products).
- Adjusting egg products adapted for industrial use which consist of either egg proteins only, or mixtures that combine them with proteins of other origins (milk, gluten, legumes, etc.).
- The search for new fractionation techniques capable of extracting certain constituents of eggs having a high added value at a competitive price.
- Looking for new storage techniques capable of retaining functional qualities more successfully.

7

Meat products

Meat technology is essentially based on producing a solution or a dispersion of myofibril products and on eliminating conjunctive tissue in order to obtain a matrix capable of forming an emulsion with water and fats. The **structural rearrangements** that influence the meat **conversion process** are of three types: modifications resulting from heat treating protein structures, reduction in size and dispersion of the myofibrils. For some time the meat products industry has been using functional ingredients in order to produce solutions for its technological problems. Intermediate food products are mainly used to prevent shrinking of muscle tissue and losses of water and fat in the cooking of meat-based products, or in order to modify their intrinsic organoleptic characteristics. Traditional ingredients have gradually been replaced by others, firstly in order to respond to consumer demand for low-calorie products and secondly in order to offer new flavours and textures.

Over 900 descriptive terms define the meats and meat-based products available on the market. Therefore, when eating 'meat' people are consuming products with a *very varied nutritional value*, principally because of the differences in the anatomical composition of the products. Even if we restrict meat to muscles alone, the diversity is only slightly reduced; we can identify 100 muscles with different structures and composition from one half carcass, and each muscle is itself not uniform.

It should be borne in mind that the progressive transition of live animals for slaughter into food products is effected in three main stages:

- During the first stage, the carcass, offal and abattoir by-products are obtained at the abattoir.

- The second stage involves further separation of the carcass into waste (bones, fats and aponeuroses) and muscles.
- Finally, during the third stage, meat products are formulated and processed, usually involving a thermal process.

We must now examine, in more detail, the composition of the carcass, the molecular and functional properties of the muscle protein, and, in conclusion, meat restructuring.

7.1 Composition of the carcass

When we refer to the composition of the carcass we consider only the three main components, which are the **muscles**, the **adipose tissue** and the **bones**. As the animal grows, the composition of the carcass changes, which means it is important to be aware of these alterations dependent on the age and mass of the animals, and according to their genotype and sex.

7.1.1 Muscles

If we take the example of cattle, animals bred for meat are likely to produce very heavy carcasses of 380–400 kg without excessive fat and with a high percentage of muscle (approximately 60%). On the other hand dairy cattle produce lighter carcasses with a lower muscle percentage (around 50%). In comparison with these variations which are significant from a commercial point of view, the differences in the way the muscle mass is distributed between animals of different genotypes or sexes are small.

7.1.1.1 Average composition

Muscle composition varies between animals and, in one animal, from one muscle to another. Although most of the water is intracellular, a significant proportion of **water** and **mineral salts** occupies the extracellular spaces, which constitute between 12 and 15% total volume. Among the mineral salts, sodium and potassium are the most abundant, and are found almost exclusively in an ionised state. Calcium and magnesium are mainly found in the form of organic complexes associated with proteins and phosphorylated compounds. These cations are activators or inhibitors of enzymatic reactions which are important in muscle contraction. Finally, in the muscle *in vivo* the movements of calcium play a determining role in the bonding between the excitation at membrane level and the activation of the contractile mechanisms in the myofibrils. As for phosphorus, it plays a determining role, in the form of organic compounds, in the metabolism of muscle energy.

Proteins which represent approximately 20% of muscle mass can be split into three categories according to their properties of solubility:

- Sarcoplasma proteins 30–35%.
- Myofibril proteins 50–55%.
- Stroma proteins (protein from the cytoskeleton and collagen).

This classification based on properties of solubility takes into account the existence of three classes of proteins corresponding to three fundamental elements in muscle structure:

- The cytoskeleton of the muscle fibre.
- The mechanisms of muscle contraction.
- The connective tissue.

In addition to the dominant proportion of water (75–80%) and proteins, muscle also contains lipids (2–4%), glycogen (1%) and non-protein nitrogen substances (1%).

7.1.1.2 Muscle structure

The base unit of the muscle tissue is the muscle fibre, a multinucleated cell several centimetres long and from 0.01 to 0.1 mm in diameter. In addition to a cellular skeleton, this cell contains contractile apparatus made up of protein filaments arranged in parallel to the cell axis (Fig. 7.1).

Myofibril organisation presents a regular axial differentiation which gives it the appearance of a series of disks or alternated bands; some, the A bands, optically dense and the others, the I bands, less dense. The I bands are split down the middle by a dark line, the Z band. The interval between two Z bands, **the sarcomere**, is considered to be the histological and functional unit of the myofibril, and is between 3 and 5 μm long when the muscle is at rest. In the fibre, the myofibrils are placed in a concentrated solution of proteins, the **sarcoplasma**, and are maintained by means of the longitudinal and transversal elements of the sarcoplasma reticulum and also by the longitudinal and transversal elements of the cytoskeleton constituted respectively by connectin, a protein having high molecular mass (10^6) and by desmin, having a molecular mass of close to 55 000 Da. These proteins, which are insoluble in solutions of high ionic strength, are only soluble in the presence of denaturing agents (urea, SDS) or extreme pHs.

The myofibrils occupy 60–70% volume of the striated muscle and represent an identical percentage of the mass. Different types of proteins can be identified in the sarcomere and are involved in muscle contraction, either directly as a motor element (myosin and actin), or indirectly as regulatory elements (troponins).

The sarcomere is made up of two types of filaments which interlock. The thick filaments of approximately 12 nm diameter are present at the level of the A band and contain myosin, whereas the fine filaments arranged on either side of the Z band contain actin. The spatial distribution of these elements is such that any myosin filament is surrounded by six of actin, and

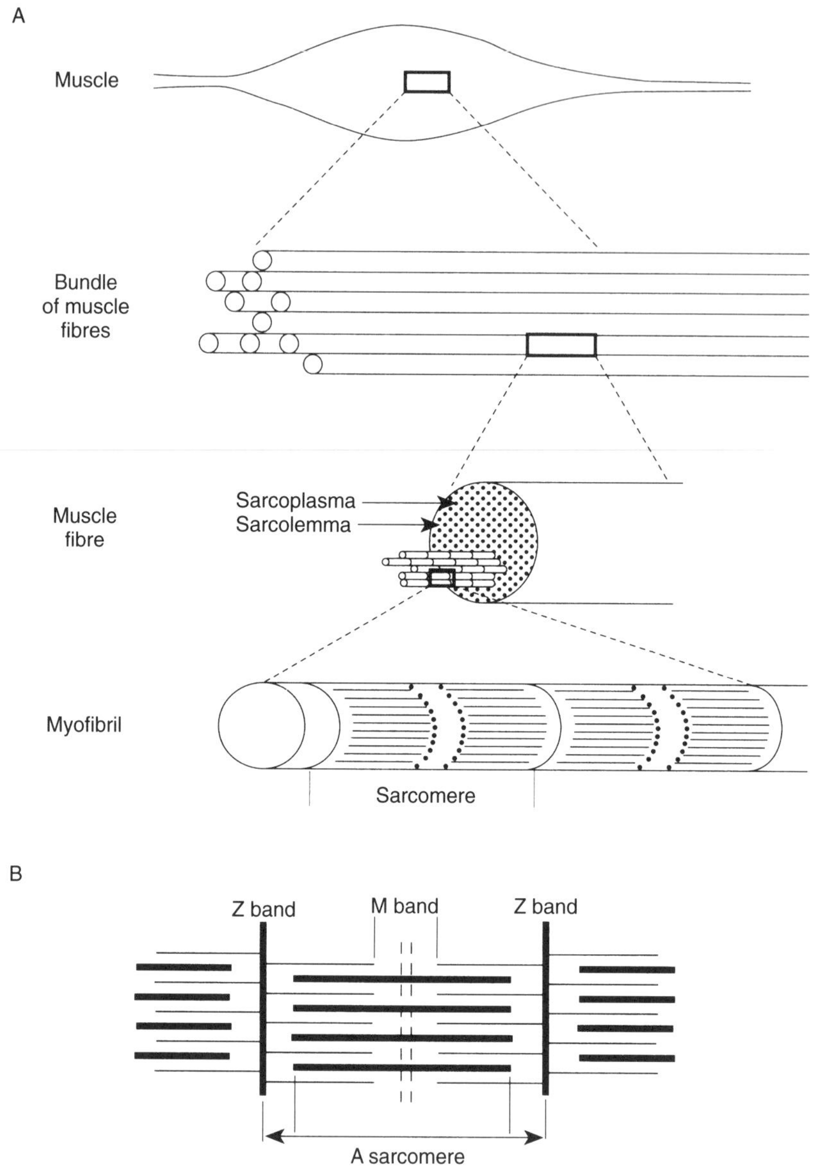

Fig. 7.1 Muscle structure: (A) general muscle organisation; (B) longitudinal section of a myofibril.

one filament of actin is surrounded by three of myosin. The molecular properties of these proteins are dealt with in Section 7.2.

The **sarcoplasma** contains five principal constituents: the sarcoplasmic matrix, the sarcoplasmic reticulum, lipid inclusions, mitochondria and Golgi bodies, with all this being separated from the external medium by a thin membrane, the **sarcolemma**, a double membrane 10 nm thick containing phospholipids and proteins.

Inside the sarcoplasma there is a very complex tubular network, the sarcoplasmic reticulum, which has two principal constituents. Longitudinal elements occupy most of the intermyofibril space. Their continuity is interrupted either at the point of the Z bands, in fish and reptiles, or at the junction of the I and A bands in mammals. A transversal tubular system extends either to the level of the Z bands, or to the A–I junction and is linked to the sarcoplasmic membrane. The sarcoplasma proteins forming the sarcoplasmic matrix constitute a very complex mixture of numerous proteins. They are extracted at weak ionic strength. These are globular proteins in a concentrated solution. In addition to a respiratory pigment, myoglobin, they consist of a mixture of numerous enzymes, in particular those of glycolysis.

7.1.2 Adipose tissue

There is an inverse ratio between the percentage of adipose tissue in the carcass and the muscle percentage. So, in cattle, adipose tissue – expressed as a percentage of carcass weight – decreases from 9% at birth to 6% in an animal of approximately 100–110 kg, then increases up to 25–30% in adult animals. The mass of adipose tissues is composed of various deposits which can be distinguished according to their anatomical localisation (Table 7.1). Generally, we can distinguish subcutaneous deposits in the carcass, inter-

Table 7.1 Lipid composition of cow and sheep muscle

Influence of position in carcass on the phospholipid and neutral lipid content of the muscles (expressed in mg/g)

Muscles	Cows		Sheep	
	Phospholipids	Neutral lipids	Phospholipids	Neutral lipids
Semitendinous	6	47	6.5	30
Long dorsal	6	78	5	50

Principal fatty acids of muscular phospholipids (long dorsal) of bovines or ovines (as a percentage of total fatty acids)

	Cows	Sheep
C16:0	19	15.5
C18:0	9	11.5
C18:1	22	30
C18:2	25	20
C20:4	11	11.5

muscular deposits, and internal deposits which line the internal face of the carcass.

The quantitative development of the adipose tissue is accompanied by a development in its composition; the lipid content of the adipose mass grows from 25% on birth to 75% in the adult, with water and proteins decreasing. Genotype and sex are the bases of biological variability in the ability to store lipids. So, heifers have a percentage of adipose deposits that is from 25 to 60% higher than that of males of the same genotype. Castrated males have between 10 and 45% more adipose deposits than non-castrated males.

Finally the differences between sexes in the distribution of adipose tissue are difficult to analyse, bearing in mind the significant variations in the overall state of fattening between different types of animal.

7.1.3 Bones

Given the low commercial value of bones in comparison with muscles, the muscle/bone ratio of carcasses is therefore a basic criterion for carcass quality.

As the skeleton portion of the live mass decreases with age, there is a progressive increase in the muscle/bone ratio. This development of the muscle/bone ratio also depends on the genotype and overall this means that values for this ratio vary from 3 (light-weight animals for processing) to over 7 for hypertrophied animals. The main industrial value of the bones is in the production of collagen (see Section 9.3).

7.2 Molecular and functional properties of muscle proteins

7.2.1 Myofibril and sarcoplasma proteins

Before examining the functional properties of these proteins, we can look at some molecular characteristics which permit us to understand their functional behaviour more clearly.

7.2.1.1 Molecular characteristics

We have seen how the sarcomere contains two types of filaments which interlock (Table 7.2). The myosin which makes up the **A filaments** is a protein with a molecular mass of around 500 000 Da. This is a very asymmetric molecule of between 160 and 170 nm in length, having an elongated shape with a globular end. Myosin is a very polymorphous protein. The differences in ATPase activities on which the various contractile types are based correspond to isoenzymes of a slow or rapid type. Although these mysosins are very similar in their general structure, consisting of two heavy chains and two pairs of light chains, they have a whole series of different

Table 7.2 Principal proteins making up the myofibrillary structure

Location	Proteins	Molecular mass (Da)	Percentage of total myofibrillar proteins
Thick filament or A	Myosin	500 000	50–55
	Protein C	140 000	2–3
M band	Protein M	90 000	3–5
Thin filament	Actin	50 000	15–20
	Tropomyosin	68 000	4–6
	Troponin T	40 000	
	Troponin I	22 000	4–6
	Troponin C	17 000	
Z band	Actinin	100 000	1–3

properties, in particular those associated with the nature of the heavy and light chains and the functional properties (hydrophobicity).

The **thin filaments** are made up of actin. Actin is a protein that can exist in two forms, G actin and F actin. G actin is a monomer with a molecular mass of 50 000 Da and can only be obtained if there are no ions in the medium. F actin is a polymer of G actin. The monomer links can involve, among other elements, Ca^{2+} and Mg^{2+} minerals and nucleotides. It is found in its polymerised form in the muscle. The I filaments are formed by two chains of F actin arranged in a spiral of 70 nm pitch with 13 monomers per turn. At the level of the Z bands the filaments of actin which, on either side of this line, have structural polarity, are divided into four intersecting strands.

Alongside these very distinctive proteins found in a high concentration in the sarcomere there is a complex mixture of globular proteins in the sarcoplasma. Their content is very low apart from myoglobin.

Myoglobin is responsible for the red colour of meat, in which it forms 90% of the total pigments; among the remaining 10% is found haemoglobin which is a tetramer resulting from the association of four units corresponding to those in myoglobin. Myoglobin concentration varies:

- according to the animal species and the muscles;
- according to age (meat becomes darker in old animals) and exercise;
- according to the iron content in the diet (iron-deficient veal is white).

Myoglobin is a porphyrinic heteroprotein. It joins the prosthetic or 'haem' groups to the protein, or globin, which is itself colourless. The structure of globin is well known: MM = 17 800; 153 amino acid residues, 80% of which are in an α-helicoidal structure, in eight segments having between 7 and 25 residues. The haem is bound by its iron atom to a histidine residue in a fold. The whole forms an ellipsoid of $4.3 \times 2.5\,nm^2$, which is stabilised by hydrogen bonds, salt bridges and hydrophobic interactions. The hydrophilic amino acids meet on the surface of the molecule, forming salt bonds (Lys, Arg → Glu, Asp). Amino acid residues with a hydrophobic side chain are arranged face to face inside the molecule; van der Waals bonds strengthen the folding.

The nutritional value of meat products is largely associated with the supply of haem iron. This form of iron, which is particularly well used in digestion, can effectively resolve iron deficiencies which mainly appear in young people, women and the elderly. It is difficult to determine the quantity supplied by meat products unless a specific quantitative analysis is carried out on this compound. In fact the quantities present are very variable from one muscle to another in the same species, and from one animal to another in the same muscle. However, it is generally thought that consuming 100 g of meat covers, on average, one-third of daily iron requirements.

Metmyoglobin is the oxidised form in which the iron is of Fe^{3+} type, and is brown in colour. Preserving the colour requires conditions that favour the reduced form: namely the presence of glucose and ascorbic acid. Nitrosomyoglobin is formed in meat in the presence of nitrite (nitrated salt or nitrited salt-preserving agents). This is the desired red colour but is not very stable. Carboxymyoglobin appears in the presence of carbon monoxide which binds with the haem iron and thus hinders oxygen transport. Heat or a low pH denatures the globin; ferro- or ferrihaemachromes are formed; the latter have a brown colour and appear when meat is being cooked. The denatured haemachrome compounds of the nitroso-myoglobulin are pink in colour (ham).

7.2.1.2 Rigor mortis

After the animal dies, the cessation of blood circulation suppresses the supply of oxygen; the muscle is then in an anaerobic condition. The ATP (adenosine triphosphate) content which allows the living muscle to conserve its elasticity, and which supplies the energy needed for muscular work, reduces progressively as a result of the action of the ATPases. After bleeding in the absence of oxygen, various mechanisms of re-synthesis will still counteract ATP depletion for a short time. However, it is hydrolysed more quickly than it is regenerated.

The most important of these reactions is **glycogenolysis**, or hydrolysis of the glycogen, which uses up the carbohydrate reserves in the muscle, forming ATP and lactic acid. This acidification inhibits the enzymes of glycolysis, which stops, again contributing to a reduction in ATP. When the ATP has completely disappeared, the muscle is 'really dead' and becomes rigid.

In practice, we can distinguish three successive phases in the establishment of rigor mortis: first of all a period of latency whose duration varies significantly according to the level of energy reserves in the muscle; then there is the period of actual establishment, in which there is a rapid reduction of size and loss of extensibility; finally there is the phase of established rigor, during which the muscle becomes completely inextensible. This occurs between 10 and 48 hours, depending on the type of animal and the conditions of slaughter (Fig. 7.2).

All these reactions result in a drop in muscle pH. This develops, in well-nourished animals which have been well rested prior to slaughter, up to a value of approximately 5.4–5.7, varying according to muscle and species. There are, however, meats known as 'high pH' meats whose final pH does not drop below 6.2–6.0. These DFD (dry firm dark) meats pose numerous problems, in particular from a hygiene point of view (their pH is more vulnerable to bacterial contamination) and from the point of view of appearance (they are dark in colour, which is unappealing to all consumers).

This phenomenon seems to be due mainly to stress, under the normal conditions for handling and transport to the abattoir. In fact, mobilising the energy reserves of an animal that has suffered severe stress results in a reduction in the level of glycogen, leading to a high final pH: on slaughter the DFD muscle has a glycogen reserve which is less than half that found in muscle under normal post-mortem conditions. All animals can produce meat with a high pH, but young bulls are particularly likely to do so. Currently the best way of preventing this phenomenon is for transport times to be short, for group stress to be avoided and for animals to be slaughtered as quickly as possible after they have arrived at the abattoir.

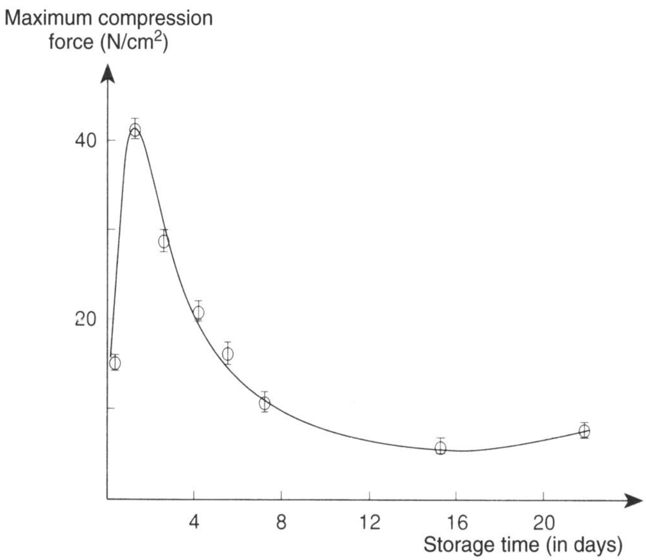

Fig. 7.2 Changes in post-mortem tenderness.

Two other frequently encountered problems are associated with abnormal corpse rigidity: these are cold contraction and pale soft exudate (PSE).

The major effect of **cold contraction** or 'cold-shortening' is an irreversible hardening of the meat. This phenomenon results from over-rapid refrigeration – imposed by regulations – of muscles whose pH has not yet reached its final value, and in particular surface muscles (on the outside of the carcass) which are first subjected to the cold. In order to overcome this phenomenon, or at any rate to avoid it, a technique has been developed: this is the technique of electrical stimulation which consists of inhibiting the contractile ATPasic activity implicated in contraction.

Exudative pork meats are another example of poor development of rigor mortis. In some pigs which are slaughtered under normal conditions, the entry into rigor is very rapid, accompanied by an abrupt drop in pH. This reaches its final value in less than an hour, sometimes in a few minutes. This rapid acidification takes place in muscles whose temperature is still very close to that of the live animal: the combination of low pH and high temperature then causes significant denaturation in the muscle proteins and alters the physico-chemical characteristics of the meat.

7.2.1.3 Functional properties

Solubility

Sarcoplasma proteins are globular proteins in water and in solutions of weak ionic strength. Because of their linear shape, their ionic strength is a determining factor in the movement of myofibril proteins into solution; for this reason they are referred to as 'salt-soluble'. Ionic strength is not the only determining factor in dissolving these proteins; another important factor is the isoelectric point of the protein in relation to the pH.

After salt is added, complexes are formed between proteins and anions, which leads to a drop in the isoelectric point of the myofibril proteins. Adding salt has practically no effect on the pH of the meat medium. The solubility of the 'salt-soluble' proteins increases as the difference between the pH of the meat and the isoelectric point of these proteins increases. It should be borne in mind that the isoelectric point of the 'salt-soluble' proteins moves from approximately 5.4 to 4.6 after adding salt.

Emulsifying power

When determining emulsifying capacity, the test that uses the measurement of conductivity under continuous current has established that the capacity of myofibril proteins is 160 cm^3 fat for 100 mg proteins and that of sarcoplasma proteins is close to 30 cm^3 oil for 100 mg soluble proteins. This difference in behaviour can be explained by the linear shape of the proteins of the sarcomere, which are more suited than the other category to forming stabilising pseudomembranes in the course of emulsification.

Different physico-chemical parameters can be significantly involved in this functional property. So, the quantity of fat emulsified is inversely proportional to the maximum temperature obtained during emulsification. The influence of the pH, for salt concentrations that vary in relation to the emulsifying capacity of the 'salt-soluble' proteins, reaches a maximum at pH 6, and beyond this it remains constant.

As sodium chloride tends to make interfacial tension vary considerably, it ought therefore to have a negative action on emulsifying capacity. In reality the opposite effect is seen; the cause no doubt lies in the fact that sodium chloride increases the solubility of the proteins and thus reduces the protein–protein interactions. This phenomenon takes precedence over the previous one and in the end dictates the behaviour of the proteins. Likewise the polyphosphates have a favourable action on emulsification. They tend to increase the speed of adsorption since in their presence the emulsion develops more quickly. By breaking the calcium bridges, the polyphosphates free the proteins from their complex. In addition, the capacity for emulsification is directly linked to the quantity of soluble proteins, and this relationship is a linear one up to a maximum, which depends on each type of protein.

Two types of test are currently in use in the meat industry for measuring the effectiveness of an emulsifying agent: emulsifying capacity and stability under heat. As far as the second type of behaviour is concerned it has been shown that the stability of emulsions generally increases with the pH of the meat. An increase in pH of approximately two units (4.5–6.5) is accompanied by a simultaneous increase of the HLB (hydrophilic lipophilic balance) of two units as well (from 11 to 13). This index value of meat proteins is close to that of the required HLB of animal fats (around 13–14), and when the protein–fat emulsifying association takes place, maximum stability is achieved.

As far as the respective roles played by the different types of protein are concerned, the following now seem established:

- Myosin has an emulsion stabilising power which is higher than all the other protein fractions in the areas of high pH.
- The sarcoplasma proteins, by arranging themselves in a concentric crown around the globule of the dispersed phase, contribute to lowering the interfacial tension between the aqueous and the lipid phase.
- The stabilising power of all these protein fractions increases simultaneously with the pH, with the exception of actin.

Gelling capacity

Myosin shows gelling capacity even at very low concentrations (0.1–0.5%). The firmness of the gels increases strongly as the myosin concentration increases. Gelling of pure myosin begins at 40 °C and the rigidity of gels increases to 80 °C; it rises as the purity of the myosin solutions increases. As with other gels, rigidity depends on the salt concentration and the acidity of the medium. The myosins that form the firmest gels are those of the rapid white muscles, at a pH close to 5.8 and in the presence of salt, which encourages the macromolecular chains to dissociate. The gels are firmer when the myosin is extracted just after slaughter (before the actin–myosin complex is formed) or when the addition of sodium pyrophosphate allows the irreversible bonds between the actin and the myosin to be dissociated.

Foaming capacity

We know that any water-soluble protein has tendency to form foams when its solution is subjected to mechanical action. Studying the influence of the physico-chemical parameters alone on the degree of aeration shows that:

- using meats with high pHs (pre-rigor meats) results in an increase in foaming capacity;
- the presence of NaCl – in a concentration of between 0 and 5% – increases protein solubility, and consequently foaming properties.

Water-holding capacity

Meat in a pre-rigor state has a high water-holding capacity. During the 24 hours after slaughter, the water-holding capacity drops considerably, and as a result the water loss from the meat during cooking increases during this time interval. In the final phase of maturation beyond rigor, the water-holding capacity increases.

The electrolyte concentration and the ATP content have an essential role to play in the drop in water-holding capacity of meat after slaughter. In fact, slight variations in pH can be enough to induce associations or dissociations of polypeptide chains. During the 24 hours after death the formation of lactic acid causes a drop in pH value; the pH is close to the isoelectric point of the proteins, the protein network tightens up following a reduction in steric repulsion and the result is a reduction in the space available for the water. Adding sodium chloride to the muscle proteins has the effect of increasing its water-holding capacity. The effect of salt on this property depends on the pH. At pH 5 salt has virtually no effect on the water-holding capacity. As a result NaCl has a more marked effect on meat in a pre-rigor state than on meat in which rigor is in the process of being established, and virtually no effect when maximum rigidity is reached. A possible explanation for the action of salt on meat in a pre-rigor state might be found in the binding of chloride and sodium ions on the polypeptide chains. This binding effect would be increased for high pH values which correspond to the maximum negative charges carried by these chains. Through the steric effect due to the Cl^- ion and especially the Na^+ ion, these bonds would inhibit the possibilities between the bivalent cations (Ca^{2+} and Mg^{2+}) and the polypeptide chains.

More specifically, some of the alkaline-clay cations (Ca^{2+}, Mg^{2+}) freed during the ATP degradation process can bind themselves on the negatively charged sites and can initiate intermolecular bridges. However, it is well known that the presence of ATP in the muscle tissue of a freshly slaughtered animal allows the muscle to retain its elasticity by inhibiting the molecular interactions between the myofibril proteins, and, as a consequence, it can retain a high water-holding capacity. The enzymatic hydrolysis of ATP into AMP (adenosine monophosphate) then IMP (inosine monophosphate) results in reduced water holding because IMP, unlike ATP, is not actually able to inhibit the actin–myosin interactions. The loose network converts itself into a tight network of actomyosin corresponding to the minimum value of the water-holding capacity. In the final process of meat maturation, certain ruptures of bonds occur in the actin filaments, resulting in increased tenderness and water-holding capacity.

7.2.2 Proteins from the connective tissue

7.2.2.1 General characteristics

The sarcolemma surrounding muscle fibre is linked to the connective tissue through a layer of interstitial substance consisting of glycosylaminoglycans, surrounded by fine collagen fibres. This body makes up the endomysium (Fig. 7.3).

The connective tissue is therefore located around the outside of fibres which it groups into bundles of varying sizes. The smallest bundles are surrounded by a connective membrane (the perimysium) and grouped into secondary bundles limited by the epimysium which delimits the actual muscle. The connective tissue finally surrounds the muscle of a membrane which links it to the collagen of the tendons used for muscle insertion. Two fundamental protein constituents should be considered: collagen and elastin. Most of the connective tissue is made up of collagen, in the order

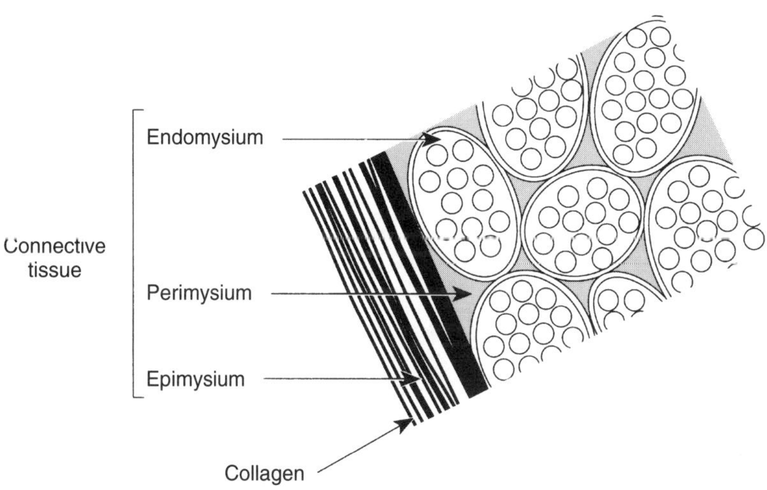

Fig. 7.3 Diagrammatical structure of connective tissue.

of 70–80% of the dry matter. Only small quantities of elastin are present, except in the semi-tendinous muscle of cattle, otherwise it is essentially located in the muscle capillaries.

The structure and properties of collagen are discussed in Section 9.3. Unlike collagen, elastin is very scarce in muscles where its very fine fibres anastomose in an interlocked network. Insensitive to heat, the structure of elastin is completely different from that of collagen. This structure has been shown to be completely non-crystalline, made up of linear chains rolled up at random and bound to each other by chemical bonds. These bridges consist of large size heterocyclic residues. These are yellow chromophores which fluoresce blue-white in ultraviolet rays. Two structures have been isolated from these bridges: desmosine and isodesmosine (Fig. 7.4).

The muscle connective tissue, and more particularly collagen, is an element in the muscle structure which participates directly in defining the toughness of the meat through the properties of rigidity of the fibres which constitute it. This basic toughness of meat, associated with its collagen content, is not particularly affected by the conversions after death which only involve the myofibril proteins and the proteins in the cytoskeleton.

From a **quantitative point of view**, it must be considered that the collagen content of muscles varies between the muscle types, in practice by a factor of 4 between tender muscles and hard muscles. This content also varies among individuals and species, sex and age. From a **qualitative point of view**, collagen is a special protein in the sense that it has a very slow speed of renewal and also because it 'ages' *in situ*. This ageing corresponds to the establishment of covalent cross-linking bonds between the fibres, resulting in a reduction of the solubility of this protein, an increase in its mechanical resistance, and, finally, in toughening of the meat. Thus, the toughness of meat depends on both the nature and the collagen content of the muscles.

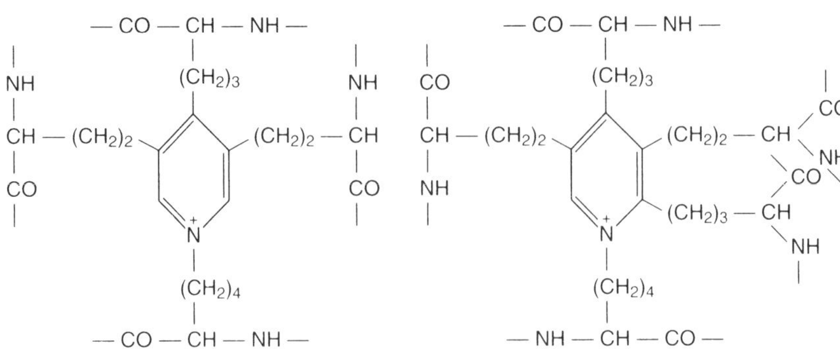

Fig. 7.4 Structure of desmosine and isodesmosine.

7.2.2.2 Functional properties of collagen

In meat products, the functional properties of collagen are a result of its special physico-chemical characteristics: composition of amino acids, molecular mass, conformation, distribution of charge and intermolecular associations (see Section 9.3). These properties depend on biological and technological factors, but we also need to consider the interactions with other muscle constituents (ions, water, lipids, carbohydrates, etc.), which largely depend on the nature and density of charge and on the environment (pH, ionic strength, temperature, etc.).

Rheological characteristics

The mechanical rigidity of collagen fibres in their natural state is the most significant characteristic. Collagen fibre must actually be considered to be an elastic structure whose modulus of elasticity is very high, with this elastic behaviour being limited to a weak field of deformation (<5%). The structure of the collagen, in particular the size of the fibres and the way they are distributed, will influence the overall shear resistance, which partly explains the variations in toughness registered for similar collagen contents. On the other hand, these fibres in the native state seem to have comparable rigidity whatever their biological origin. Technological processes (acidification, heating, enzymatic action) are potential ways of tenderising meat.

Binding capacity

In its natural state, collagen is poorly hydrated, which contributes to its properties of mechanical resistance. Bound water (0.15–0.30 g/g protein) in mono- and multimolecular layers is bound by the ionised amino acid residues. The mobility of free water is largely dependent on the spatial structure of the protein, variations in electrostatic force of the ionic groups and the number of hydrogen bonds. Variations in the pH of the medium and in ionic force, by modifying the conformation of the protein, affect the potential sites for water bonding, with variations in charge resulting overall in water entering the fibres.

After thermal denaturation of the protein (above 60 °C), the rupture of the hydrogen bonds that stabilise the spiral structure considerably increases the reactivity of gelatin with water. When the intermolecular bonds are broken by thermal processing, the soluble gelatin possesses a remarkable water-binding capacity (see Section 9.3).

Hydration capacity

In comparison with the normal swelling of the collagen in meat, three types of swelling of the fibrous structure are traditionally differentiated:

- **Osmotic swelling** is due to a drop in pH; the side chains lose their negative charges, and the free negative ions remaining in the polar zones produce local osmotic swelling accompanied by slight longitudinal contraction of the fibres and water absorption. This phenomenon is reversible and is noticeably reduced in the presence of salt, which cancels the free charges when it binds on the protein. In this strongly hydrated structure, the spiral arrangement of the protein chains is retained insofar as the intermolecular bonds are not destroyed; this increase in hydration through a modification in protein charges very slightly reduces the heat stability of the protein and is therefore the basis of the technique for tenderising meat by means of marinating in an acid medium (see Section 8.2.4).

- **Lyotropic swelling** is induced by neutral salts which mask the ionised groups of the protein and reduce electrostatic attraction. In this way they allow the side chains to relax completely. The increase in intra- and interfibre volume increases the likelihood of water absorption. As a general rule, this swelling is irreversible and corresponds to chemical denaturation of collagen. Moreover, it is less in size than the osmotic swelling. The cations (Na^+, K^+, NH_4^+, Mg^{2+}, Ca^{2+}, in decreasing order of occurrence) and the anions (I^-, NO_3^-, Cl^-, SO_4^{2-}) seem to exercise an independent action. Among the organic anions, acetate produces more marked swelling than citrate and tartrate.
- **Swelling on heating**, associated with the thermal denaturation of collagen (60 °C), is more complex in nature. This swelling is slowed down by the number and stability of the cross-linkage bonds and conditioned by the level of denaturation achieved (temperature and heating time) and by the level of thermal solubility; beyond 80 °C the fact that an increasing proportion of gelatin dissolves reduces swelling.

Before concluding this paragraph, we can add that pre-treatments (partial hydrolysis, reduction into very fine particles) allow the collagen to contribute to the emulsifying capacity of the muscle proteins, but its effectiveness remains more limited than that of the myofibril and sarcoplasma proteins.

7.3 Meat restructuring

Texturisation aims to reproduce a texture identical to or different from that of meat. This is why texturisation processes initially involve varying degrees of disintegration of the structure of the raw material. After this **denaturation stage** which involves the unwinding of molecules and the destruction of molecular structures by rupturing low energy bonds without damaging the peptide bonds and, if possible, the side chains of the amino acids, the protein molecules go through an **orientation** and/or **organisational stage**. Finally comes the **binding stage** of the new structure, obtained when all the intra- and intermolecular bonds which are broken in the course of the first stage are randomly reformed.

The principal restructuring techniques have been dealt with in Chapter 3.

7.3.1 Manufacturing restructured meats

The preparation of restructured meat consists of at least two phases:

- The fragmentation phase, involving numerous chopping, cutting or slicing processes.
- The restructuring phase, which allows a sufficiently strong bond to be obtained between the pieces cut up in this way, as a result of the proteins dissolving, and through the action of binding agents, temperature and pressure.

7.3.1.1 Raw materials and additives

The raw material generally used for restructured meat is beef that is too tough to be eaten in its natural state. Different muscles and various types of meat as well as binders can be incorporated in these products. Certain meats other than skeletal muscles can be used in variable proportions. Elements from offal can be incorporated, in proportions that can vary from 10% for the oesophagus to 40% for the tongue. Likewise, mechanically separated meats (top-quality poultry products) are often incorporated. Moreover, plant proteins have also been used in restructured meats. These proteins have a particular effect on texture and water holding. Soya isolates and gluten yield very similar products but they have poorer cohesion, are less juicy and require a higher fragmentation force in comparison with a product containing a texturising agent (NaCl, polyphosphate). Gluten is less detrimental to flavour and produces textural characteristics that are closer to muscle than soya.

In order to reach an acceptable level of juiciness, restructured meats must have a higher fat content – usually between 12 and 25% – than meat in its natural state. Obviously, the type of fat incorporated influences the quality of the end product. Sodium chloride is a virtually essential additive in the manufacture of restructured meat. It actually allows the proteins to dissolve so that the product has better cohesion for shaping. Polyphosphates are currently being used because they allow better water holding and therefore improve juiciness. These two additives (sodium chloride and sodium tripolyphosphate) have been shown to have an effect on water-holding capacity, the thiobarbituric acid (TBA) index (measurement of rancidness) cooking losses and shearing force of the restructured meat.

7.3.1.2 Conditions of manufacture

The processes of fragmentation, crushing or slicing mean that pieces of varying sizes can be obtained. A size that is neither too thin nor too thick is generally preferable in terms of appearance and colour. When the size of the pieces increases, tenderness reduces and fat loss is higher. Likewise, mechanical measurements indicate that the shearing force increases and the elasticity index reduces.

In order to be able to use parts from the offal, it is necessary to reduce the quantity of collagen which can be recognised in the end-product. This aim can be achieved by means of mechanical or enzymatic tenderising. The first process referred to allows a better bond to be obtained but this treatment can result in problems with colour stability. Tenderising the meat first using cathepsin improves the textural characteristics of the restructured meat.

Among the other parameters studied that have an important effect on the quality of structured meats are the factors of temperature and mixing time. Temperature is a very important factor at the time the meat fragments; this temperature actually needs to be fairly low to produce satisfactory texture and appearance. The mixing time (from 5 to 20 minutes) has no effect on the TBA acid index. On the other hand, loss of mass on cooking reduces as mixing time increases.

To sum up, in order to obtain restructured meat that resembles meat in the natural state, large pieces are preferable, but the product is tougher – which is why it needs to be tenderised first – and it has poorer cohesion. What is more, in this case the lipid inclusions are more visible and the loss of mass on cooking is higher.

7.3.1.3 Different types of product

Depending on the country, the meaning of 'restructuring' can be either broad or restricted. The manufacturing technology can consist of a simple reforming of muscle parts until all the structures have been completely destroyed by means of a mechanical process, or of solubilisation followed by spinning or extrusion in order to imitate the muscle model or to create new networks. Specialists within this industry distinguish the following three broad categories of product:

- Formed meats: products of a given shape obtained by means of mechanical processes, which macroscopically retain their native structure. Nothing is added to these products.
- Restructured meats or those that have been further texturised: meat products that have undergone various treatments (mechanical and chemical) which modify their structure (tenderising, chopping) with a limited addition of non-meat products which improves their appearance and their use (cohesion, colour, etc.).
- Reconstituted meats: mixture of meat of different kinds with addition of ingredients normally present in meats (fat), imitating certain cuts of meat.

7.3.2 Properties of structured meats

In this section, we will be examining only the properties associated with the lipid composition and with the quality and quantity of collagen in raw materials. The protein content of the muscle tissue is in fact similar in the various species; in the order of 20 to 22%. In addition the similarity of the amino acid profiles is very high between beef, pork and lamb.

As far as lipid composition is concerned, the differences among species depend on diet and the actual metabolism of the lipids in each species (Table 7.1). In meats we can distinguish first of all intra- and intercellular lipids and, secondly, extracellular lipids or fatty deposits. Intracellular lipids contain a high proportion of phospholipids, located inside the fibres where they mainly constitute part of the different cellular structures. The intercellular fat contains the lipids, mainly triglycerides, located between the muscle fibres.

Although the fat composition of muscle tissue is fairly constant, that of the 'fatty deposits' varies widely among animals. As this variation among species has a very high biological importance and huge consequences on meat technology, animal 'fatty deposits' can be classified in three categories:

- One group whose composition depends almost exclusively on the diet (as in fish).
- A category of fats whose composition is influenced by the lipids in the diet and by endogenic factors (as in pork and chicken).
- The last group is that of ruminants whose lipid composition is not influenced by diet. Most polyunsaturated fatty acids which are present in the diets of this type of animal are saturated in the rumen.

As a major limiting factor in toughness of meat, collagen lies at the very basis of processes for breaking down meat prior to restructuring. A very low collagen content (<0.5%), corresponding to muscles that are not suitable for restructuring, would result in an over-soft texture after fragmentation. The temperature at which product properties are determined is particularly critical in the case of collagen. In fact, when cooled after cooking, the partial renaturation of the natural spiral structure of the collagen will noticeably increase the product consistency (elastic component) and this happens once the temperature is lower than 30 °C. The conditions of refrigeration (air velocity, temperature) and the storage time of the cooked product will also influence rheological properties. When the pre-cooked product is reheated, the gelatin will liquefy when the temperature reaches a critical threshold, which depends on the molecular composition of the gelatin and the characteristics of the medium (pH, salt); this threshold is situated at around 45–50 °C for beef muscle at pH 5.6. Gelatin liquefaction can then cause the structure to break down (loss of binding power) with losses of lipids and water. In order to understand the contribution made by collagen to the final quality of a restructured product, all the processes ought to be taken into account, in particular thermal processes (heating, cooling, re-heating) including the conditions when the product is eaten.

7.3.3 Meat emulsions

7.3.3.1 Definitions

Meat emulsion is a diphasic system, formed by the solid dispersing in a liquid in which the solid is not miscible. This definition must always be viewed with some reservations because although, for example, it takes into account the behaviour of particles that are insoluble in the aqueous phase, this is far from being the case in terms of fats. This type of biochemical constituent forms a structure that is closer to that of an emulsion than that of a dispersion.

Within the area of chopped products grouped together under the name of meat emulsions, we can define fine paste as a mixture composed essentially of lean meat, 'fat' and water, whose homogeneity is such that the grain of the added constituents is no longer visible to the naked eye. Mortadella, frankfurters and liver pâté are some of the products typical of this category. Their basic ingredients are meat (all kinds), adipose tissue, water, salt and spices, all finely ground during manufacture.

7.3.3.2 Structure of fine pastes

In analysing the structure of fine pastes, some scientists emphasise the emulsifying properties of the soluble muscle proteins. In fact, when meat is chopped, the bundles of fibres are separated and their membranes are broken. By rupturing the sarcolemma and progressively freeing the myofibrils, water capture and swelling of the actomyosin system increase. The matrix obtained consists of two phases: a solid phase consisting of insoluble proteins, muscle particles and connective tissue, dispersed in a liquid phase, which is an aqueous solution of salts and carbohydrates. The fat is then dispersed within this matrix. After chopping, the paste obtained is a multiphase heterogeneous system.

For other scientists, the gelling capacity of the myofibril proteins plays a fundamental role. In fact, as the chopping temperature does not exceed 15 °C, not all the fat needed for the formation of droplets can be liquefied. Small lipid particles disperse in the aqueous phase of the paste, and in the cavities of the matrix network. The raw paste, formed of a network of filaments of actin and myosin, swells to a greater or lesser degree after water is incorporated. This colloid progressively moves into a sol state then, where appropriate, into the gel state, with the ionic strength determining whether or not sol or gel is obtained.

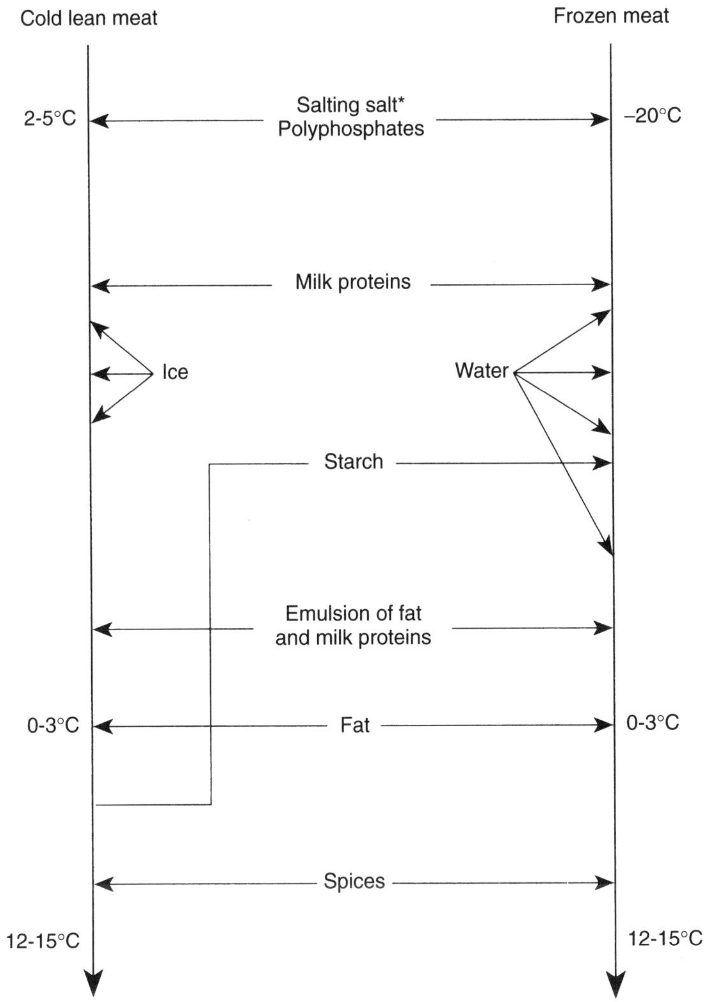

* salting salt: 99.4% NaCl, 0.6% $NaNO_2$.

Fig. 7.5 Chopping in order to manufacture fine-paste products. Order of adding ingredients at the temperature indicated.

7.3.3.3 Manufacturing technology

Manufacturing meat emulsions comprises two essential stages:

- The first one consists of making a dense mixture of the fat and muscles of finely ground offal.
- The second stage consists of obtaining a solid product with a stable structure by cooking the liquid mix.

During manufacture, the mixture is far from being in a stable state. As a consequence, chopping, illustrated in Fig. 7.5, is critical in relation to the desired result. In this respect the most important factors are as follows:

- Addition sequence.
- Emulsification point.
- Ratio between proteins, lipids and water.
- Types of lipid present at all stages of manufacture.

The protein binders (caseinate or isolate) are generally added at the solution or gel state and dispersed at the same time. Adding caseinate to finely textured meat products makes chopping less critical, especially from the point of view of temperature. Adding whey derivatives to fine-paste products is also current practice. Either whole whey is used, or demineralised whey with lactose removed.

8

Products from the sea

Life began in the oceans around 3 billion years ago, whereas the first land species appeared approximately 400 million years ago. This explains the enormous diversity of living marine forms. The richness of the biochemical potential of marine organisms must be viewed in relation to the physico-chemical conditions of this medium: strong salinity, wide range of temperatures, high pressure, etc. These properties are currently arousing industrial as well as academic interest. In addition, in line with the current socio-cultural trends (convenience, fitness, health) the seafood sector is flourishing.

8.1 Structure of fish flesh and seaweed

The biochemical composition of both marine and freshwater fish flesh is very similar to that of land animals. The main chemical constituents are water (between 66% and 84%), proteins (15–25%), lipids (0.1–22%), mineral substances (0.8–2%), a low quantity of carbohydrates (0.3% of glycogen) and vitamins.

If fish flesh can be compared to meat in its protein composition, mineral salts and vitamins, this is not the case as far as its extremely variable content of fats is concerned. In addition, it has a high content of polyunsaturated fatty acids. The relationship between its various constituents has a marked influence on the texture of the flesh. To summarise this from a practical point of view, the flesh of small size fish (250–600 g) contains few lipids and has relatively neutral characteristics, dominated by the constituents of the muscle myofibrils. The flesh of larger size fish (2–4 kg) is characterised by both myofibril and lipid constituents.

8.1.1 Fish muscle

The muscle is the most useful part in terms of food. When fillets are removed on the production line it is essentially the muscles from the trunk that are being cut out. They represent approximately 35–40% of the weight of the whole fish (whiting, cod) to 65–70% (herring, carp, anchovy, pollack).

8.1.1.1 Biochemical characteristics

Fish muscle contains the same protein constituents as the skeleton muscle of mammals (Table 8.1), but it has some different characteristics. From a structural point of view, these muscles have a metamere structure which is not often found among mammal tissues. The principal part of fish flesh is formed of long muscles divided into 'leaves' of a more or less conical shape whose tip is directed towards the head. These numerous segments, which are also called **myotomes** (lamellae with short myofibrils whose length is ⩾3 cm) are piled on top of each other but separated from each other by partitions of connective tissue: the **myocommas**. When the connective tissue separates from the myotomes, the leaves separate from each other and this is referred to as 'gaping'.

In order to allow the intense contraction required during brief periods of flight or hunting to take place, white muscle, rich in glycolytic enzymes, is preponderant. Brown muscle (between 5 and 30% by weight according to species) is particularly well developed in oily fish such as tuna and salmon. It is in fact relatively rich in lipids, myoglobin and mitochondria. This type of muscle is sensitive to cold contraction (after death) and is responsible for deterioration in colour and odour (oxidation of lipids catalysed by the haem-proteins).

Moreover, collagen, present in lower proportions than in mammal muscle, has a lower thermal resistance; it gelatinises from 35–40 °C instead of 60–65 °C in the case of bovine collagen. As far as protein composition is concerned, the respective proportions of the various myofibril proteins are

Table 8.1 Comparison between the composition of fish muscle and the skeletal muscle of mammals

	Fish muscle	Mammal skeletal muscle
Water*	70–80	65–72
Proteins	15–26	15–23
Lipids	1–10	4–15
Carbohydrates	0.3–1.0	0.5–1.0
Minerals	1.0–1.5	1.0–1.3
Sarcoplasma proteins**	20–35	30–35
Myofibrillary proteins	60–75	50
Stroma proteins	2–5	15–20

* Expressed in g per 100 g muscle.
** Expressed in g per 100 g total protein.

similar in fish muscle to skeletal muscle in mammals; however, it must be stressed that the myosin molecules are more sensitive to proteolysis as well as to heat denaturation.

8.1.1.2 Changes after death

In fish, the phenomena connected with the appearance and relaxation of corpse rigidity are rapid and on average occur 5–22 hours after death respectively when immediately stored at 0 °C (Fig. 8.1). The *post-mortem* hardening and the drop in pH remain moderate. The pH generally drops from 7 to 6.5–6.0 in the case of non-oily fish and to approximately 6.0–5.6 in the brown muscle of oily fish. This lowering of muscle pH is generally not sufficient to inhibit or slow down microbial development. However, fish viscera and skin are rich in psychrotropic bacteria and the muscle contains many compounds that can be used as nutritive growth substrates.

In addition to having a proteolytic activity capable of rapidly softening the tissues, enzymes such as lipases and phospholipases remain remarkably active when cold and promote the oxidation of free fatty acids. Decarboxylase activities, in particular, result in the formation of histamine from histidine, while bacterial reductases convert trimethylamine oxide into trimethylamine, a smelly compound characteristic of fish that is no longer fresh. So, it is clear that with an enzymatic and microbiological potential of this type, fish must be refrigerated immediately after death and stored at 0 °C for less than 24 hours.

To summarise the muscle changes in fish from a practical point of view, the most important point is that the pH does not drop as low during rigor mortis as it does in fresh meat. So, for fish flesh, the problem of tenderisation does not exist. In contrast, real decomposition (drop in protein solubility, liberation of fatty acids) begins earlier. The pH rises fairly quickly, disagreeable flavours appear, and the fish is very susceptible to bacterial contamination.

8.1.1.3 Functional properties of myofibril proteins

Myofibril proteins from fish possess a very good **gelling capacity**. They are actually capable, at a protein concentration in the region of 20%, in the presence of 0.5 M NaCl, of forming an elastic gel, smooth in appearance, whose firmness is superior to that of coagulated egg white. Under these conditions these proteins undergo structural modifications. The myofibril proteins are partially dissolved. The formation of actomyosin can be observed. The denaturation temperature measured by DSC drops by between 5 and 10 °C.

It should be appreciated that a gelled elastic pre-network can be produced at low temperature: within 24 hours at 0 °C; 1–2 h at 20 °C; 30 min at 40 °C. The heavy chains of myosin are those that seem to aggregate by preference during this pre-gelling stage. Although myofibril fish proteins possess a pre-gelling capacity that is higher than the muscle proteins of

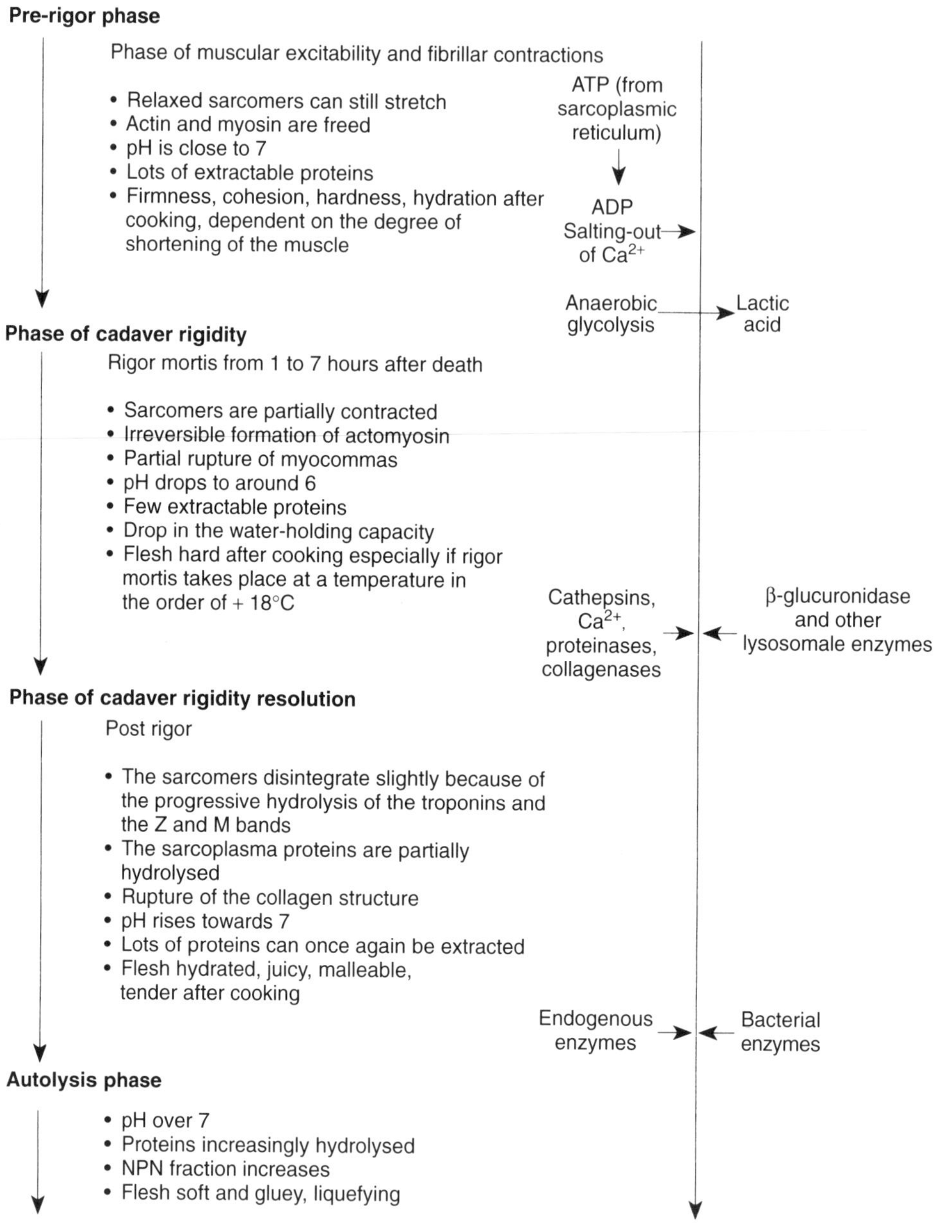

Fig. 8.1 Changes in fish muscle after death.

other animal species, cold water fish have a pre-gelling capacity that is higher than other species of fish.

After being heated to 60–70 °C, the gelled network is, however, clearly weakened. This phenomenon is probably due to the activity of the endogenous proteases in the muscle and/or to protein denaturation. As a result of increasing hydrophobic interactions, subsequent heating (or direct heating, without pre-gelling) to 90–100 °C clearly reinforces the firmness of the gel network. This functional property is obviously used to good effect in the technology of *surimi* (see Section 8.4). As the gelling capacity and the pre-gelling power at low temperature depend on the species and the freshness of the fish, various processes for improving these properties have been developed. These include, for example, adding starch, soya, gluten or egg white proteins, or using a transglutaminase from a *Bacillus* source. This enzyme catalyses the formation of isopeptide bonds between the protein chains in accordance with the following general reaction:

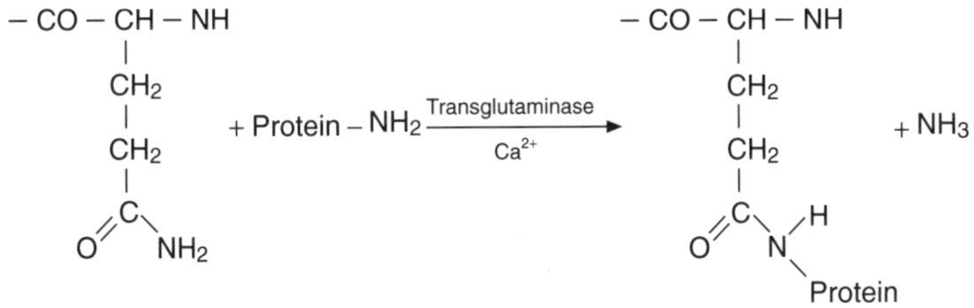

It should be borne in mind that these bonds are broken by the intestinal enzymes and so using transglutaminase does not affect the nutritional value of the protein preparation. Not only is the behaviour under heat of the myofibril proteins of fish relatively atypical, but these proteins also undergo progressive denaturation in the course of frozen storage. This phenomenon can be explained by the aggregation of the protein chains, which occurs when the formation of pure ice crystals concentrates the remaining unfrozen aqueous phase, and brings together the protein chains. Biochemical studies have shown that the aggregation takes place as a result of the SH groups of myosin oxidising to intermolecular disulphide bridges and also as a result of hydrogen bonds and hydrophobic interactions. Moreover, protein denaturation is accentuated when the fish muscle – in particular that of non-oily fish – contains a high amount of trimethylaminoxidase demethylase (TMAO demethylase), an enzyme that is active in the frozen state. Now, this biocatalyser frees formaldehyde from trimethylamine oxide, and this can bind itself onto the free amino groups of proteins, causing shrinkage of the polypeptide chains.

8.1.2 Seaweed

Seaweed is a lower form of plant life which, with fungi, makes up the group of Thallophytes. Unlike fungi, however, seaweeds exhibit photosynthetic activity. These plants can be subdivided into two main categories: micro-seaweeds, which are unicellular organisms, and macro-seaweeds, which are macroscopic plants generally growing on the ocean bed. A wide variety of pigments allows a distinction to be made between green algae (Chlorophyceae) and brown algae (Phaeophyceae).

The term micro-seaweed groups together all the photosynthetic eukaryote and prokaryote micro-organisms. These seaweeds represent a very diversified group of about 25 000 species, only about fifty of which have been studied in detail. Macro-seaweeds are the subject of industrial exploitation based either on their nutritional properties (food seaweed industry), or on their glucan content (colloid industry) (see Chapter 12).

The technological exploitation of the micro-seaweeds covers various fields:

- The production of protein substitutes, vitamins and amino acids for animal and human food.
- The extraction of base products such as pigments, glycans, lipids and glycerol.
- The extraction and purification of high added value biological substances: antibiotics, antifungal agents, enzyme inhibitors.

In conclusion we should add that micro-seaweed can be cultivated in order to produce energy, as a significant proportion of the biomass grown can be converted into methane or alcohol.

8.2 Preservation technologies

8.2.1 Fundamental principles

Fish decomposes so quickly that elementary methods of conservation are obviously essential: drying in the sun, smoking over wood fires, salting and curing. Other techniques have gradually been added to these age-old methods, and have contributed to the success of the fishing industry: dehydration in a drying chamber, freezing and industrial quick freezing.

Because of its high water content, fish was therefore traditionally preserved and stored using methods that reduced its water activity. Water acts as a solvent for the soluble components making up fish flesh; every process intended to eliminate water, such as **drying**, concentrates the soluble components in the remaining aqueous phase. Rather than eliminating water it is also possible to add solutes that will immobilise part of the water. This is what happens during **salting**, and salting and drying are often combined in order to preserve foods of animal origin. Traditional salting carried out using dry salt makes water migrate from the fish to the exterior and the resulting brine penetrates and diffuses into the flesh; some of the water can be eliminated by subsequent drying, either in the open air or in a drying chamber. Salting can also be followed by **smoking** (when the two operations are combined this is referred to as curing) either cold (20–35 °C) or hot (65–90 °C).

8.2.2 Drying

Drying must be carried out quickly to avoid the fish decomposing, but not too quickly, otherwise the flesh will toughen. The technique most frequently used is forced air drying, in which a current of hot air – as dry as possible – spontaneously establishes a difference in temperature and partial water vapour pressure between the fish flesh and the surrounding air. This results in a transfer of heat from the air towards the fish and a transfer of water in the opposite direction, with the air acting both as a heating fluid and as a vehicle for the water which has been removed. An equilibrium is reached between the partial pressure of the water vapour in the air and the liquid phase of the fish. This apparently simple situation is the result of complex physical phenomena: humidity of air–water vapour mixtures, enthalpy of humid air and adiabatic saturation of the air.

Drying was initially carried out in the open air and this is still the current practice in many countries. However, in this case, the ambient temperature cannot produce more than a given quantity of water vapour, depending on the temperature, resulting in excessively long drying times. In industrial drying, air circulation is accelerated by means of suitable ventilation and by using equipment that allows the temperature and humidity to be regulated. In a ventilated drying chamber, unlike the sequence of events during air drying, the water vapour from the fish flesh is removed by the circulated air, the atmosphere is never saturated, and evaporation continues up to the desired limit.

From a physical point of view, drying is the result of water migration in the following sequence: diffusion within the flesh → migration to the surface → evaporation into the surrounding air → reduction in moisture close to the surface. This transfer of water takes place as a result of the heat transferred in the opposite direction: supply of energy → transfer from air onto the surface of the fish → conduction through the flesh → contribution of latent vaporisation heat to the water in the fish. The rate of drying, the output from the drying chamber, the quality of the fish and the cost of the process depend on the values of the parameters controlling these various sequences (product temperature, surface area, humidity, temperature, pressure and velocity of circulation of surrounding air).

The density, which depends on the lipid content/water content ratio, the thermal conductivity – lower in oily fish – and the moisture content of the flesh are the main physical properties that will affect the extent and speed of water migration in the fish. In fish, the water diffusion properties are practically isotropic and almost all the non-oily species have the same water-diffusion coefficient. This remains more or less constant until the water content drops to around 0.1 g water per g of dry matter.

As the composition of the fish is modified by reducing its water content, drying does not permit the initial sensory characteristics to be regained on reconstitution. If it is carried out successfully, it will produce a different end-product, which can be stored for a long time (except in humid conditions) especially if it has been salted first. The effects of drying on the biological and organoleptic properties are essentially due to the drop in the a_w, but also to the direct action of heat: lipid and pigment oxidation, partial denaturation of proteins, Maillard reaction and destruction of vitamins (especially B_1). Problems with flavour can result: rancid, soapy, bitter, sharp or sour flavours.

Loss of aroma observed during drying depends on the molecular mass of the aromas, their vapour pressure and their solubility in water. However, the loss of aroma is less than that which might be expected, given its volatility at the drying temperatures; in fact their diffusivity through the fish flesh decreases as drying proceeds.

The texture of the fish also undergoes modifications as a result of drying. This is essentially due to the irreversible changes the proteins undergo. From a practical point of view, dehydration results in a shrinkage of flesh which increases in accordance with how slowly the drying is carried out. If drying is rapid the flesh will tend to remain more porous, therefore more suited to rehydration, although more sensitive to oxidation.

8.2.3 Salting

Salting begins as soon as contact is made between the fish and the salt, and ends when the salinity of the fish is sufficient for it to be unsuitable for contaminating bacterial proliferation and when it has acquired the specific flavour, odour and consistency characteristic of salted foods ready for consumption. Between these two stages, the salt has penetrated into the flesh under the influence of a certain number of physical and physicochemical factors such as capillarity, diffusion, ionic strength and osmosis, associated with the chemical changes in various constituents (proteins, in particular) in the fish. Clupeins (herrings, sardines, anchovies) as well as gadids (cod) are particularly suitable for salting. As stated in Section 8.2.1, there are some mixed techniques that combine salting with drying and smoking. The term 'curing' applies to the various types of salting, smoking or drying used alone, or in combination. The salted product is both an end-product and a starting material suitable for further processing such as smoking and drying.

When salting begins, the water bound to the proteins cannot take part in this exchange mechanism. While the salt content is between 2 and 5%, the water remains bound to the proteins which swell, and this is known as turgescence. At this low concentration this phenomenon is due to the adsorption of Cl^- ions on the surface of the proteins, thus increasing their number of negative charges at neutral pH and, as a consequence, causing an increase in the forces of repulsion inside and between the polypeptide chains. In the space created there is an increase of protein hydration water since an additional quantity of water is needed to hydrate the new negative charges.

As the concentration of NaCl increases in the flesh, the myofibril proteins gradually become soluble, reaching a maximum solubility at between 3 and 12%, depending on the temperature and species of fish. Then, as the salt concentration continues to increase (above 12%, for example, for cod), proteins precipitate in the tissues (coagulation of albumins, in particular). At salt concentrations close to saturation, virtually all the proteins precipitate. In fact, the Na^+ and Cl^- ions and the muscle proteins have a strong affinity for water; at high salt concentration, however, these ions compete with the proteins for the relatively low quantity of water present, and, as they generally attract the water molecules in preference to the proteins, these dehydrate, and are 'denatured' proteins.

What happens in terms of the water-holding capacity (WHC) of fish flesh during salting? At the pH_i of proteins, the space between the myofilaments is restricted, and hydration is at its minimum level. On both sides of the pH_i, the increase in the number of charges causes intra- and interchain repulsion, and the WHC increases. Adding a neutral salt (as with salt for salting) displaces the pH_i of the proteins and, in highly salted fish, causes a drop in the WHC through protein dehydration and consequently a shrinkage of tissue and reduction of volume of the fish.

Finally, the high fat content of some fish delays salt penetration and makes the fish susceptible to oxidative rancidity. This oxidation is due above all to the oxidising power of the Cl^- ions, but also to the catalytic effect of the haem pigment of proteins such as haemoglobin, myoglobin and cytochrome C. Moreover, in the flesh of cod, although the rate of lipolysis differs between lightly and strongly salted cod, the quantities of free fatty acids are the same in both cases (approximately half the total fatty acids) at the end of salting. The formation of free fatty acids is accompanied by a proportional drop in the phospholipid content at least up to a certain salt content of the flesh (17% for cod). Above this, salt inhibits the enzymes responsible for lipolysis. The unsaturated fatty acids are, as we know, very susceptible to oxidation and the resulting substances, ketones and aldehydes, are responsible for the specific flavours and taste of salted fish, as well as for its colour.

After salting, the fish undergoes maturation. The duration of this depends on the tissular and digestive enzymes of the fish itself, on the bacterial enzymes and the bacterial fermenting action of the micro-organisms. During maturation, the most obvious modifications concern proteins: increase in the content of free amino acids and other forms of non-protein nitrogen (NPN); and at the same time the appearance, consistency of the flesh, smell, and taste develop in different ways depending on whether the fish is an oily or non-oily type. Figure 8.2 shows the biochemical changes of the nitrogen compounds during maturation. During the first phase, protein solubility reduces strongly, especially that of soluble proteins of high ionic

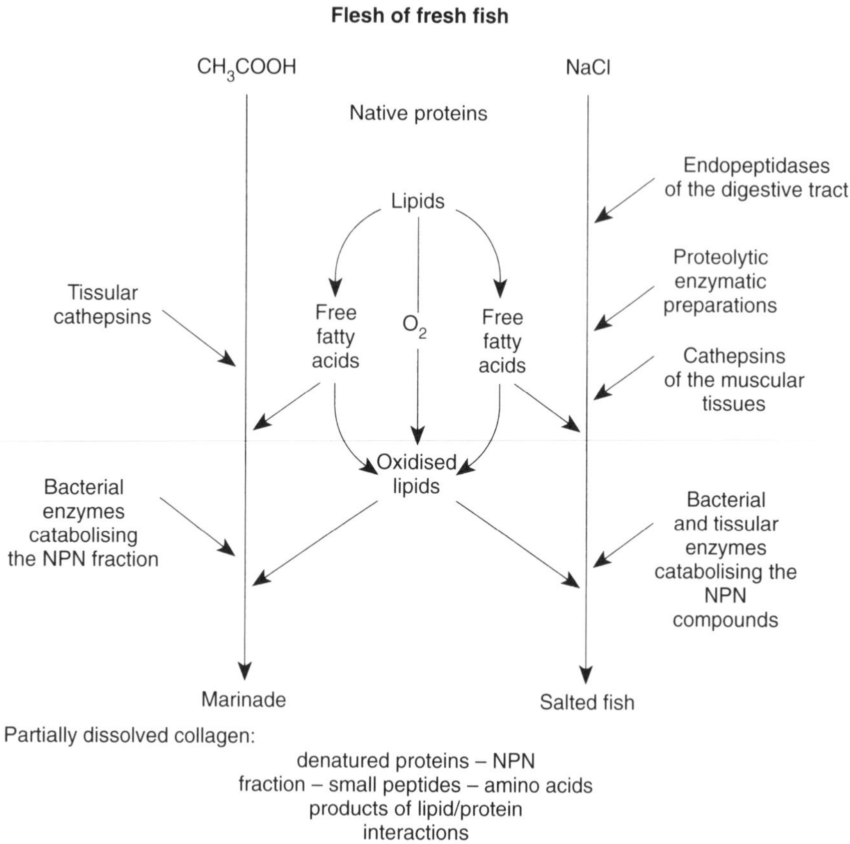

Fig. 8.2 Biochemical changes in nitrogen compounds during maturation of marinated or salted fish.

strength (actomyosin). The insolubility of actomyosin (denatured by the high salt content) makes the fish tough and dry. Then the progressive degradation of the actomyosin moves on to the small peptides and the soluble amino acids of low ionic strength. These soluble compounds contain, alongside their protein fraction, a fraction that cannot be precipitated by trichloracetic acid (NPN) which increases over time. The content of amino nitrogen also increases.

Maturation must take place at temperatures that are determined according to the type of salting used, but above all depending on the species. All the factors likely to modify the enzymatic action or the development of the micro-organisms (temperature, moisture content, salt content, relative humidity) obviously influence maturation. So, well-salted cod can be kept for four to six months between 10 and 15 °C and for one year at +4 °C; whole 8–10% salted herring is stored at between 2 and 4 °C, to ensure that it undergoes a slow favourable maturation.

We can determine the degree of maturation of fish by analysing the quantities and proportions of various nitrogen fractions. For example, during maturation of the herring, the quantity of soluble nitrogen is lower than 20% of the total nitrogen in fresh herring and over 30% of the total nitrogen in over-matured herring. In addition, the proportion of free basic amino acids moves from 40 to 15% during maturation.

8.2.4 Marinating

Marinating is a conservation process which consists of reducing the water content of fish through the action of salt, and of inhibiting the growth of micro-organisms through acidification of the medium. Marinades can involve whole fish or portions, fresh or frozen, processed by means of the combined action of vinegar or dilute acetic acid (in some countries) and salt.

In the course of **cold marination** of raw fish, the preservation properties and the taste of the marinade are solely due to the action of the vinegar and the salt, with the flavour being refined by means of spices.

After normal preparation (removing heads, gutting, possible filleting), followed by short-term 'firming' in saturated brine and 'blanching' (removing blood) in a vinegar solution, the process is carried out in two stages. In the first stage, by immersing the fish in a 'finishing bath' (actual marinating) consisting of vinegar (5–10% acetic acid) and salt (10–15%) for approximately a week, the product acquires its characteristic texture and flavour, while being protected from inopportune bacterial growth. In the second stage, after washing and draining, the fish is immersed for three months in a cold (0–2 °C) marinade containing vinegar (1–2% acetic acid) and 2–4% salt, flavoured with various spices and sometimes with the addition of sugar.

Hot marinating is obtained by first of all cooking the pre-salted fish in brine for 10 minutes at 85 °C in a bath known as a 'blanching bath' consisting of 2–4% vinegar and 6–8% salt. After cooling, the pieces of fish are placed in a marinade containing 2% acetic acid, 3% salt and between 4 and 5% gelatin. This type of marinade is generally less acidic than fried marinades or cold marinades of raw fish.

It should be pointed out that salt penetrates the fish flesh more slowly than vinegar.

During marinating, the proteins are partially hydrolysed by tissue enzymes (with proteolysis being limited by the presence of salt) with the liberation of peptides and amino acids (Fig. 8.2). The autolytic process is rapid for the first two to three days at 10 °C and, at the end of marinating, the quantity of amino acids resulting from proteolysis is approximately the same as that pre-existing in fresh fish. However, some of the amino acids, resulting from the proteolysis, migrate into the finishing bath marinade, but this leaching loss is small as a percentage of the total. If contaminated by acid-tolerant micro-organisms, the amino acids are de-carboxylated and the pH rises owing to the formation of biogenic amincs.

In comparison with cold marinating of raw fish, a higher number of species (herring, lamprey, whiting, haddock, smelt) are fried before marinating. The principle of this technology consists, after immersion in brine at 10% salt for one or two hours and draining, of coating the fish twice with wheat flour and sifted rye so that it is rich in gluten. This flour coating is followed by frying in a mixture of oil which must be at a temperature of between 160 and 180 °C. After cooling, the fried fish is processed under its protective marinade coating.

8.2.5 Smoking

The principle of smoking consists, after pre-salting in most cases, and a degree of drying, of impregnating the fish flesh with the smoke resulting from the slow combustion of wood shavings or sawdust. In practice, smoking is mainly used for organoleptic purposes, rather than as a means of preservation, at least in industrialised countries. Because of the combined use of salting (or brine pickling), drying and smoking, the term curing is the correct name for the process applied. The most important modifications which the fish flesh undergoes take place during the salting and drying process, as the action of the smoke only takes place afterwards.

There are two variations on the technique of smoking:

- **Cold smoking** lasts from several hours to several days. The fish must remain raw, and is dried at approximately 30 °C during the first phase; it is smoked at 24–27 °C during the second phase.
- **Hot smoked** products are cooked, most often after drying at a temperature in the order of 30–40 °C for between 30 and 90 minutes; fish is smoked at a temperature which rises progressively towards 70 to 80 °C for 1–2 hours.

The optimum conditions for producing 'food' quality, non-toxic and effective smoke are as follows:

- Combustion temperature between 350 and 400 °C.
- Oxidation temperature of 375 °C.
- Air velocity in correlation with the humidity and size of wood pieces (sawdust, shavings, etc).
- Use of hard wood, preferably at 17–20% moisture content.
- Hot smoked products are cooked, usually after drying at a temperature in the order of 30–40 °C for 30–90 minutes; fish is smoked at a temperature that moves progressively to between 70 and 80 °C for 1–2 hours.

The composition of wood smoke is extremely complex, as over two hundred constituents have been identified. They can be classified into phenols (the largest group), organic acids, alcohols and carbonyl compounds. The smoke contains numerous polycyclic aromatic hydrocarbons, but these constituents are small in quantity in terms of the amounts found in smoked fish (in the order of several parts per billion); however some of them (in particular 3,4-benzopyrene and dibenzanthracene) attract attention because of their possible carcinogenic effects.

The reactions that occur during smoke production, its transfer and being deposited on the surface of the flesh, and the interactions among the constituents of smoke and the proteins, lipids and other biochemical molecules in fish are not yet precisely understood. We will mainly refer to the effects of smoke on the organoleptic characteristics of fish.

The colour of smoked fish is essentially due to Maillard type reactions. Smoked products marinated in vinegar have little coloration because the pH is too low. On the other hand, a very dry fish can result in a darker colour because the dehydration reaction is accelerated.

Much of the 'smoky' flavour is attributed to the phenol fraction, in particular phenols with a low to average boiling point. The carbonyl compounds, organic acids, vanillin, diacetyl, although only involved in a minor way, also contribute to the typical aroma. The smell of smoke is mostly associated with syringol and 2,6-dimethoxy-methylphenol whereas guaiacol and eugenol mainly contribute to the smoky flavour. However, because of the interactions between the constituents of the smoke and those of fish itself, the flavour of the smoked product does not only depend on the compounds in the smoke.

Smoking also has an effect on texture. First of all there is the action of heat during the pre-drying stage, then the drying stage resulting from the smoking process; then the actual constituents of the smoke which slightly modify the texture. That of cold smoked fish is soft and tender, that of hot smoked fish is tougher and drier. It is, however, possible to obtain hot smoked fish which only harden slightly on the surface and remain soft on the inside. The essential changes: water loss, melting of fat, denaturation of proteins in the connective tissue (gelling of the subcutaneous layer during hot smoking) are mainly the result of the temperature, especially as it is acknowledged that fish proteins are among the most sensitive to heat. However, the shiny, glossy brown film that then forms on the dry surface of fish flesh during smoking is a result of the action of the formaldehyde from the smoke which coagulates the proteins with the help of volatile acids. Certain specialists attribute the shiny, glossy film to the desiccation of the actomyosin solution formed during pre-salting.

To conclude, the biochemical action of smoke can be summarised in a few words. In the case of cold smoking, lysine losses from proteins are in the area of 20%, whereas they can exceed 55% during traditional hot smoking.

Apart from the partial denaturation of proteins and a C-nitrosation of the phenols, the important role of smoke is, above all, its anti-oxidising effect, due to the phenols, on the fish lipids: they inhibit the stage of propagating auto-oxidation. Whereas in hot smoking it is mainly the heat that destroys the micro-organisms, in cold smoking it is the smoke that is the essential element. Although for many years formol was thought to be the principal reason for the anti-bacterial action of smoke, it has now been established that it is the phenols with low boiling points that are the most active. Their action is selective, but this selective effect is itself very diversified for different types of smoke.

8.3 Hydrolysates: economic development of the protein fraction

Hydrolysis can be effected by the tissue or digestive enzymes of the fish itself, where an **autolysate** is produced, or by exogenic enzymes, where a **heterolysate** is obtained. The chemical hydrolysates strictly involve acids or alkaline substances, and must not be confused with mixed hydrolysates whose natural enzymatic action is accelerated by modifying the pH. Another type of mixed hydrolysate can be obtained by means of lactic fermentation.

The principle of preparing enzymatic fish proteins consists of 'dissolving' the proteins using a technique, for example hydrolysis, then isolating the proteins from the ambient medium by precipitating them selectively, either by isoelectric precipitation, or by salting them out by modifying their ionic strength, or by heat coagulation. These techniques are 'positive' and milder than those used to obtain fish protein concentrates (FPC: extraction by solvent intended to eliminate fats and odorous compounds). In addition, the functional properties of the proteins of these isolates are generally well retained (see Chapter 3).

8.3.1 Traditional products

Broadly speaking, the proteolytic enzymes of the fish itself, and the micro-organisms, intervene, in the presence of high salt content. Fish is processed in this way mostly in the regions of south-east Asia to produce sauces, pastes and matured salted fish. An infinite variety of techniques are used. Microbial activity is generally slow, owing to the halophilic bacteria largely responsible for the characteristic flavours and aromas typical of the sauces.

Sauces based on hydrolysed fish are usually prepared from small tropical fish mixed with salt in a fish–salt ratio in the order of 3/2 or 3/1. The typical autolysis established generally lasts for several months (with a few exceptions). Although the principal action is certainly associated with the endogenous enzymes in fish, bacteria are nevertheless involved, although their number reduces rapidly over the first month. A liquid which floats on the surface is formed, and its volume increases with time. The content of soluble nitrogen from the fish proteins increases and the final sauce appears in the form of a brown liquid. *Nuoc-mam*, for example – undoubtedly the best known and most important of all the sauces made from hydrolysed fish – must have an ammoniacal N/amine N ratio lower than 0.5 in order to be of good quality.

Hydrolysed fish pastes usually contain partially dried fermented products, namely the residues from sauce manufacture to which carbohydrates are often added in the form of fermented rice, bran, flour and toasted grains. Generally the pastes have a higher nutritional value than the sauces, as a large proportion of the total nitrogen is soluble (essentially free amino acids).

Matured salted fish – again called hydrolysates of whole fish – is prepared from fish that have been gutted through the throat. They are salted (25% salt) and matured anaerobically. The fish partially dehydrates. On maturity, namely after two or three months, the flesh is soft, pasty and bright red, with a pH of 6. This preparation contains some volatile fatty acids and some amines: it becomes brown as a result of the oxidation of the lipids and the typical odour is attributed to the methylketones.

8.3.2 Industrial hydrolysates

Taking inspiration from traditional techniques, the industrialisation of hydrolysate manufacture means that the time required for the process can be reduced while retaining the desirable organoleptic characteristics. Hygiene qualities can be improved, as can the procedures and the yield, and so new products can be produced, such as the enzymatic isolates of proteins. Obtaining proteins for human foodstuffs presents some difficulties: rapid contamination of fish, odours and flavours which are difficult to avoid or eliminate, loss of some of the functional properties of the proteins during drying. Enzymatic proteolysis allows the functionality of the proteins to be retained. By selecting the appropriate enzymes and by controlling the conditions of hydrolysis, processes have been designed to recover proteins at different stages of fish decomposition.

8.3.2.1 Starting material

Any type of edible fish can, in principle, be considered as a starting material that can be used for manufacturing a hydrolysate. Its physical appearance is not important, as the essential factor is that the fish is fresh or in a good state of preservation. However, different processes must be used for oily fish than for non-oily fish, because above 1% of fats the lipid fraction must either be eliminated or must be stabilised using anti-oxidants. On the other hand, although oxidation products are small, they can contribute to colour and aroma by reacting with the amines and the peptides.

It should be emphasised that fish that is not used commercially or the waste from the fish-processing industry can constitute the raw material which can be used to make hydrolysates. For example, a process of selective extraction of krill flesh has been developed. This produces an important waste product which contains large quantities of proteins (between 50 and 65% of the total weight) which are very good substrates for preparing an isolate after hydrolysis.

Among the three main categories (sarcoplasma, myofibril and connective proteins), only the sarcoplasma proteins are water soluble. The myofibril proteins are influenced mostly by the salt content of the medium whereas the proteins in the connective tissue are insoluble in salt solutions, whatever the salt concentration.

Some fish constituents contribute to the flavour of the hydrolysates:

- The **amines** produced from the amino oxides or the amino acids of fish release the odour of 'fresh' fish; then from 'fresh' to 'putrid' (pyridine, pyrrolidine, piperidine, indole, skatole, putrescine, cadaverine). The ratio of the amine/amino acid content determines the quality of the sauces: a ratio that is too high is of poor nutritional value and the sign of an overactive decomposition by bacterial action. On the other hand, ammoniac exists in almost all these products.
- Butyric and isovaleric **acids** formed from amino acids by the bacteria produce very pleasant '*sushi*' flavours (marinades of raw fish) whereas citric, succinic and lactic acids are responsible for the characteristic flavours (for example the 'sea-food' flavour is associated with disodium succinate).
- The **neutral constituents**, aldehydes, ketones, alcohols and esters are the origins of certain flavours referred as 'fruity' (esters), 'creamy butter' (aldehydes), bitter or greasy (scission of the long chains of oxidised fat), 'cheesy' (methylketones in *nuoc-mam*, for example).
- **Sulphur compounds** produce unpleasant odours at high concentrations, but an aroma that is pleasant at low concentrations (H_2S, methylmercaptan, dimethylsulphurs and diethyl-sulphurs).

8.3.2.2 Proteolytic enzymes

Proteolytic enzymes which are involved in the manufacture of hydrolysates can be classified as follows:

- Endogenous fish enzymes: tissue enzymes (cathepsins, alkaline proteinases, dipeptidases) and digestive enzymes (trypsin, chymotrypsin, pepsin).
- Enzymes from other sources added during processing: enzymes of plant origin (papain, bromelain) or enzymes of microbial origin (pronase, subtilinase, alkalase).

Plant enzymes and microbial enzymes have a wider spectrum of activity than the digestive enzymes of fish; these are much more specific and produce breakdown products having a higher molecular mass. This is why the action of the endogenous enzymes of fish is supplemented, to a greater or lesser degree, by that of other added enzymes, although this can vary according to species of fish; thus, mackerel and sprats require only a few exogenic enzymes for complete hydrolysis to occur. In all these processes, increasing the quantity of enzymes beyond 1% does not produce any noticeable improvement in hydrolysis yield.

8.3.2.3 Parameters of the manufacturing process
Apart from its anti-bacterial action, increasing the salt content beyond 10–15% results in a reduction in the soluble nitrogen content. However, the solubility of the proteins increases with the ionic strength in the region of neutral pH (for a salt content lower than 20%). It reduces if the ionic strength increases in the regions of extreme pH. The pH of the medium must be adjusted to the level of the values of optimum activity for the enzyme being used. However, extreme pHs allow a larger quantity of proteins to dissolve and prevent bacterial contamination. The ideal temperature is that corresponding to the optimum activity for the enzymes involved; it must be sufficiently high to prevent bacterial growth but not too high to prevent the thermal denaturation of the enzymes and excessive production of inhibitors.

8.3.2.4 Properties and applications of hydrolysates
The **autolysates**, which are the result of proteolysis due to the unique action of the endogenous enzymes in fish, are obtained by an industrial process that accelerates this natural process. Manufacture involves a digester, an industrial dryer, a centrifuge, a filter, an evaporator and a spray drying tower. The end-products are packaged in a liquid form, in concentrates or as a powder. The concentrated autolysate contains less than 45% water, over 6% nitrogen (that is, approximately 40% proteins) and less than 0.5% lipids. The dehydrated concentrate from the spray dryer has the same appearance as the protein concentrates of fish, but with better preserved functional properties and a higher protein efficiency coefficient; the autolysate powder has a pH $\leq$5 and moisture <7%.

Hydrolysis of chopped fish flesh, whether or not this has been salted, or whether or not it is raw or pre-cooked, is carried out by adding enzymes of animal, plant or microbial origin, and by avoiding all contaminating bacterial action. The actual **heterolysate** (aqueous phase) is separated from the lipid phase and the undigested sediments; it may be concentrated and dehydrated (Fig. 8.3). The heterolysate can be produced from fish flours or fish protein concentrates, in order to obtain a product with varying content of nitrogen fractions and with better functional properties (solubility, dispersability, etc.) intended for human foodstuffs. Continuous processes have been developed at an exploratory stage, as have the use of membrane techniques.

The composition of proteins and amino acids of the heterolysates is close to that of the original fish. The powders from these hydrolysates contain

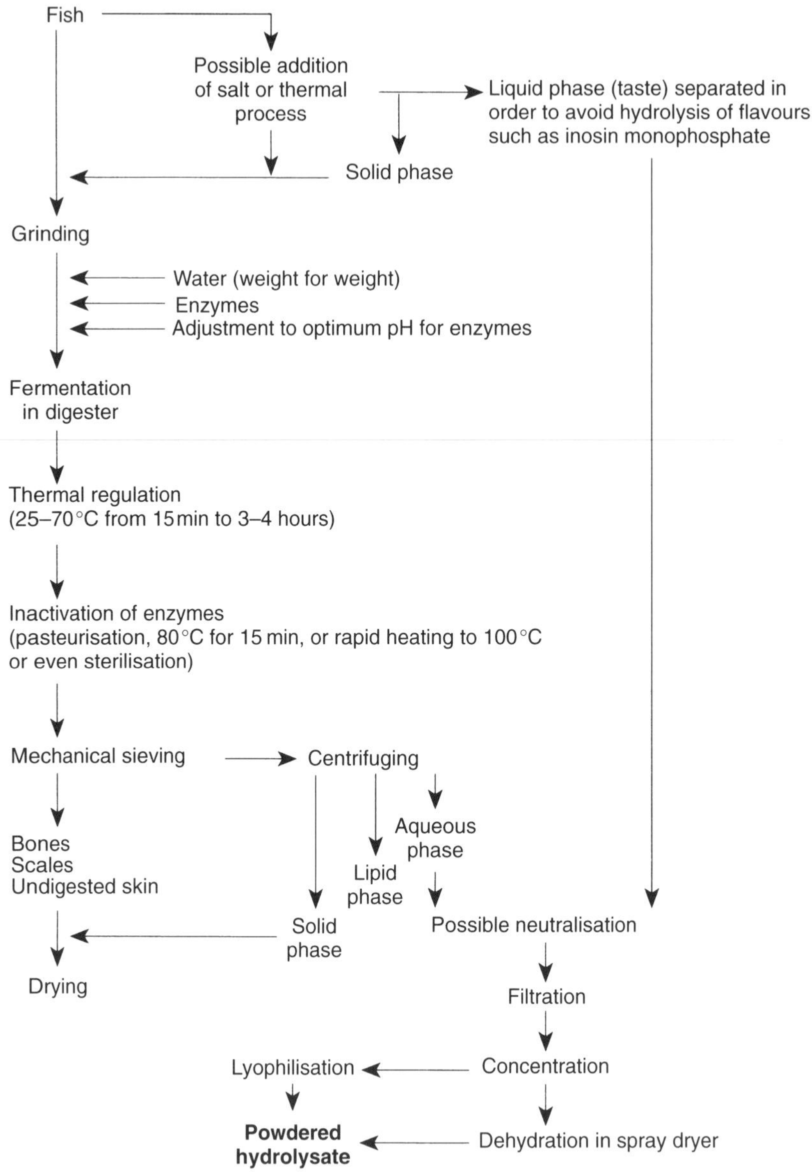

Fig. 8.3 Production of an enzymatic heterolysate.

between 75 and 85% of protein, for a moisture content of between 3 and 8%.

As **chemical hydrolysis** causes racemisation and decomposition of certain amino acids, its uses tend to be reduced. In addition, the yield of soluble proteins is lower than that from enzymatic hydrolysis. Certain mixed hydrolysates involve the autolysate or the enzymatic heterolysate combined with an addition of chemical components; this is 'silage making'. Acid silage is produced by traditional hydrolysis from chopped fish with the addition of an organic acid (usually formic) or, less frequently, an inorganic acid (HCl, for example). Organic acid and mineral mixtures are also used. Other hydrolysates are mixed: '**biological hydrolysates**' employ hydrolysis which is provided by the endogenous enzymes of fish and lactic fermentation favoured by adding fermentable carbohydrates and by a pH from 4 to 6 depending on the lactic acid bacteria present. Mycological fermentations (yeasts and lipolytic and proteolytic moulds) have also been used.

The functional properties of the proteins that have been isolated in accordance with the techniques described above can be further improved by chemical modification of these partially hydrolysed proteins. In fact, unlike plant proteins, fish proteins are unstable, as they are sensitive to changes in temperature and pH, salts and contact with polar solvents. This sensitivity results in a substantial loss of functional properties of fish proteins when they are prepared in the form of FPCs and even isolates. This loss can be alleviated if the proteins are chemically modified. In this new form they are often referred to as 'modified proteins' and 'modified hydrolysed proteins' if hydrolysis has taken place at a specific time during the process. Acylation (processing by means of acetylation or succinylation) which takes place between the amino groups and the acylating agents, is one of the reactions most frequently used. Succinylation of myofibril proteins results, for example, in a powder consisting of modified proteins with improved dispersability, higher emulsifying capacity and higher heat stability; if this type of protein undergoes gentle hydrolysis (using bromelain, for example), its foaming properties are improved.

We have already emphasised that the nitrogen composition of the hydrolysates is comparable to that of the original fish and their dietary value is approximately identical. Hydrolysis does not result in any improvement in the protein efficiency coefficients (PECs); on the other hand, the digestive utilisation coefficients (DUCs) can be improved by hydrolysis and oil extraction.

To sum up, the uses of hydrolysates in human foodstuffs, although very diversified, remain small in terms of quantities. On the other hand, the demand for 'silage' intended for feeding young animals is booming.

8.4 *Surimi* and by-products

Surimi, which is the name for fish flesh that has been ground and washed, and *kamaboko*, which is a gel of myofibril proteins of fish, obtained by adding salt to *surimi*, followed by thermal coagulation, are foods that have been prepared in Japan for centuries.

8.4.1 Preparation of *surimi*

The aim of the conversion process is to obtain a stable product from fish pulp, by means of a succession of washing processes and by adding cryoprotective substances. *Surimi* can be prepared from various species of non-oily or oily fish. The choice of fish depends not only on economic reasons but also on its good pre-gelling and/or gel-on-heating properties. The manufacturing process, in which the temperature is maintained at 0–5 °C to reduce the risks of microbial growth, involves the following operations (Fig. 8.4):

- **Opening up the fish and de-heading.** This very mechanised stage must be as early and rapid as possible in order to avoid the possibility of intestinal proteases penetrating the fish flesh and reducing the gelling properties of the muscle proteins.
- **Obtaining fish flesh or pulp.** The production of fish flesh is effected by means of simple mechanical separation. The flesh is partly destructured and passes through the perforations of a continuous mechanical separator, while 90% of the bones and the majority of the skin are retained and eliminated. This pulp, which has been subjected to lipid oxidation or protein denaturation, is stored at –40 °C in the presence of antioxidising agents (α-tocopherol, ascorbyl palmitate) and cryoprotective agents such as sorbitol.
- **Washing operation.** The aim of this stage is to eliminate lipids and brown muscle (in oily fish) and especially the soluble compounds of low molecular mass (amines, trimethylamine oxide, urea, creatine, nucleotides, etc.), the undesirable sarcoplasma proteins (myoglobin, oxydoreductases, etc.).

  As a general rule, the first wash is done using 0.5% solution of $NaHCO_3$ which increases the pH and buffers it; the second wash is carried out using water and the third with a solution of 0.1–0.5% NaCl so as to reduce the water content of the washed pulp by means of osmosis. Washing thus allows a concentrate of myofibril proteins (actomyosin) to be obtained, which possesses excellent gelling properties, is relatively stable in terms of lipid oxidation, is very light in colour and often odourless. On the other hand, the quantity of protein lost represents approximately 30% of the flesh weight of the processed fish. About 80% of the proteins present in the wash-water are proteins which are soluble in water.
- **Drying and refining.** The washed pulp is dried, for example in a press. It then passes through another piece of apparatus of the same type (mechanical separator) which allows the residual fragments of skin, connective tissue, black intestinal membrane, small bones and scales to be retained and eliminated.
- **Addition of cryoprotective substances.** Using a mixer and at a temperature of less than 10 °C, cryoprotectors are incorporated in the washed

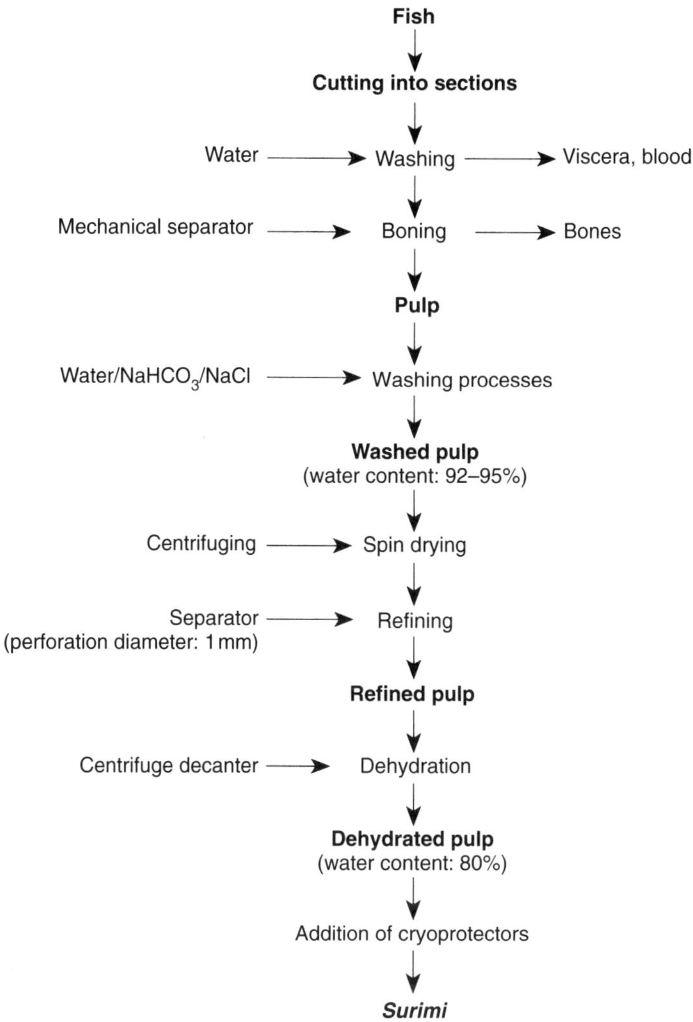

Fig. 8.4 Principal stages in preparing *surimi*.

flesh, in particular carbohydrates, sorbitol and polyphosphates which have the effect of limiting protein denaturation. Adding salt can also limit the deterioration of the *surimi* during storage, but its presence causes the proteins to gel, and because of this the *surimi* will be less suitable for conversion into more processed products.

The quality of the *surimi* is not only linked to the amount of residual water but also to the freshness of the fish, the composition of the cryoprotective agents used and the method of storage, which must be below –20 °C. There are a certain number of variations within *surimi* technologies. We can look at the one which is currently employed in Japan: the fish pulp is finely ground in order to eliminate the undesirable soluble compounds more effectively. Preparing *surimi* involves a stage during which the pulp is finely ground using colloidal grinders or special homogenisers, and refining is replaced in some cases by filtration. To sum up, this concentrate of myofibril proteins, *surimi*, has a very high nutritional value. It is actually a base material without any collagen, low in both lipids (<1%) and in cholesterol. However, the washing operation reduces the vitamin content (B_{12}) and mineral content (K^+).

8.4.2 Preparation of *surimi* by-products and seafood substitutes

The following processes take place using *surimi* as a starting material:

- The formation of gel using myofibril proteins which are dissolved by salt, using a cutter, in the presence of salting ingredients, at a given temperature and for a certain length of time. These are all variables that directly influence the strength of the gel, depending on the type of fish.
- The addition of binders, colourants and aromatic preparations. Ingredients such as starch from potatoes, wheat or maize, can be used, as can powdered egg white, soya proteins or gluten (between 1 and 5%). Adding protein gelling agents improves the texture of gels prepared with poor quality *surimi*. One method of firming up the texture consists of adding calcium salts, for example, in the form of ground fish bones. Likewise, adding starch modifies firmness and increases water holding on cooking. In addition, in order to reduce the final salt content, first of all dissolving the myofibril proteins by adding 3% NaCl, then diluting the salt by adding various ingredients, is recommended. After this stage the mixture has the consistency of a smooth, very viscous paste with a shiny appearance.
- The actual manufacture of the various products obtained by simple extrusion or co-extrusion (without texturisation), by spinning (texturisation), by compound extrusion and finally by emulsification.

The traditional products based on *surimi* are numerous and are differentiated mainly by their method of cooking: steaming in the case of *kamaboko*, baking in the case of *chikuwa* and *yaki-ita kamaboko*; cooking in boiling water for *hampens*, *naruto* and *tsumire*; frying in oil for *satsumage*. More recently, imitation shellfish and seafood products have been developed. Manufacturing these new seafood substitutes combines the various methods of cooking referred to in the previous paragraph. Broadly speaking, we can distinguish two groups of products according to the method of production. The first method consists of producing a thin film of gelled *surimi*, then rolling it into a cylinder. This is, for example, how 'crabsticks' and imitation scallops are obtained. The second method consists of manufacturing filaments or ribbons of gelled *surimi*, then introducing them into a mould with a thermocoagulable binding agent, in order to produce the desired shape. This is the method used for producing shrimp and lobster tail substitutes. The principle of preparing these products is very simple: a block of *surimi*–NaCl mixture is prepared and thermocoagulated by cooking. The block is then cut into fibres, shavings, scraps, filaments or ribbons. These fragments are coated in a thermocoagulating fluid (binding agent), possibly after being modified, then introduced into a lobster tail-shaped mould and cooked again. In this way a fibrous heterogeneous texture quite similar to that of the natural product is obtained.

9

The exploitation of by-products

9.1 Whey

9.1.1 General characteristics of whey

Whey must be considered as a derivative rather than a by-product of cheese manufacture, or casein. Two types of whey can be distinguished: that which results from the coagulation of non-acid milk, by means of rennin, which is known as 'sweet whey' and that which results from either the manufacture of fresh curd or soft cheese, or from the manufacture of lactic or acid casein, which is known as 'acid whey'.

9.1.1.1 Average composition

The average composition of these wheys is given in Table 9.1. It can vary, because it depends on the originating milk. The presence of constituents that have a high nutritional value and good functional properties, as well as molecules with a high added value (lactoferrin, lactoperoxidase) are arguments in favour of the commercial exploitation of this 'co-product' of the dairy industry.

However this strategy has some drawbacks, including the following:

- Significant reduction of the dry matter (1/15 as opposed to 1/8 in whole milk).
- Fairly high salinity (between 8 and 10% of the dry matter (DM)).
- Variability of shelf-life (depending on the type of manufacture and the care taken with storage procedures).
- The predominance of lactose (between 70 and 75% of the DM).
- A vulnerability to pathogens; it is a fairly good culture medium for numerous micro-organisms whose biological oxygen demand is 4–5 × 10^4 mg/l.

Table 9.1 Average composition of whey serum (% dry matter)

	Milk	Whey		Permeate (of sweet whey)
		Sweet	Acid	
Water	87.6	93.0	93.5	94.3
Dry matter	12.4	7.0	6.5	5.7
Fats	3.4 (27.5)	0.4 (1.5)	0.1 (1.5)	0
Casein	2.6 (21.0)	Traces	Traces	Traces
Soluble 'proteins'*	0.7 (5.6)	0.9 (13.0)	0.7 (10.5)	0.15 (2.3)
Lactose	4.7 (38.0)	5.0 (71.0)	4.5 (69.0)	4.9 (86.2)
Salts (ashes)	0.9 (7.3)	0.6 (8.6)	0.7 (10.5)	0.5 (8.8)
Lactic acid**	–	0.1 (1.5)	0.6 (0.9)	0.15 (2.6)

* Including non-protein nitrogenous substances.
** Whey contains some citrate, depending on its origin.

9.1.1.2 Composition of the protein fraction

Although proteins do not constitute the major fraction of whey, it is the one that is most important from an economic and nutritional point of view, and as far as potential uses are concerned. The nitrogen substances of milk are retained in whey, with the exception of the caseins. If we look at Fig. 5.1 showing an overview of cow's milk, we can see that the nitrogen substances are the so-called soluble proteins which are found in whey, together with non-protein nitrogen substances. In whey from milk to which rennin has been added, we also find 'caseinoglycopeptide' which results from the enzyme reaction with the κ-casein.

The structural characteristics of these proteins are covered in Section 9.1.3.1. We must stress the nutritional properties of this fraction. The nutritional value of whey proteins is actually higher than that of the caseins, because their amino acid content is more balanced. Their ease of digestion and biological value are both high, and close to those of the proteins in eggs (Table 9.2).

9.1.1.3 Minerals and vitamins

The mineral content of whey is a problem in terms of its use in human or animal foods, as well as in technological processes, especially those concerned with the preparation of pure lactose and proteins. It is therefore advantageous to remove these minerals, either wholly or partially, using physico-chemical techniques (filtration, reverse osmosis, ion exchange, etc.).

Whey contains most of the water-soluble vitamins present in milk; it is particularly rich in riboflavin (which gives it its greenish colour) and the vitamin B content (per litre) meets an appreciable proportion of the daily human requirements, especially in B_2, B_5 and B_6 (Table 9.3).

Table 9.2 Comparative nutritional properties of whey proteins, casein and egg proteins

Composition in essential amino acids (g/100 g proteins)

	Whey	Egg	Recommended food and agriculture organization balance
Threonine	6.2	4.9	3.5
Cysteine	1.0	2.8	2.6
Methionine	2.0	3.4	2.6
Valine	6.0	6.4	4.8
Leucine	9.5	8.5	7.0
Isoleucine	5.9	5.2	4.2
Phenylalanine	3.6	5.2	(Phe + Tyr) = 7.3
Lysine	9.0	6.2	5.1
Histidine	1.8	2.6	1.7
Tryptophane	1.5	1.6	1.1

Biological parameters

	Whey proteins	Casein	Egg proteins
Protein efficiency ratio (PER)	3.2	2.5–3.0	3.9
Net protein utilisation (NPU)	95	79	94
Biological value (BV)	100	85	94
Digestive utilisation coefficient	97	97	97

Table 9.3 Vitamin content of whey

	Concentration (mg/ml)	Daily requirements
Thiamin (Vitamin B_1)	0.38	1.5 mg
Riboflavin (Vitamin B_2)	1.2	1.5 mg
Nicotinic acid (Vitamin B_3)	0.85	10–20 mg
Pantothenic acid (Vitamin B_5)	3.4	10 mg
Pyridoxine (Vitamin B_6)	0.42	1.5 mg
Cobalamin (Vitamin B_{12})	0.03*	2 μg
Ascorbic acid (Vitamin C)	2.2	10–75 mg

9.1.2 Industrial products, both modified and non-modified

9.1.2.1 Raw whey

Traditionally, whey was used for pig feed, spread as fertiliser or discharged into the river. The disappearance of pig farms, the fight against river pollution and the concentration of cheese production into larger and larger units have resulted in the cheese industry finding new outlets for whey. These obviously involve concentrating and drying whey (see Section 5.2.2.1).

The operations involving final concentration and drying depend significantly on both the quality of the starting wheys and the processes they undergo. The most significant reductions in quality are, firstly, lactic fermentation which converts some of the lactose into lactic acid and results in a significant increase of the total flora and, secondly, the degradation and denaturation of the whey proteins, essentially associated with heat processes. The methods of dealing with these two risks are opposing. A compromise always has to be found.

Good quality **sweet whey**, correctly collected, cooled and stored, does not present any difficulties as far as concentration is concerned, but direct drying after exiting the evaporator often leads to clogging in the drying chamber and produces a hygroscopic powder. On the other hand, cooling to 20 °C, followed by several hours' storage, brings about partial crystallisation of the lactose, and satisfactory drying is obtained.

We might assume from this that the presence of crystallised lactose encourages drying in the chamber, but this phenomenon appears not to be the only one, with the state of the whey proteins changing during storage. Slight flocculation, which is reversible as it does not involve denaturation, is observed and this also appears to encourage drying.

In whey that has been **acidified** by biological means, which is normally the case, part of the lactose has been converted into lactic acid, a small portion has also been hydrolysed, and, finally, mineralisation has increased. These three factors contribute to inhibiting crystallisation. In addition, the bacterial population has increased, so we need to have a more severe heat process with a low pH, which leads to the partial denaturation of the proteins which become more hydrophobic.

These considerations partially explain the fall in production of raw whey in powder form. These wheys are mainly incorporated in animal foodstuffs (veal, pigs, poultry, etc.).

9.1.2.2 Modified whey

Demineralised whey

Removing minerals from whey has extended the use of this product. The demineralisation operation consists of removing some of the metal ions and lactic acid from sweet whey of very high bacteriological quality containing very little lactic acid and very few minerals. This operation is usually carried out by means of electrodialysis or by ion exchange. The result is a product that contains few minerals and whose organoleptic qualities have been improved. In addition, demineralisation followed by acidification to pH 4.5 allows 99% of the bacteria and 90% of the residual lipids to be removed. The elimination of residual lipids improves storage stability, foaming and emulsifying properties. Demineralisation allows the quantity of whey that can be incorporated into bread to be increased and makes it a particularly useful ingredient in the manufacture of dry products. Demineralised whey powder is used in infant formula, dietary foods, confectionery and cake making and can partially replace skimmed milk powder in ice-cream.

Whey with lactose removed
Whey powder whose lactose has been removed partially or totally is in fact a by-product of lactose manufacture. In order to extract lactose, the whey is concentrated by evaporation, cooled and centrifuged. The proteins have therefore been partially denatured by the heat process and the mineral content of the dry extract is very high. This product is not manufactured with a view to incorporating into specific foodstuffs and is normally used in animal feed. The mineral content can, however, be significantly reduced (from 24% to 1–3%) by means of dissolving and washing at pH 3.5.

Whey protein concentrates, whose lactose has been hydrolysed, are also produced. They are used to reduce lactose crystallisation and to lessen the effects of lactose intolerance. They can also be used as sweetening and stabilising agents in frozen desserts, as nutritional additives or browning agents in bread or even to extend the storage time of whey and to prevent it from drying out.

9.1.3 Whey proteins

Whey proteins are labile and their functional properties can be modified as a result of the processes applied to the whey. Using these proteins to advantage in the foods in which they are incorporated requires a good knowledge of their structure, the way their functional properties depend on various factors (varying complexity of the medium, processes, etc.) and their behaviour during the preparation operations used in various industries: biscuit-making, confectionery, cooked meats, soups and sauces, ice-creams, desserts, etc. Table 9.4 provides a general outline of these properties.

9.1.3.1 Structure and stability factors of whey proteins

Structural characteristics
The two major proteins, β-lactoglobulin (50%) and α-lactalbumin (22%), are compact globular molecules with a primary sequence that is distributed uniformly between polar (both charged and uncharged) and hydrophobic residues. They can fold over to mask the hydrophobic residues in the centre of the molecule, so that intermolecular associations are unlikely. Their three-dimensional structure is stabilised by disulphide bridges (between 1 and 2 for β-lactoglobulin, 4 for α-lactalbumin). The spatial structure of β-lactoglobulin resembles a very compact cone (Fig. 2.8) whereas that of α-lactalbumin, very similar to lysozyme, has an almost spherical shape. These are small molecules (MM of 18400 for β-lactoglobulin and 14200 for α-lactalbumin) in comparison with other whey proteins (serum albumin, enzymes and immunoglobulins, proteose-peptones).

Table 9.4 Applications of whey proteins

Products	Functions
Biscuit industry	Protein supply, water holding, gelling, texture (interaction with gluten)
Pasta	Protein supply, texture
Cake-making (meringue, genoese sponge)	Emulsifying, foaming, water holding, gelling agent
Confectionery (caramel, nougat, etc.), milk chocolate	Emulsifier, flavour, texture, dispersibility
Soups, sauces	Thickener (interaction with starch), emulsifier
Ready-made meals	Thickener, emulsifier, water holding
Lacteal flours	Protein supply, solubility
Lacteal or fruity drinks	Soluble when hot and/or at acid pH, thickener
Dietary and baby foods (enteral feeding)	Protein supply, solubility, thickener
Natural and processed cheese	Emulsifier, thickener, gelling agent
'Imitation cheeses, dips', spreads, coffee whitener, ice-cream	Emulsifier, thickener
Creams, desserts, flans, yoghurts	Emulsifier, thickener, gelling agent
Meat products (sausage, pâtés, hamburgers)	Emulsifier, thickener, binder, gelling agent, water holding and fats

Stability factors
The structure of β-lactoglobulin depends on the pH. At the pH of milk, a dimer forms whereas at a pH lower than 3.5 and higher than 7.5 the complexes dissociate into unfolded monomers.

The structure and functionality of these two proteins are very strongly influenced by the presence together of thiol and disulphide groups. The reactions of oxidation–reduction and of the heat processes in particular result in exchanges of disulphide bridges which modify most of the properties, starting with solubility. Thermal denaturation takes place between 50 and 75 °C and results in the unmasking of the SH groups and in the molecule unfolding; depending on the pH, reassociation can take place in an intramolecular (precipitation or coagulation in a slightly acid medium) or intermolecular (gelling in a neutral, alkaline or acid medium) way.

9.1.3.2 Concentration and fractionation of proteins
Proteins are preferred to raw or modified whey in many processes. In the areas of dietary or therapeutic products, for example, it is essential to use concentrates or protein isolates.

Thermal coagulation
Traditionally nothing is simpler and cheaper than to precipitate whey proteins by means of heating in an acid medium. Fromage frais has been

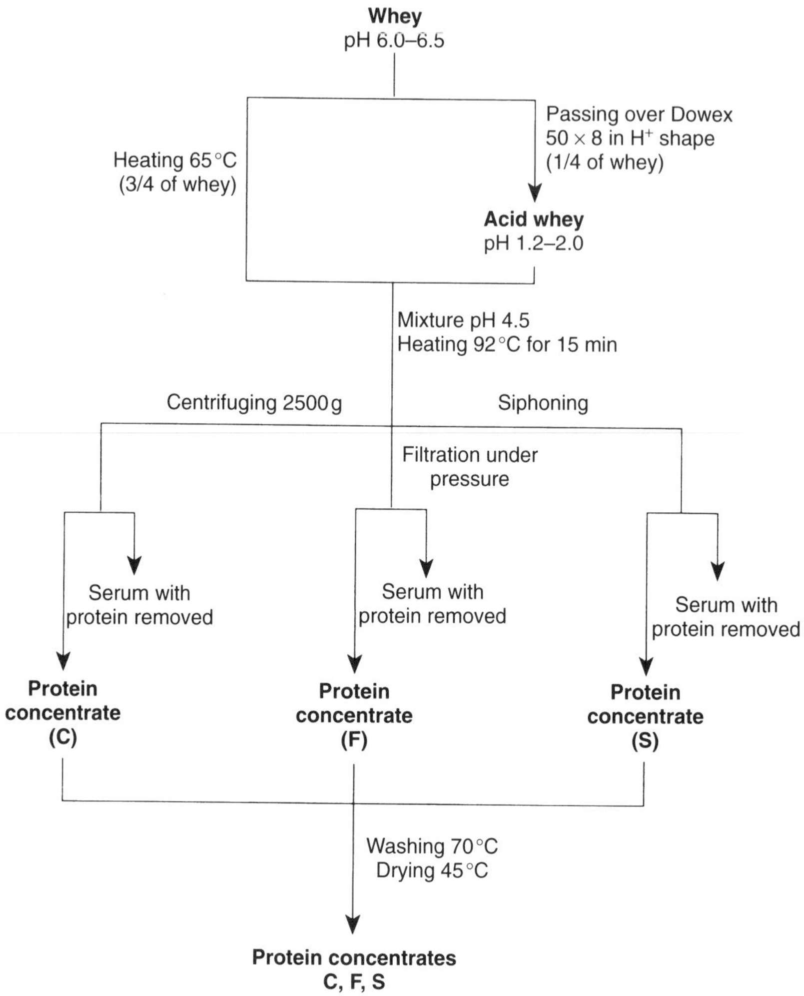

Fig. 9.1 The technology of whey protein thermal coagulation.

produced in this way for centuries, as have other sorts of cheese made from buttermilk; a 'protein milk' is recovered and is reincorporated in the milk used for cheese-making ('Centri-Whey' process) (Fig. 9.1).

The pH of the thermal process obviously has a strong influence and numerous variations of this process have therefore been suggested. However, the maximum recovery yields are obtained at around pH 5.0–5.5. Precipitation at a much lower or higher pH allows proteins that are more soluble or more easily dispersed to be produced. Heating to pH 7.5, for example, produces whey proteins with good structuring properties;

however, it should be borne in mind that a process carried out in a slightly alkaline medium encourages undesirable modifications of essential amino acids (see Chapter 3).

Finally, lowering the pH by passing whey over an ion exchanger resin slightly complicates preparation but has the great advantage of reducing the salt content.

Other processes

If a heating process is not used then the proteins are separated more or less intact, but these processes are obviously more costly in terms of investment and running costs.

Ultrafiltration is the process most often used. In order to avoid membrane fouling, various pre-treatments have been recommended: filtration, the addition of a calcium sequestering agent, demineralisation and moderate heat treatment. The industry produces concentrates (retentates) with a variable protein content (between 40 and 80% of the DM).

Ion exchange chromatography is used for the industrial production of concentrates (90% and over). The 'Spherosil' process uses spheres of porous silica onto which the exchanger groups have been grafted (Fig. 9.2). Fractions enriched in any of the constituents of whey proteins can be obtained. In the case of sweet whey, the process chosen consists of implementing an anion exchanger Spherosil column (Spherosil QMA) and a weak cation exchanger Spherosil column (Spherosil C). In fact, at the pH of sweet whey (pH = 6.6) most of the proteins are in an anionic form and are adsorbed on anion exchangers, and a small proportion (between 7 and 10%), essentially consisting of immunoglobulins, is in a cationic form and is adsorbed on a cation exchanger. With acid whey, the process used consists of implementing a cation exchanger Spherosil column (Spherosil S); in fact, at the pH of acid whey (pH = 4.6) all the proteins are in a cationic form and are adsorbed on cation exchangers.

Cold precipitation by polyelectrolytes has been investigated on numerous occasions. Carboxymethylcellulose has mainly been recommended (see Chapter 3).

9.1.3.3 Main functional properties of whey proteins

Whey proteins, in the forms in which they are found, can differ widely in terms of functional properties: hydration properties, and texture and surface properties.

Solubility

This property is often a criterion for powder quality; in addition, it is often associated with other properties such as viscosity, gelling, emulsifying or foaming capacity. It depends on numerous factors: separation pretreatment, methods of concentration and drying, pH, ionic strength,

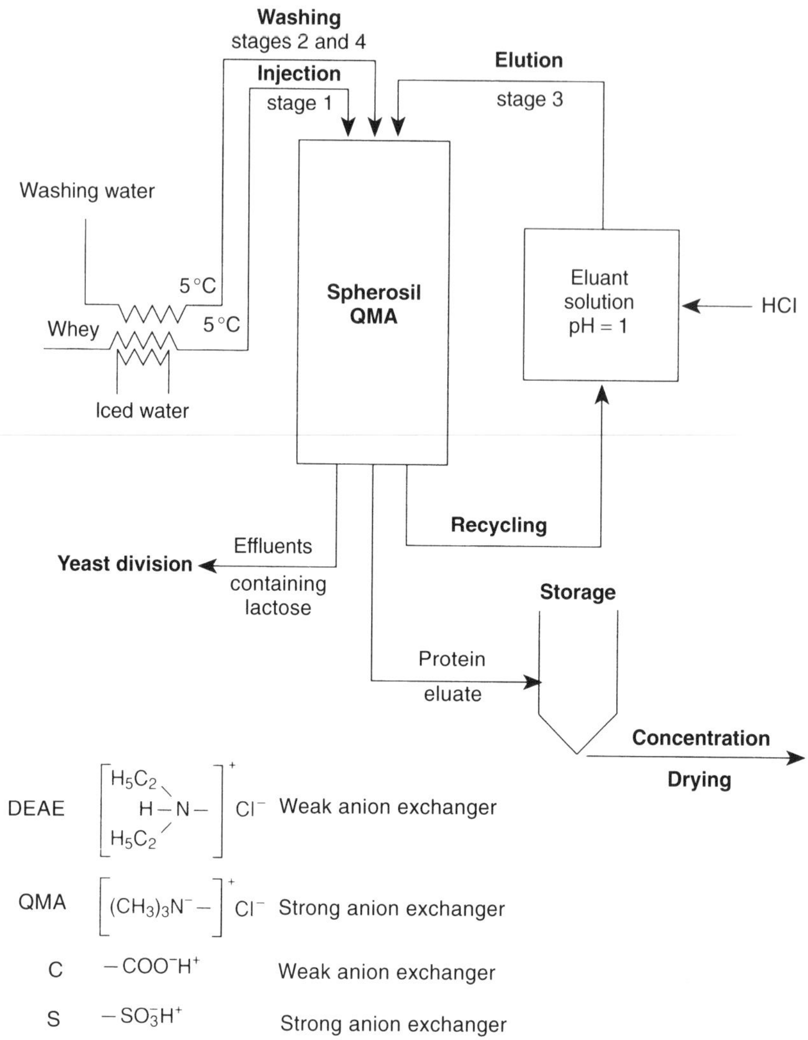

Fig. 9.2 Ion exchange extraction (Spherosil process).

temperature, whether or not Ca^{2+} ions or other complexing salts are present. Solubility, in particular, is greatly reduced at pH_i following denaturation.

Absorption of water – swelling

Whey proteins do not absorb much water (0.5 g/g) but thermal denaturation (80 °C for 45 s) improves the binding capacity (useful in cake-making and cooked meats).

Coagulation – gelling

Coagulation can be considered to be a disorderly aggregation, produced by the thermal denaturation at the average pH_i of whey proteins (4.9–5.2).

Gelling involves the formation of a more or less orderly continuous network of peptide chains after unfolding; gel formation requires protein concentration of at least 8%. The firmness of the gels depends on the following:

- The heat process: firmness increases with temperature.
- The pH: gels are firm in an acid and alkaline medium, and granular at a neutral or slightly acid pH.
- The protein concentration: gels are stable at high concentrations.
- The carbohydrates present: whey gels are very stable in the presence of sucrose.

Emulsifying properties

These properties arise from an ability to reduce the interfacial tension between hydrophilic and hydrophobic constituents; they are often associated with the solubility of the protein in water.

The emulsifying capacity (quantity of oil emulsified per g of protein prior to phase inversion) and emulsifying stability increase with the protein concentration (optimum at 2–3%) and are often minimal at pH_i.

Foaming properties

These properties, which are a result of unfolding at the water/air interface, are highly valued in the baking industry (cakes, meringues, soufflés, etc.).

Maximum swelling (or foaming capacity) and the stability of whey protein foams are excellent if the proteins are purified (as in the case of β-lactoglobulin) and if the pH is close to neutral. Thermal denaturation is an improving factor.

9.1.3.4 Improvement processes

Processes that change the composition of the preparation

Purifying proteins improves all their qualities, especially if the process involves little denaturation (as in the case of extraction by means of ion exchange). Likewise, eliminating salts (electrodialysis and ultrafiltration) has a favourable effect (see Section 5.2.3.1).

Heat processes

These processes are used as means of extraction in a thermocoagulated form or as an improvement treatment, either alone or combined with other pre-treatments (ultrafiltration, dialysis, etc.).

- **At acid pH:** these processes, whether or not they are associated with ultrafiltration, allow the gelling properties and the stability of foams to be improved.

- **At pH_i:** if this process is followed by heating in an alkaline medium, the gelling and foaming properties are considerably increased.
- **At neutral pH:** combined with clarification and demineralisation, this process above all produces excellent emulsifying and foaming properties.

Enzymatic or acid hydrolysis

Limited proteolysis by pure protease or in the form of mixtures, improves the solubility of the concentrates obtained via thermal coagulation and more specifically close to the average pH_i (4.8–5.2). Unlike limited hydrolysis of polymeric proteins (such as those in legumes and cereals) which improve foaming and emulsifying properties, limited hydrolysis of the native proteins in whey damages these properties. The only positive results have been obtained on concentrates that are already not very soluble, denatured and strongly polymerised.

However, heat processes carried out at low pH (0.5–1) hydrolyse the β-lactoglobulin very weakly and also result in partial polymerisation which increases the stability of the emulsified systems.

Chemical modifications

Whey proteins have been used as model substrates in numerous trials on chemical processes for the purposes of improving their functional properties.

By means of **succinylation** or **acetylation**, the α and ε-amino groups, whose charge is strongly positive at neutral or acid pH, are converted into anionic groups which carry negative or zero charges respectively. Modifying electrostatic repulsion by means of creating negative charges or reducing positive charges results in a stretching or unfolding of proteins, leading to a reduction in oil–water and air–water surface tension. For example, succinylation of whey proteins that have been precipitated at the isoelectric point improves all functional properties except foaming. In the bakery trade, the use of succinylated whey protein concentrates reduces dough rising time. Succinylated whey protein concentrates can replace 20% of the sodium caseinate in imitation cream and 20% of egg yolk in salad dressings (improved emulsion stability and viscosity). However, essential amino acid residues that have an $\varepsilon\text{-}NH_2$ group, such as lysine, are involved in covalent links which cannot be hydrolysed. Nutritional value is therefore affected, the metabolism of these products is unknown, and, as a consequence, the use of the products obtained using these methods conflicts with legislation.

One of the advantages of **esterification** is that it acts on non-essential amino acids such as aspartic acid and glutamic acid. Esterification of whey proteins increases hydrophobicity and gelling power. In the esterification of β-lactoglobulin, the emulsifying activity is slightly lower than that of the native protein, but the stability of the emulsion is significantly increased.

As in the case of succinylation, **phosphorylation** creates new electrostatic repulsion within the polypeptide chain by means of increasing the number of anionic hydrophilic groups. Chemical phosphorylation is obtained by means of covalent attachment to the phosphorus atom of the protein coming from phosphorus oxychloride on several possible sites: serine, threonine, tyrosine and lysine. Internal links can also be established via phosphate bridges. Phosphorylation of the β-lactoglobulin increases dispersability, emulsifying and gelling capacity, emulsion stability and viscosity. These properties of β-lactoglobulin make the use of whey proteins a possibility in sauces and mayonnaises requiring ingredients with good emulsifying properties.

9.1.3.5 Uses of whey proteins

Table 9.4 summarises the principal industrial applications of whey protein concentrates. Whey protein concentrate (approximately 60%) heated to 80–100 °C for 10–30 min has the appearance and consistency of hard egg white. When mixed with skimmed milk (1.5% protein concentrate) – that is, half the content required for egg white – a gel can be obtained after heating for 5 min at 85 °C. Thermal gelling allows the whey proteins to act to improve texture and water retention (meat products, cheese products, sauces, etc.).

Whey proteins are also well-known for their foaming properties. In cake-making, for example, meringues that look exactly like those obtained using egg white can be obtained by replacing 50–100% of the egg yolk or the whole egg with whey proteins depending on the type of product, provided that the recipe and the cooking conditions are slightly modified. The colour and the texture obtained are very acceptable, but the typical flavour of egg is lacking.

Whey proteins are used in the baking industry for their protein contribution, water retention capacity, texture, ability to brown bread and enhance flavour and nutritional value, but they reduce the volume of the dough. The use of reducing agents (iodoacetamide) or thermal denaturation in the presence of bromate improve the raising powers of the dough but these processes are not permitted.

The gelling, thickening, emulsifying and water retention properties of whey proteins are used in meat products.

Because of their solubility at acid pH, these protein concentrates can be used at up to 3% in order to fortify drinks and fruit juices with proteins. The product remains clear and stable, provided that it has been clarified first.

Whey protein concentrates, both with and without hydrolysed lactose, are used in infant formulae and as a protein supplement in dietary and pharmaceutical products because of their nutritional value.

In association with sodium caseinate, whey proteins contribute to the formation of a stable emulsion, bind the water and give butters and 'light' spreads the texture and spreadability required.

Finally, these proteins can be used as binding agents in foods that contain a mixture of textured proteins. It is possible to incorporate up to 20% of dried sweet whey or protein concentrates with rice, maize, wheat flour or whole flour without altering the product's expansion capacity, its consistency or its colour. Likewise, protein fibres prepared from whey protein concentrates can be used in meat products.

9.2 Blood

9.2.1 Average composition

Blood is composed of a liquid, plasma, in which the red and white corpuscles are immersed. These are essentially erythrocytes or red corpuscles, leucocytes or white corpuscles and blood platelets. After an anticoagulant has been used, centrifuging allows the following to be collected:

- plasma (60–65% of the total);
- cruor (35–40% of the total).

Although blood is a universal element within organisms, its composition can fluctuate depending on a wide variety of factors: race, age, diet, physiological condition, etc. The compositions given (Table 9.5) are average values: 80% water, 18% proteins and 2% carbohydrates, lipids and mineral salts. Blood can be broken down, firstly into plasma, mainly containing white corpuscles, circulating proteins and salts, and secondly into cruor, which contains the red corpuscles.

The composition of plasma proteins is as follows: 44% albumin, 14% α-globulins, 11% β-globulins, 31% γ-globulins and 0.6% fibrinogen.

It should be pointed out that plasma also contains other substances in solution: e.g. lipids, carbohydrates, amino acids and mineral salts, most of which are nutritional substances or metabolic waste in transit through the blood.

Cruor, made up of erythrocytes, leucocytes and platelets, contains 30% of dry material made up of a large fraction of proteins, one of which, haemoglobin, is dominant.

Table 9.5 Composition of bovine blood, plasma and cruor (g/100 ml)

	Blood	Plasma (60%)	Cruor (40%)
Water	80–85	90–92	70–78
Proteins	15–18	6–8	25–29
Lipids	0.15	0.5–1	0.2
Carbohydrates	0.1	0.08–0.12	–
Mineral salts	1	0.8–0.9	Traces
Other substances	0.55	0.2–0.3	–
Dry matter	15–20	8–10	22–30

9.2.2 Nutritional and functional properties of blood

9.2.2.1 Nutritional value

The amino acid composition of blood proteins is close to that of proteins which are considered well balanced from a nutritional point of view. This is the case with the essential amino acids. However, the Ile and Met concentration is lower than that of milk proteins, whereas the Lys and Thr concentrations are higher.

Table 9.6 shows the nutritional characteristics of blood and plasma proteins and that of some other animal proteins. The PECs of bovine blood and plasma are slightly lower, but on the other hand the digestibility percentage is good. Obviously the nutritional value of blood proteins does not just depend on their initial composition. The type of processing they undergo can change their intrinsic value.

From an economic point of view, the nutritional value of blood proteins is not a conclusive argument for their use, partly because of the low amounts used, and secondly because of the abundance of other top-quality proteins within industrialised countries.

9.2.2.2 Functional properties

Blood by-products are used as protein additives mainly because of their functional properties and preference is given to plasma proteins.

Solubility

Plasma solubility is good over the whole pH range: 100% when neutral, dropping to 75% at pH 4.8, which is the minimum solubility point; spray drying at 160 °C or 193 °C reduces this solubility by approximately 20%; adding lactose provides a degree of protection.

Table 9.6 Nutritional characteristics of some animal proteins

	Limiting amino acid	Chemical index*	PEC	DUC
Bovine blood	Met	23	2.0	99
	Ile	27		
Bovine plasma	Met	29	2.2	
	Ile	56		
Beef (meat)	Met	66	2.8	93
	Phe	77		
Casein	Met	80	2.5	97
Whey proteins	Met	51	3.2	97
Fish flour	Phe	68	2.6	95
	Met	82		
Egg		100	3.9	98

* Chemical index calculated on the most limiting amino acid.

Emulsifying properties

Blood concentrate has an emulsifying potential double that of all the other tissue concentrates including muscle concentrate. Blood-based emulsion also has good thermal stability of the same order as that of muscle-based emulsions, so blood is a potential source of top-quality emulsifiers.

The emulsifying capacity of plasma is virtually independent of the pH, the spray drying temperature (160 or 193 °C) and the presence of lactose during spray drying. This allows plasma to be used as an emulsifying agent in meat products (cooked sausage, cooked ham, liver sausage, etc.).

Binding properties

On cooking, blood proteins coagulate by binding the elements with which they have been mixed, while remaining involved in lipid and water holding. These properties have not been closely studied but can be put to best advantage together with emulsifying properties in meat products. They contribute to the stability of emulsions during cooking.

Properties of cruor

In the 'purée' collected after centrifuging blood, haemoglobin is responsible for the red colour which turns to brown under the effects of oxidation $HbO_2(Fe^{2+}) \rightarrow$ Met Hb (Fe^{3+}). This brown colouring capacity is so intense that it precludes the use of blood in human foods, as we will see later, and greatly restricts the use of blood in non-human food; cruor contains two-thirds of the blood proteins and its use would allow the collection and storage costs to be spread over a wider basis than those for plasma. We will return to this point Section 9.2.3.2.

9.2.3 The use of blood within the food industry

9.2.3.1 Collection and utilisation of whole blood

The hygienic collection of blood requires a collection system that avoids contamination during slaughter by the products of regurgitation: in some abattoirs removal by trocar through a vein is used. In addition, if a veterinary inspection has not authorised the use of carcasses for human consumption, the blood cannot be collected in a mixing tank; in practice, collection and storage must be separated prior to mixing.

Blood can be used fresh, coagulated or shaken. It is principally intended for human foodstuffs, where it is involved in the composition of certain cooked meat products (in particular black pudding, and also sausages), but this use is limited to pigs' blood collected in the main over the winter period.

Frozen blood is mainly used for domestic animal feed but the costs of freezing are high, which limits expansion of this market.

Powdered blood is mainly intended for animal feed, in particular domestic animal feed, and for producing fertilisers for market gardens or flower fields. Drying, effected either in a cooker, or in a spray dryer, is a costly process, which makes slaughterhouse butchers increasingly reluctant to heat-process blood, especially as this powder quickly reabsorbs moisture from the surrounding air, thus leading to risks of microbial activity.

9.2.3.2 Utilisation of cruor and plasma

Centrifuging separates the plasma from the cruor and thus allows a more profitable use of blood, because of the functional properties of the plasma which are currently highly valued in salted foods (Fig. 9.3). It can be used either in the form of powders or in a frozen form.

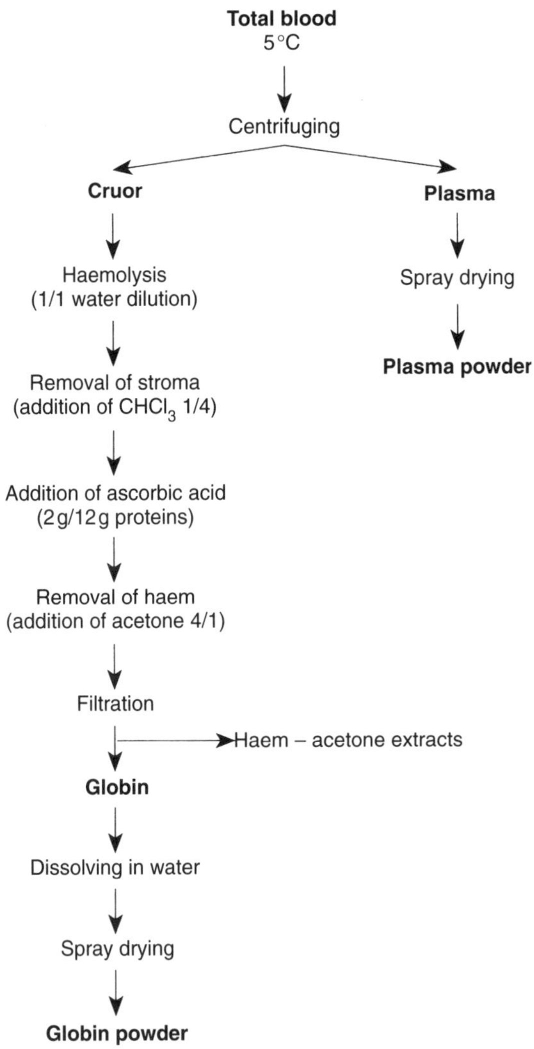

Fig. 9.3 Fractionation of blood.

The use of the residue from water extraction, or cruor, is more delicate, bearing in mind the colouring of haemoglobin. In fact the bright red colour of fresh blood turns to brown under the effects of oxidation where the oxyhaemoglobin converts to methaemoglobin. The brown colouring is so intense that it renders the use of cruor in human foodstuffs impossible and considerably limits its use in foodstuffs for dogs and cats.

Three solutions have been proposed in an attempt to use cruor in human foodstuffs in spite of its very intense colour: colour stabilisation, colour masking or decoloration.

As far as **stabilising** the colour is concerned, numerous reducing agents have been tested for their ability to stabilise the bright red colour of fresh blood so that haemoglobin can be used as a colourant (in salted foods, for example). The tests that have been carried out have shown that substances such as ascorbic acid were more likely to slow down the oxidation of haemoglobin and that only compounds such as carbon monoxide or nitrites produce stable combinations with haemoglobin.

As far as colour **masking** is concerned, various techniques have been developed to allow either cruor or blood to be used in certain special products. The Smirniskaya process, in particular, involves heating mixtures of cruor and skimmed milk at 95 °C for 5 min in the presence of $CaCl_2$. The protein precipitate is pressed until its moisture content reaches between 70 and 75%. It can be used to replace up to 15% of beef in sausages.

Numerous methods for cruor **decoloration** have also been studied and are now patented. Some of these are applicable not only to cruor but also to whole blood.

All the techniques for blood decoloration are based on the principle of allowing the haem–protein link to break within the haemoglobin molecule so that these two fractions separate. In the techniques under consideration, the link can be ruptured either chemically or enzymatically; the separation of the two fractions is based either on their differences in terms of molecular mass or on their differential solubility in the various solvents. This involves uncoupling the pigment from the globin and separating the two constituents, with the difficulty lying in preserving the functional properties of the molecule without which they are virtually valueless. Several processes have been put forward but none has reached the industrial stage.

The best-known process is summarised in Fig. 9.3. After haemolysis of the blood by means of simple 1/1 blood dilution in water or intense homogenisation, this involves treating the product obtained with acidified acetone: acidification reduces the affinity of the protein for the haem which, when freed, dissolves in the acetone while the protein precipitated undergoes decoloration; alcohol can act in the same way as acetone, but is less effective. This process was subsequently applied with various modifications. Some researchers begin by converting the haemoglobin into cholemethemoglobin using ascorbic acid before extracting it with acetone; others have suggested starting by concentrating the haemoglobin solution by precipitating it with NaCl prior to treatment with acetone, in order to reduce the quantities of acetone needed.

The globin obtained by the acetone process has useful functional properties and consequently good prospects for use. As far as the solubility, which influences the other properties, is concerned, globin has its own particular characteristics in comparison with plasma proteins and most proteins, in that it presents a maximum at low pH; solubility remains at over 90% between pH 2.5 and 6, but falls to around 20% when close to neutrality. This particular characteristic can be used to advantage in order to add proteins in solution to a product. As far as the emulsifying capacity of globin is concerned, this is good when solubility is good; at pH 4, the maximum volume of the oil phase is approximately 50% for a protein concentration of 0.2 g/100 ml and 83% for a concentration of 0.8 g/100 ml; drying with or without lactose makes little difference to the emulsifying capacity. However, the use of this long-standing process is not widespread, probably because of the quantity of solvent required and the problems with food safety posed by the use of products of this kind, and also no doubt because of the rather disagreeable flavour of the product obtained.

Haemoglobin can also undergo decoloration via a process that uses hydrogen peroxide. The method consists of shaking it with hydrogen peroxide for 30 minutes at 70 °C and of destroying the excess of hydrogen peroxide by means of catalase; under these conditions the haemoglobin is converted so that the haem is released. The protein obtained is insoluble and almost flavourless; its functional properties are poor, but it can be used as a protein charge in some meat products. However, current regulations do not allow the use of this process.

Lastly, a final method is currently arousing interest; this is the method that consists of effecting the **partial enzymatic hydrolysis**, of the haemoglobin, with varying degrees of thoroughness, which liberates the pigment, associated to a greater or lesser degree with the peptides; the separation of peptides and the pigment associated with the small peptides is then carried out by ultrafiltration, or by adsorption on activated carbon.

Hydrolysis slightly modifies some functional properties: the emulsifying capacity goes from a maximum to a low degree of hydrolysis (DH); foaming capacity does likewise.

In addition, the hydrolysates almost always have a bitter flavour, which can, however, be improved by employing certain treatments. They can then be used to enrich various foodstuffs with proteins: soups, dietary products, even hams, subject to the country's regulations.

9.3 Collagen and gelatin

Gelatin is obtained by the partial hydrolysis of the collagen contained in the skin, the connective tissue and the bones of animals. Collagen is a fibrous protein found throughout the animal kingdom. In terms of quantity, it is the most important compound in the connective tissue. It is therefore the most widespread protein in the animal kingdom. It is the major structural element in multicellular organisms.

From time immemorial humans have attempted to exploit the large quantities of natural collagen available, either as a foodstuff in the form of a jelly, or as an industrial product (wood sizing, for example). For a long time these preparations were made in the household or by craftsmen, but over the past few decades a gelatin industry has come into being and this has allowed increasingly pure gelatins to be produced on a regular basis, for use in widely differing purposes: in the food industry (confectionery, dairy products, salted foods, etc.), in the pharmaceutical industry (capsules), and in various industrial techniques (photography, etc.).

9.3.1 Origins, structure and manufacture

9.3.1.1 Origins

Gelatin originates in the skin and bones of animals. These are moist and contain, alongside the collagen that will produce the gelatin, other proteins, fats and mineral salts.

In reality, the supply is more or less limited to the following four types of raw material:

- Fat and moist bones of local origin.
- Dry bones, whose fat has been removed, from hot countries.
- Cold stored pig skins.
- Scraps recovered from tanneries when sheep and cattle skins are cut up.

9.3.1.2 Collagen structure

The term collagen covers several isotypes, in all of which at least part of their molecules have a characteristic structure, known as 'triple helix' (Fig. 9.4). This structure consists of three polypeptide chains (known as α chains), each of which form a left-hand helix. These coil round to form a 'super' right-hand helix. This spatial arrangement is possible because the primary sequence of the α chains consists of a succession of Gly-X-Y triplets. As one amino acid from each chain must be found at the centre of the triple helix, out of all three residues, only glycine can occupy this position because of its small size. The X and Y amino acids are proline and hydroxyproline, respectively, in 30% of cases.

The collagen forms fibrils, filaments or networks within the extracellular matrix of the connective tissues. Each molecule of collagen can have up to three different α chains, and there must be over 25 encoding genes for collagen α chains. Among these different isotypes, we can distinguish either fibrillar collagen or non-fibrillar interstitial collagen.

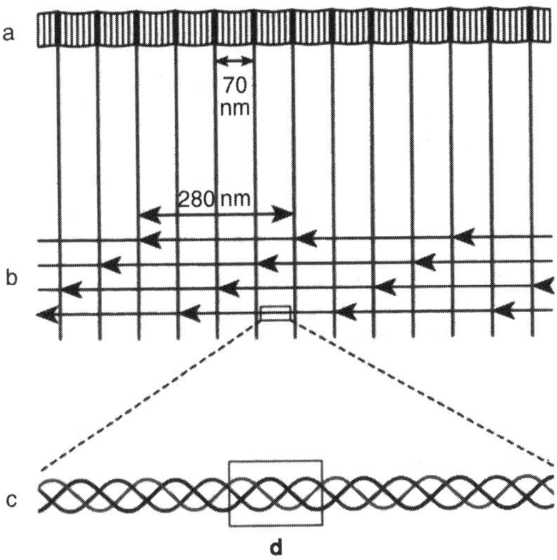

(a) Diagram of collagen fibres (enlargement 100 000 times)
(b) Molecule of tropocollagen represented by an arrow (equivalent to four stria of 280 nm)
(c) Triple helix of tropocollagen
(d) 'Super helix' – with three cords by twisting to the right of three left helices

Fig. 9.4 Collagen structure.

The fibrillar collagen is synthesised in the form of pro-collagen whose globular N and C-terminal parts are eliminated sequentially, to form fibrils by means of side aggregation of the molecules.

The collagen fibrils, of several dozen nanometres in diameter, are therefore composed of aggregates of collagen molecules along a parallel axis, but whose extremities are evenly overlapped to produce bands every 70 nm. The aggregation of the molecules into fibrils is spontaneous in purified solutions, at physiological pH, temperature and ionic strength.

The collagen molecule, 280 nm in length, with a molecular mass of 300 000 Da consists of a triple helix stabilised by hydrogen bonds as well as by intramolecular bonds which are established between lysine residues of the telopeptides which make up the non-spiral sections of the chains at the N and C terminal extremities (representing somewhere in the region of less than 5% of the molecule). The cohesion of molecules within the fibre is assured by intermolecular bonds of the same type as the intramolecular bonds but which are established between telopeptides of adjoining molecules and between telopeptides and the spiral part of the adjoining molecules.

9.3.1.3 Manufacturing gelatin

The general principle of the process for extracting gelatin consists of converting the insoluble collagen into soluble gelatin. The alkaline or acid processes must cut the intermolecular bonds and all or some of the intramolecular bonds. Gelatin manufacture involves a whole series of stages which vary according to the raw materials and the type of gelatin being manufactured (Fig. 9.5).

The skins are treated directly by gelatin factories, after having been simply cut up and washed. The fat bone, however, must have its fat and flesh residue removed by means of washing in hot water. Dry bones contain a mineral fraction (principally calcium phosphates) and an organic fraction (collagen). By means of a treatment employing hydrochloric acid, the phosphates pass into solution and are precipitated by the lime in the form of dicalcium phosphate, removed, washed, dried and marketed for animal feed.

The organic matter in the bone, which is insoluble in acid, retains its shape by becoming more flexible: this is ossein or bone collagen, from which the gelatin will be extracted.

So, the processes for obtaining gelatin, whether starting from skins or from ossein, are very similar. The collagen contained in these raw materials is found in an insoluble form. Two processes are used to obtain the controlled breakdown of the organised structure of the collagen in order to obtain soluble gelatin:

- **The alkaline process:** the ossein or scraps are immersed in an alkaline medium, generally lime, for several weeks, at room temperature.
- **The acid process:** the ossein or pigs' skin is left to soak in an acid bath for one day, at room temperature. The alkaline or acid reagent is eliminated by washing followed by adjusting the pH.

Extraction takes place via cooking: the hot water digests the collagen, resulting in its hydrolysis, then the gelatin passes into solution, which is referred to as 'liquor'. The collagen is processed at a strictly controlled temperature, by means of successive extraction procedures, until the source material is completely used up. In the case of pigs' skin, the grease is separated from the gelatin liquors and recovered at this point.

The liquors identified are carefully filtered to eliminate impurities in suspension, traces of grease and albumin-like substances which have coagulated under the influence of heat. This process produces clear gelatin solutions.

These solutions are concentrated in evaporators under vacuum to achieve a concentration of 30–40%, depending on the raw materials and the products extracted. The concentrated solutions are sterilised at 140 °C, then cooled sharply to obtain a jelly extruded in the form of 'noodles' which are spread out on the drying chamber conveyor.

The various extraction products are subsequently ground and sieved to the selected particle size depending on the intended use of the product.

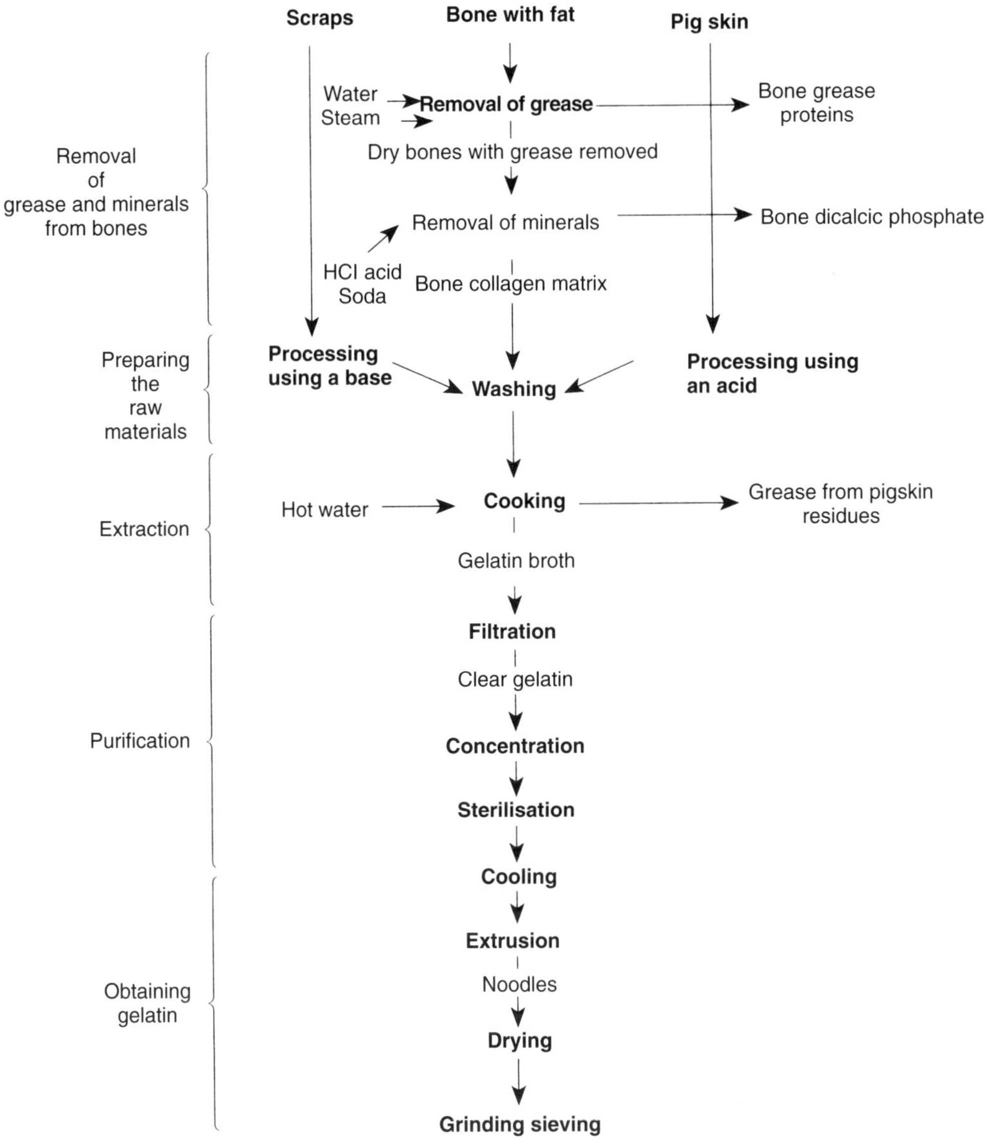

Fig. 9.5 Principal stages of manufacturing gelatin.

9.3.2 Physico-chemical and functional characteristics of gelatins

9.3.2.1 Various types of gelatin

The amphoteric properties of gelatin allow two types of gelatin to be distinguished, depending on the manufacturing process used and the value of the isoelectric point:

- Type A: pH_i of between 6.3 and 9.5.
- Type B: pH_i of between 4.5 and 5.2.

In addition, the various uses of these gelatins has resulted in the definition of several families of gelatin:

- F gelatins having a low gelling capacity; they have properties similar to the protein hydrolysates and do not form gels under the normal conditions of use.
- Gelatins having a strong gelling capacity.
- Chemically modified gelatins whose essential use often lies in their increased insolubility at the isoelectric point (out of phase in comparison with that of the initial gelatin) as in the case of phthalylated and phenylcarbamylated gelatins used in the photographic industry.

9.3.2.2 Physico-chemical and functional characteristics

Gelatin actually consists of several polypeptide chains which are either free or bound to each other. It is a polydispersed macromolecule whose **molecular mass** is between 10^4 and 10^6 Da. The molecular distribution determined by HPLC, depends on the raw materials and the extraction process used; it is possible to identify monomers (α chains: molecular mass of approximately 100 000 Da), dimers (β chains) and trimers (δ chains) as well as lower polypeptides.

Gelatin is characterised by its **pH** and its **isoelectric point,** which are generally measured in a 1% gelatin solution. The viscosity of the gelatins is strongly dependent on this. Viscosity is lowest around the isoelectric point and increases when the pH moves away from this zone. The reason for this lies in the repulsion due to charges which are identical all along the chain when the pH is reasonably distant from the isoelectric point.

The **solubility** of gelatin in water is very good, especially at concentrations of under 10%. When plunged in cold water, the gelatin swells, absorbing between 5 and 10 times its volume in water. When heated to 50–60 °C the swelled gelatin dissolves to give a solution, the 'sol' which converts into a 'gel' on cooling. For this to take place the gelatin concentration must be adequate and the cooling temperature must be lower than the melting point (approximately 40 °C). This 'sol–gel' conversion is reversible.

Unlike most hydrocolloids of polysaccharide origin, the gelling of this protein is independent of the pH, and does not require the presence of other reagents (cations, carbohydrates, etc.). As in the case of other gelling agents, the gelling mechanism of gelatin can be explained by the formation of a three-dimensional network. The junctions between macromolecules are mainly due to electrostatic interactions or hydrogen bonds. As these bonds are quite well spaced out, the gels have remarkable mechanical properties.

Gelatin, a high polymer, produces solutions whose **viscosity** is generally between 1.5 and 7.5 mPa s, measured by taking the time for a solution of gelatin of 6.67% concentration to flow through a viscometer pipette at a temperature of 60 °C. As viscosity varies with concentration, temperature and pH, measurements must be taken under specific conditions.

Gelatin solutions can form films during the reversible sol–gel changes in state, characterised by three measurements which correspond to the following transition zones:

- Melting point measured by the speed of fall of a steel ball in a 10% preparation.
- Setting point measured by the immobilisation of air bubbles in a preparation at the same concentration.
- Rotatory capacity.

The **jelly strength**, or Bloom, expressed in grams, is associated with the mechanical elasticity of the gel. It allows gelatins to be classified. The standard measurement is taken using a gelometer. This corresponds to the force required for a standard cylinder of a given depth to penetrate a gel at 6.67% protein, previously prepared at 10 °C for 16–18 hours. This value of between 50 and 300 Bloom allows the properties of the gels at the product storage temperature to be defined. The jelly strength increases with time: this is the 'maturing' of the gel, which varies inversely with temperature.

The measurement of turbidity, expressed as a nephelometric unit of turbidity (NUT), determines the quantity of light diffused at an angle of 90° by a gel placed in a test-tube lit along its axis. It indicates the quality of the filtration carried out; its value goes through a maximum at the isoelectric point.

9.3.3 Uses within the food industry

The food and non-food applications of gelatin are constantly developing. As a result of its protein nature, it is considered by most legislation to be a food and not an additive – except in the case of cooked meats where its content is limited. The multifunctionality of gelatin is obvious when its numerous uses are reviewed, shown in Table 9.7 according to the various industries involved.

9.3.3.1 Meat products industry

Gelatin of animal origin can be considered to be a natural ingredient of meat-based proteins. It is often incorporated in the following products:

- Ham cooked in moulds, cooked shoulder.
- Tinned meats and semi-preserves.
- Meat emulsions (cooked meats, pâtés).
- Jellies.

During preparation, hams cooked in moulds are sprinkled with gelatin prior to being scored, in the place where the bone has been removed; this is also carried out to restick flesh or rind whose excessive fat has been

Table 9.7 Food uses of gelatin

Industrial sector	Gel strength (Bloom unit)	Viscosity			Gelatin concentration (%)
		High	Average	Low	
Meat products					
Ham	150–250	+	+		1–2
Coating	150–250	+	+		5–20
Jellies	150–250	+	+		3–15
Dairy products					
Gelled milks	150–250		+		0.2–2
Cream desserts	150–250		+		0.2–1
Yoghurts – fermented milks	150–250		+		0.2–2
Ice-creams – sorbets	150–200		+		0.2–1
Confectionery					
Gums	100–150		+	+	10–15
Chewing gums	100–150		+		0.5–2.5
Caramels – toffees	100–150		+		0.2–1
Meringue products	75–125		+		2–5
Extruded aerated products	100–125	+			3–7

removed. On coming into contact with the meat, the gelatin powder absorbs water. During cooking, it forms a film which, after cooling, welds the flesh.

Gelatin gels the salted juices that exude during cooking and contains them in and around the ham. It reinforces the jelly obtained directly during cooking by the connective tissues. The main result is better cohesion on slicing.

Because of their high fat and water content, cooked meat products present particular problems of stability, which result in the exudation of water or fat, or in problems in texture after cooking. In addition to having a stabilising effect on the emulsion, incorporating gelatin creates a better water bond and gives the pâté a homogeneous texture.

Gelatins with average jelly strength (150–250 Bloom) are used for these purposes.

Ready-to-use jellies are employed in the following ways:

- Glazing, an operation that consists of coating a preparation, for example cooked meat, with jelly.
- Coating ham noisettes, salamis, etc.
- Filling up tinned meats or semi-preserves.

Some of these jellies are prepared with a mixture of gelatin and carrageenan. The addition of this glycan reduces the time required for the gel to set, increases its strength and raises its melting point; for example, adding 10% carrageenan raises the melting point by approximately 20 °C.

9.3.3.2 Dairy and associated industries

The industry uses stabilisers in the preparation of milk and its by-products, milk desserts, fermented products, ice-creams, etc. Among the most widespread uses of gelatin are the following:

- Milk jellies: using gelatin in association with other stabilisers produces an end-product with a smooth, flexible gel structure, avoiding syneresis.
- Cream desserts where gelatin offers the same advantages even when there are significant changes (freezing, for example).
- Yoghurts and fermented milks: when legislation authorises this, gelatin is incorporated for its anti-syneresis effects. In some countries, the shelf life of 'yoghurts' or fermented milks is extended by additional pasteurisation after incubation. Adding a mixture of gelatin (0.5%) and modified starch (between 1 and 1.5%) before heat processing allows a good texture to be obtained and the effects of syneresis to be controlled.
- Ice-creams, sorbets: gelatin, in association with other stabilisers (pectins, xanthans, carrageenans, etc.), is used for its emulsion stabilising capacity and/or its swelling properties.

9.3.3.3 Confectionery

There is no doubt that this is the industry in which all the functional properties of gelatin are used to advantage. These multiple uses can be explained in particular by the following:

- The ability of gelatin to disperse in an aqueous solution and to absorb between five and six times its own weight in water.
- The wide range of possible viscosity values.
- The protective colloidal power which allows crystal formation to be controlled.
- The foaming capacity which varies according to the type and quantity of gelatin.
- The film-forming property put to advantage for coating nuts when manufacturing sugar almonds.
- The binding capacity used in manufacturing tablets.

Gelatin is used in association with other stabilisers or gelling agents in the manufacture of numerous types of confectionery: gums, chewing gum, caramels, toffees, nougats, soft centres, spreads, meringue products, extruded aerated specialities and sugarless confectionery products based on polyols.

Hydrolysed gelatin obtained enzymatically has significant outlets within the confectionery industry (chewing gums, for example), in cake-making (aerated milk desserts) and also in oenology for clarifying wines and fruit juices.

This quick look at the food applications of gelatin does not include the ways in which this by-product has been developed in other industrial sectors. To mention only the best known uses, gelatin is still a basic product within the photographic industry, for producing capsules within the pharmaceutical industry, and for surfacing in the paper and cosmetic industry.

Part Two – Extraction and modification of biomolecules

10

Sugar chemistry

10.1 Definitions and functions of carbohydrates

It is easy to define carbohydrates in the chemical sense of the term; they are carbonylic substances formed from one or several units of polyhydroxy-aldehyde – or ketones. On the other hand, carbohydrates – often incorrectly termed sugars – and their by-products, constitute a body of carbon substances that is difficult to define as a whole. In fact, these compounds constitute a heterogeneous unit in terms of their chemical structure and they fulfil various functions (nutritional, organoleptic, preservational) yet they have a shared identity of a virtually cultural kind associated with their sweetening power.

New products are constantly appearing alongside a well-established 'sugar' industry, thus underlining the enormous wealth of this type of molecule (Fig. 10.1) in plant raw materials.

The industrial preparation of carbohydrates from legume or cereal grains is an operation widely used in the case of **starch**, for both food and industrial purposes. The economic development of starch and its by-products is covered in Chapter 11. The recovery of parietal carbohydrates now falls within the field of biotechnological economic development and will be covered in this chapter.

Alongside this technological innovation, sugar chemistry has long been involved in the exploitation of the essential product of the 'sugar-beet' industry: **sucrose**. Sucrose, which is stable, cheap, constant in quality and easy to store, is used directly in many food processing operations, and, as a consequence, there is no advantage in selling it in the form of intermediate products, apart from a few specialised products involving only small tonnages.

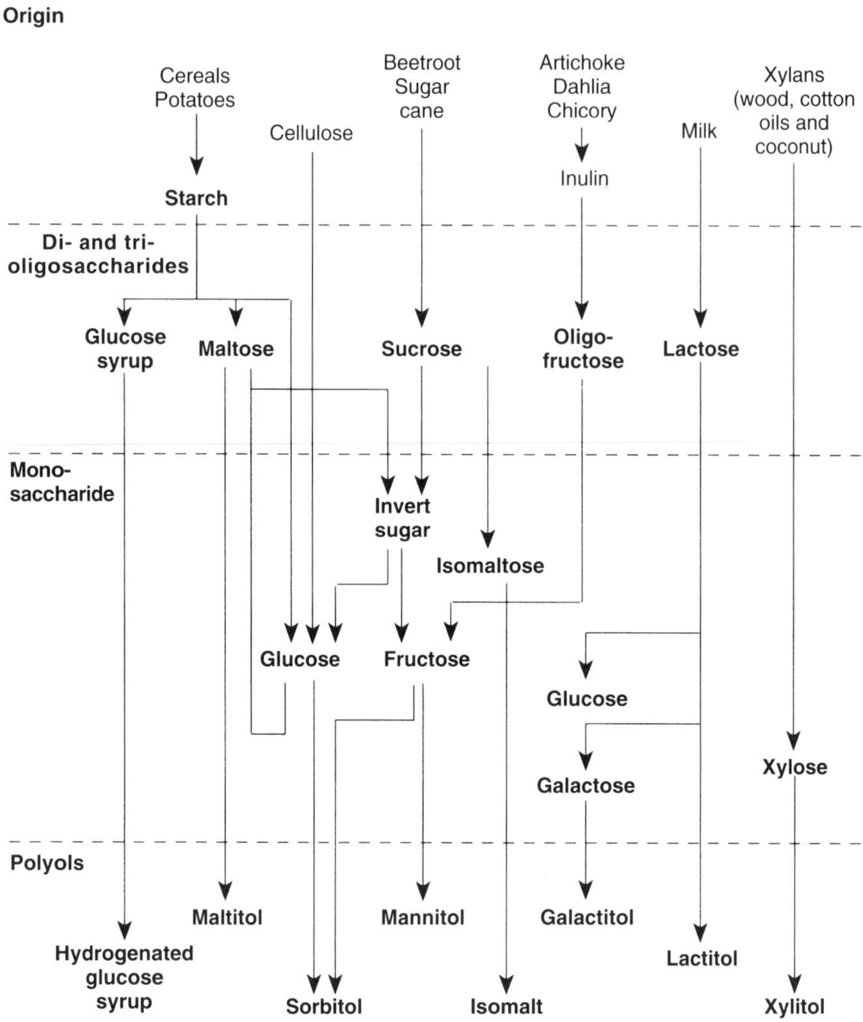

Fig. 10.1 Principal chains of sugar chemistry.

Almost all the **lactose** contained in milk (approximately 50 g/l) is found after coagulation in whey. If we consider world production of whey, the theoretical possibilities of using lactose and its by-products are absolutely enormous. Unfortunately the actual state of affairs is quite different and any possible uses must take into account legislation, price levels, potential substitutes, tradition, sweetening efficacy and the technical difficulties involved in hydrolysis.

New openings have appeared in the sugar chemistry sector in relation to the production of organic and biological molecules, for both human and animal foodstuffs and for non-food uses. This discussion will restrict itself to dealing only with the food applications of these products.

In this chapter we will be examining in turn the properties and uses of sucrose, lactose, parietal carbohydrates, plant oligosaccharides, polyols and intense sweeteners of both carbohydrate and non-carbohydrate types.

10.2 Sucrose

Sucrose is easily the most widespread natural simple carbohydrate and this is why the term 'sugar' in the singular is the legal definition for disaccharide.

Sucrose is found in all plants that contain chlorophyll. Whether extracted from cane or sugar-beet, 'sugar' manufacture uses only very simple purification and extraction processes, without using any additives or synthetic products. The 'sugar' industry produces the largest tonnages of the purest products in the world and obtaining this purity of sucrose, which currently exceeds 99.9%, involves only the use of fairly simple chemical processes.

10.2.1 Structure and principal properties

Sucrose is a carbohydrate food which is very important for the populations of developed countries; moreover it is the only pure crystallised food consumed by humans. Sucrose is known as ordinary cane or beet 'sugar' and its hydrolysis gives a glucose and a fructose molecule. Its scientific name gives a precise definition: α-D-glucopyranosyl-(1→2)-β-D-fructofuranoside.

Commercial sucrose takes the form of a shiny white crystalline substance (rhomboidal prisms) which is not hygroscopic. It has no smell or characteristic taste. Its moisture content is very low (in the order of 0.05%) and its stability when stored is very good.

Sucrose comes in various granular forms suited to the many requirements of the user industries, and is very soluble in water, especially as the latter's temperature rises; at 20 °C solubility is 67 g per 100 g solution. When heated dry, sucrose melts at around 160 °C then converts into caramel before 'burning' at around 190 °C. Its specific dextro-rotatory power is: $\alpha_D^{20°} = 66.5°$.

10.2.2 Invert sugar

Sucrose hydrolyses easily in an acid medium. Two hydrolases are responsible for enzymatic hydrolysis: α-glucosidase and β-fructosidase; the latter has been called 'invertase' since the mixture of the monoses, because of the strong laevo-rotatory nature of the fructose, retains this property (20°). This mixture is still known as 'invert sugar'. Invertase is extremely widespread and is found principally in the human intestine.

From an industrial point of view, invert sugar is produced either by enzymatic catalysis or by acid catalysis. Enzymatic catalysis is particularly suitable for producing invert sugar with a very high degree of hydrolysis. The traditional acid method produces syrups that are strongly mineralised (after neutralisation of the acid) and highly coloured (coloration due to the extreme conditions of the reaction).

The method which uses cation exchanger resins is, on the other hand, a very well-controlled process in comparison with the above. In addition, this process can be continuous and can be fully automated. It is possible to obtain the desired inversion percentage by varying the temperature and the flow rate through the column.

10.2.3 Uses of sucrose and invert sugar

Invert sugar in solution is much more fluid than liquid sucrose (at 20 °C the viscosity of a sucrose solution is three times as high as the equivalent solution of invert sugar). It lowers a_w more strongly than sucrose; in other words it increases the water-holding capacity. This property is used in order to improve product shelf-life from both a texture and microbiological point of view. In addition, sweetening power is higher.

Mainly because of its moisture-providing properties, invert sugar is used in moist confectionery (items incorporating gelatin, fruit jellies, etc.), industrial cake-making (gingerbread, madeleines, brioches, cakes, fillings, etc.). Because of its physical properties (sweetening power, lowered freezing point, etc.) it is used in ice-creams, sorbets, fresh dairy products, drinks (crème de cassis, etc.).

Sucrose is one of the ingredients in flavouring caramels. It is interesting to note that – as in the case of other carbohydrates – when the partial solubilities of glucose and sucrose are added together in mixed solutions, it is possible to obtain total concentrations of as high as 75%, considerably higher than the concentration of one of these carbohydrates alone. In this way syrups can be manufactured while avoiding crystallisation or fermentation.

In its dry form sucrose can absorb only small quantities of water in the form of vapour and can contribute to keeping a stored mixture dry. Sucrose in concentrated solution has the best food preservation properties.

Traditional applications produce high tonnages within the jams and syrups sector. In cakes, sucrose also binds water, thus extending the shelf life of products.

Sucrose has texturising, lubricating and bulking (role of viscosity) properties, as well as modifying and flavour-enhancing properties (even in very low quantities which allow acid and bitter flavours to be 'eliminated', etc.), and dispersive and caramelising agents.

10.2.4 By-products

The sucrose molecule has a very rich potential. It is a non-reducing, stable, pure and cheap polyol on which the traditional reactions of organic chemistry can be carried out (oxidation, reduction, ammonisation, substitution, etc.).

Esterification is a very productive method which is used extensively, either in a solvent medium or in an emulsified medium (starting with an anhydride or an acid chloride or by trans-esterification). Certain esters have been used for many years, such as 'sucrose' diacetate hexaisobutyrate (SDH) used as a dispersing agent in essential oils and in the food industry, formulation of hotmelts, coatings, etc, in which the well-known plasticising sucrose octobenzoate (Velsicol) is also found.

Sucrose esters obtained with fatty acids are being extensively developed. These biodegradable, non-toxic products have considerable potential as non-ionic, emulsifying and stabilising surfactants in the food and cosmetic industries. In addition, by acting on the hydrophilic lipophilic balance (HLB) specific applications can be envisaged.

Sucrose polyesters or **sucro-esters**, which are a hexa-, hepta- and octa-ester mixture of sucrose with long chain fatty acids (generally between 8 and 12 atoms of C) are not hydrolysed by any enzyme in the digestive tract (Fig. 10.2). They have been subjected to clinical trials, producing the following results:

- Their efficacy: weight loss (200 g/J) in obese patients, 20% reduction of blood cholesterol, reduced absorption of cholesterol contained in foods.
- Their innocuity: they do not pass into the blood (unlike synthetic sweeteners such as aspartame) and have no notable side effects apart from a slight loss of vitamin E (liposoluble vitamin).

Sucro-esters, whose appearance, taste and consistency are identical to those of a vegetable oil, can replace up to 25% of the fat in cooked foods, without

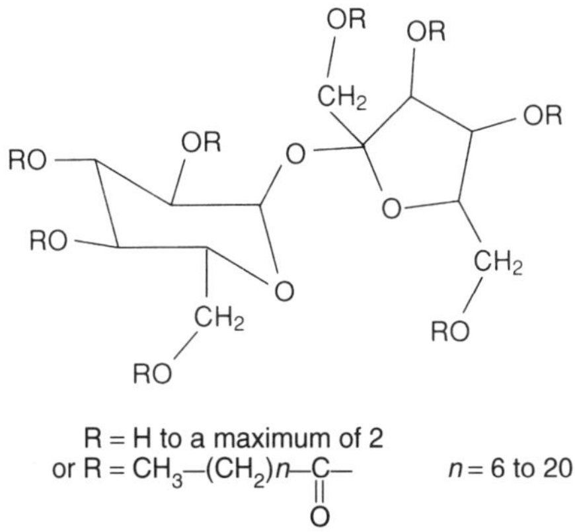

Fig. 10.2 Structure of the sucrose polyesters or sucro-esters.

any modification of recipes, and up to 70–75% of lipids in cooked dishes of suitable composition.

Sucrose can be replaced by other carbohydrates (glucose, fructose, etc.) or polyols (xylitol, erythritol, sorbitol, etc.). Sorbitol, in particular, is the starting point for synthesising various esters. When heated to a temperature of between 220 °C and 250 °C, in the presence of an acid catalyser and fatty acids, sorbitol simultaneously undergoes intramolecular dehydration, with the formation of the corresponding anhydride (sorbitan), and the esterification of one or several alcohol groups produces mono- or polyesters of sorbitan (see Fig. 13.4).

By means of the controlled addition of molecules of ethylene oxide, under pressure, at 140–160 °C in the presence of an alkaline catalyser, sorbitan monoesters can produce polyoxyethylene by-products. These products are known as **polysorbates** and vary in terms of the number of fixed oxyethylene groups and the nature of the fatty acid. The use of these compounds in human foodstuffs is not authorised in France, although it is in other countries.

Sucroglycerides are produced as a result of the trans-esterification of a natural triglyceride (suet, palm oil, etc.) by sucrose; a mixture of sucrose mono- and diesters and mono- and diglycerides is formed (see Chapter 13). These substances are used as non-ionic surfactants; for example in the manufacture of fat-enriched milks (using suet), for calves, or in other neutral or acid foods. They lower the surface and interfacial tensions significantly, are not toxic and can be assimilated.

The essential structural characteristic of these two by-products of sucrose is that two spatially distinct fields are found in their molecule: the apolar one is made up of hydrocarbonated chains of fatty acids, and the other polar one corresponds to the non-esterified hydroxides of sucrose, as well as glycerol in the case of the sucroglycerides. As the relative size of these two fields is quantified by the HLB, the sucroglycerides have HLBs that are midway between those of the sucro-esters and those of the monodiglycerides.

Because of their amphiphilic nature, adding sucro-ester or sucroglyceride molecules allows the water/oil interfacial tension to be lowered and a stable emulsion with a lower energy supply to be obtained. Depending on whether the continuous phase is aqueous (emulsion of an oil-in-water type such as in dressings) or fatty (emulsion of a water-in-lipid type such as margarines), an emulsifier will be chosen which is either more hydrophilic (sucro-esters) or more lipophilic (sucroglycerides), respectively.

Sucroglycerides also exercise a significant anti-bleaching action in chocolate-making and an anti-crystallising action in fillings for confectionery products. They prevent re-crystallisation from taking place and give the end-product greater flexibility. They are incorporated hot in syrups.

These sucrose by-products also have potential uses in cereal products. Sucroglycerides improve the performance of flours and dough during kneading and cooking. In particular they improve the extensibility of the dough and its cohesion during cooking.

In addition, the chlorination of sucrose increases its sweetening power 600 times, double that of saccharin and almost four times that of aspartame, yet at the same time its caloric value remains very low. However, unlike these two sweeteners, this compound, known as **sucralose**, is very stable to heat (see Section 10.7.4.2).

10.3 Lactose

10.3.1 Structure and general properties

This is the carbohydrate found in milk and nowhere else; it is a reducing agent: β-D-galactopyranosyl-(1→4)-D-glucopyranose in the α or β forms (Fig. 10.3). It is predominantly found in the milk of ruminants, at a level of 5% in cows' milk.

The two forms, α and β, have very different properties (Table 10.1) and both have practical applications. The α form crystallised with a water molecule is ordinary lactose; this can be dehydrated. The β form crystallises, without binding water, from a concentrated solution, at above 90 °C, which is the critical crystallisation temperature.

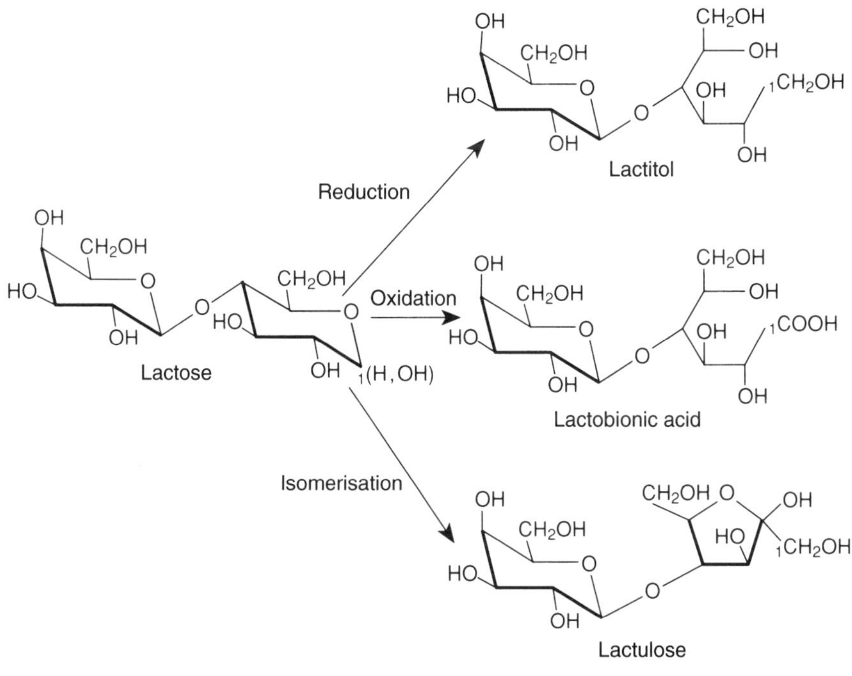

Fig. 10.3 Structures of lactose and by-products.

Table 10.1 Properties of lactose isomers

	α (H_2O)	β	Equilibrium (20 °C)
Rotatory power $(\alpha)_D^{20}$	+89.4	+35	+55
Melting temperature (°C)	202	252	–
Solubility (g per 100 g water)			
15 °C	7.3	50	17
100 °C	70	95	–
Concentration in solution at 20%	38	62	β/α = 1.63
Sweetening power (sucrose = 1)			0.2

Anhydrous lactose and mono-hydrated lactose have little affinity for water, so this disaccharide can be used in many formulae without any risk of clotting.

Lactose is not very soluble in water, approximately four times less than sucrose at ambient temperature (17 g per 100 g). The difference is less at 100 °C, because lactose solubility increases considerably. This disaccharide is therefore fairly easy to separate by crystallisation; but obviously thick syrups or jam cannot be obtained with lactose. In addition it does not have a great deal of sweetening power (1/6 that of sucrose); milk is not very 'sweet'; milk diets can be tolerated because they are not sickly.

Lactose is an interesting example of crystalline polymorphism; during slow crystallisation various forms can be observed, which result in a hard, long, complex crystal. Certain substances inhibit crystallisation; for example riboflavin (vitamin B_2), but at a concentration higher than that found in milk.

Lactose is extremely effective in binding flavours and pigments. This is why this carbohydrate is used in tomato juices or concentrates, sauces and products enriched with carotenoids, as in the case of the majority of tinned vegetables and most ready cooked dishes.

10.3.2 Lactose hydrolysis

Hydrolysing lactose offers nutritional, technological and economic advantages:

- Increased digestibility for people with lactose intolerance.
- Improvement of sweetening power.
- Increased non-enzymatic browning capacity.
- Products of hydrolysis much more soluble than lactose, as they avoid the crystallisation phenomenon observed in a concentrated medium.

Two methods are possible: chemical hydrolysis or enzymatic hydrolysis. Chemical hydrolysis is the most traditional process. It is carried out at high temperatures, after adjusting the pH to a high acid value. During the initial stage, the juice has its cations removed and is then hydrolysed by being passed over a strong cationic resin at high temperature (95 °C) and at acid pH (1.5).

In the second stage, the juice which has been 80–90% hydrolysed is then cooled to 20 °C and has its anions removed. After a final filtration, the level of dry matter is increased by evaporation from 6 to 78%, in order to obtain a lactolysed syrup.

In order to carry out enzymatic hydrolysis, commercial lactases, which are all of microbial origin, are used. This β-galactosidase is present in most yeasts and moulds; it is less common in bacteria and is a distinctive metabolic feature typical of certain groups such as lactic bacteria and coliform bacteria.

The lactase from *Aspergillus niger* can be distinguished by its acid pH_0. There are several alternatives with regard to the process of enzymatic hydrolysis:

- Using the free enzyme in a vessel.
- Hydrolysis by the immobilised enzyme.
- Inclusion in cellulose-type fibres.
- Immobilisation on porous glass.
- Hydrolysis by the free enzyme in a membrane reactor.

The last technique is the one currently used most frequently. It allows continuous hydrolysis of the lactose-enriched juice alongside enzyme retention, with the latter being retained in the hydrolysis circuit by an ultrafiltration membrane.

Hydrolysis of the whey lactose, which may be associated with protein purification and demineralisation, is an interesting technological process, as it produces a product with good functional properties which is therefore very useful for the secondary processing industry (biscuit-making, confectionery, etc.). Numerous uses have resulted: toffees, spreads, fondants, Danish toasts, cakes, ice-creams, etc. Lactose hydrolysate can therefore replace all or part of the sucrose in numerous products.

10.3.3 Lactose by-products

10.3.3.1 Lactitol

Lactitol consists of a molecule of galactose bound to a molecule of sorbitol (Fig. 10.3). Depending on the conditions of preparation, lactitol crystals can be anhydrous, monohydrated or dihydrated.

Lactitol presents the general properties of alditols. Crystals of hydrated lactose are non-hygroscopic and stable. The hygroscopy of lactitol is very low and identical to that of maltitol. This property is used to advantage in the manufacture of chocolate confectionery products, chewing-gum and cakes. Obviously this means that lactitol cannot replace sorbitol as a water-holding agent.

Lactitol's solubility in water is good (149 g of lactitol per 100 ml at 25 °C).

The viscosity of the aqueous solutions of this polyol is comparable to that of sucrose, as is the crystallisation diagram. The chemical stability of lactitol in an alkaline medium is higher than that of lactose. For the same reason – absence of the carbonyl group – lactitol does not participate in Maillard reactions.

The non-cariogenic effect of lactitol can be compared with that of sorbitol. Lactitol is neither hydrolysed nor absorbed in the small intestine and, like a number of alditols, presents a non-dependence towards insulin. In addition, its use by people with diabetes is reinforced by its low caloric value.

Lactitol's activity as a laxative depends on the direct osmotic effect of the molecule, if it is administered at a high dosage. On the other hand, when lactitol is administered at moderate doses, the laxative effect depends on the enzymatic breakdown of the molecule by the bacterial flora in the intestine. At moderate dosage lactitol behaves as a fermentable fibre.

10.3.3.2 Lactulose

Lactulose, which was produced synthetically over 50 years ago, is a disaccharide derived from lactose: β-D-galactopyranosyl-(1→4)-D-fructofuranose (Fig. 10.3). Lactulose synthesis is based on the isomerisation of lactose by strong alkalis. It is produced commercially in the form of an odourless crystalline powder, which is hydrosoluble (67 g per 100 ml) and white in colour.

Lactulose is not absorbed in the small intestine and when digested in the colon produces little energy. Lactulose was originally offered under medical prescription to combat metabolic disturbances caused by proto-systemic encephalopathy, and has been used more generally as a laxative. We are now fully aware of the way in which the lactulose acts in the digestive tract, as a result of our knowledge of intestinal physiology and the ecosystem of the colon.

Lactulose prescribed at moderate doses (20–40 g/day) acts in the colonic tract in the same way as glycans in the diet or in the mucilages. It is differentiated from these by its low molecular mass, absence of viscosity and hygroscopic power and the consistency of bacterial hydrolysis in the right colon.

10.3.3.3 Galacto-oligosaccharides

Human milk contains galacto-oligosaccharides which are present as trace elements and have the ability to reduce the incidence of digestive and respiratory pathologies in breast-fed babies compared with babies fed on substitute milks. This observation stimulated the research into, and synthesis of, molecules related to these galacto-oligosaccharides.

This section will consider the properties of the galacto-oligosaccharides obtained by enzymatic synthesis from whey, which is still called transgalacto-oligosaccharide (TOS).

The TOS is synthesised by means of the action of the β-galactosidase of *Aspergillus oryzae* on lactose in solution. It appears in the form of a mixture of tri-, tetra-, penta- and hexa-oligosaccharides, in which the bonds between the different monosaccharides are indicated below in decreasing order:

$$\begin{array}{cc} \multicolumn{2}{c}{\text{Gal—(Gal)}_n\text{—Glc}} \\ \beta\text{-}1\rightarrow 6 & \beta\text{-}1\rightarrow 4 \\ \beta\text{-}1\rightarrow 4 & \beta\text{-}1\rightarrow 2 \\ \beta\text{-}1\rightarrow 3 & \beta\text{-}1\rightarrow 6 \end{array}$$

This initial enzymatic reaction can be followed by a second one which implements the β-galactosidase of *Streptococcus thermophilus*. It is thus possible to increase the sweetening power of the mixture and to obtain a product that is more suitable for certain food uses.

The sweetening power of these oligosaccharides depends on their richness in mono- and disaccharides. It is approximately 50–80% lower than that of sucrose. Likewise, the water activities in the solutions of the different TOSs are similar to those of the reference disaccharides, or slightly higher.

In an acid medium, the TOS is more stable than sucrose at all the temperatures studied. It does not break down when stored for several months at pH 2, between 5 and 37 °C.

The TOS, which cannot be digested by the enzymes in the intestinal tract, is fermented by the microbial flora in the colon. The use of this oligosaccharide is recommended in a wide variety of food products: baby foods, drinks, breads and cakes; like many synthetic compounds it possesses both the properties of sweeteners (sweetening power and anticariogenic action) and some properties attributed to dietary fibres (action on the intestinal tract and on the composition of the digestive microflora).

Lactobionic acid (Fig. 10.3) obtained by oxidation of the aldehyde function on the carbon 1 of the glucose is another of the by-products of lactose. This product has not yet been extensively used. It is a good souring agent and an active conveyor of calcium, iron and certain antibiotics (erythromycin).

10.4 Parietal carbohydrates

10.4.1 Fractionation and properties of the products obtained

The preparation and use of parietal carbohydrates on an industrial scale has not progressed to match the developments made with starch, since these fractions had not been recovered previously. During fractionation, extraction and purification of plant proteins in a liquid form from legumes (peas and beans) and cereals (wheat), starch and parietal carbohydrates have been the subject of recovery studies. The techniques used are based on their structure and properties. The overall composition of the parietal carbohydrates of wheat and barley, for example, essentially represented by glucose, xylose and arabinose monomers, corresponds to the different types of polymers. In fact, wheat is rich in arabinoxylans, whereas barley is rich in β-(1→3)(1→4)-glucans. Their structures and properties differ according to their histological location. So, wheat flour or endosperm is mainly made up of arabinoxylans, weakly substituted and highly soluble (commonly known as pentosans) and β-(1→3)(1→4)-glucans, which gives them a certain water-holding capacity. On the other hand, wheat bran or pericarp with 30% cellulose and 70% strongly substituted and insoluble arabinoxylans presents higher water absorption properties.

Fractionation by air classification and by enzymatic and chemical aqueous fractionation of the carbohydrates from wheat bran produces fractions enriched with hemicelluloses and pentosans. Their physico-chemical properties show a direct correlation between the expansion and water-holding capacity and their content in arabinoxylans. This confirms the accelerating action on intestinal transit attributed to wheat bran, owing to the phenomenon of water absorption, and allocates responsibility for this to the multiramified portion of arabinoxylans.

The same properties have been observed on the cellular walls of peas and beans. The exploitation of pea and bean by-products, enriched in starch and cellular walls, after protein extraction, has allowed starch and 'parietal carbohydrates' to be purified simultaneously by sieving. The 'hemicellulose' fraction has been shown to have a water-holding capacity in direct relation to its content in pentoses. Recently it was shown that the presence of uronic acids in this fraction was responsible for its chelating properties, one of which is the absorption of cholesterol esters. It may be that these polymers with a heterogeneous (neutral monosaccharides and uronic acids) and ramified structure, which gives them their chelation and water-holding properties, will not be involved significantly in the near future in the nutritional–medical and pharmaceutical fields.

Because of their composition and chemical structure, low content in the grains and properties of solubility, water absorption and chelation, these heterogeneous polyosides are not easy to isolate in the native state. However, laboratory preparation of purified fractions and the study of their properties for capturing various metabolites (water, salts, cholesterol and lipids) suggest a certain future for nutritional–medical and other uses. The latter fractions have the advantage of coming from natural plant products in comparison with the chemical products currently used for similar properties.

To sum up, the recovery and recent exploitation of parietal carbohydrates is still posing numerous problems, in terms of their preparation on an industrial scale and their possible uses. However, there are already some industrial uses in existence.

10.4.2 Industrial exploitation of pentoses and their derivatives

These are four in number: ribose, arabinose, xylose and lyxose. They are present in extremely variable quantities in plant structures. The two corresponding ketopentoses, ribulose and xylulose, which are very rare in the natural state, occupy the essential role of encouraging the isomerisation of the corresponding aldopentoses.

As in the process for preparing xylose, obtaining pentoses involves the prior extraction of oligomers of variable size followed by their hydrolysis. Three methods are possible:

- Solubilisation in an alkaline medium (not often practised industrially).
- The manufacture of cellulose paste in the presence of organic solvents.
- The destruction of plant fibres under high pressure in the presence of steam.

The subsequent recovery and purification stage requires successive extractions, using azeotropic water/ethanol and isopropanol mixtures.

Pentose is extracted by crystallising the resulting liquor, which has been concentrated, dehydrated or purified by chromatography over ion exchangers from demineralised extraction juices.

The principal economic developments of xylose involve converting it into xylitol, ethanol, furfural and furanic polyols.

Hydrogenating xylose syrup to produce xylitol is dealt with in the following section. The conversion of D-xylose into furfural and the chemistry of the latter is a prototype example indicating what the biomass as a raw material could contribute to the chemistry of the future.

Because of its direct applications, its principal by-products already available and new molecules which are reaching the pilot stage, furfural should undergo significant development over the next few years. Sources of raw materials are abundant and cheap, and, in addition, this polyfunctional molecule is capable of generating monomers similar to those produced by petrochemistry.

Furfural was used initially for its ability to act as a selective solvent. Its principal by-products – furan, furfurylic acid, tetrahydrofuran and tetrahydrofurfurylic alcohol – have now started to be developed at a rapid rate within this industrial sector.

These molecules are used either as solvents or as intermediate products of synthesis for C_4 synthons, di-functionalised at the end of the chain, or saturated heterocycles, both functionalised and di-functionalised. So, for example, tetrahydrofurfurylic alcohol can, by reacting with ammonia at 600 °C, produce pyridine, or, by reacting with nitric acid, produce an ester used as an additive which increases the octane index of gasoil.

The following additional molecules are some examples which are found in smaller quantities and are also worthy of mention, without being exhaustive:

- Furfurylamin, nowadays used essentially within graphic arts, is a potential base for cationic furannic surfactants.
- Furoin, produced as a result of the dimerisation of furfural in the presence of cyanide anions, is a well-known plasticising agent and an intermediate product in solvent synthesis.
- Methyl-2-furan, which is a by-product of the hydrogenation of furfural into furfurylic alcohol, is a very good solvent for resins and paints, and can also be used in the formulation of flavours.
- Furoic acid, after esterification, is a useful flavouring agent. These esters are also very good solvents and intermediate compounds for further chemical synthesis.

10.5 Plant oligosaccharides

Numerous plants contain high concentrations of oligosaccharides. These reserve substances often consist of sucrose, with one or several monosaccharide residues lying along one or other side. An example of this is the family of α-galactosides which contains one or several α-D-galactoses bound by an α-1→6 bond to the sucrose's glucose unit. These oligosaccharides are indigestible by humans.

However, there are other oligosaccharides which are beginning to become involved significantly as ingredients in IFPs. Some of these are found in certain natural foods; for example, melezitose in honey, oligosaccharides of fructose in chicory, garlic and Jerusalem artichoke. Others, such as maltotriose and maltotetraose, are prepared from maize starch.

10.5.1 Inulin

This glycan is a product of the condensation of around a hundred units of D-fructose, in a furannic form, by means of β-2→1 oside bonds, with a few D-glucose units at the end of the chain (Fig. 10.4).

Inulin is a useful product for more than one reason:

- Because of its gelling capacity it can improve the stability of emulsions and expanded foods.
- Because of its non-degradability in the organism it behaves in the same way as a dietary fibre and arrives virtually intact in the colon where it is hydrolysed by bacteria, in particular the bifidobacteria and some lactobacilli.
- Because of its low caloric value of between 4 and 10 kJ/g depending on the length of the chains.

By means of special technology the functional properties of inulin have been enriched recently by a new development: a fat substitute for manufacturing low-calorie foods. The process consists of mixing water (60%) with

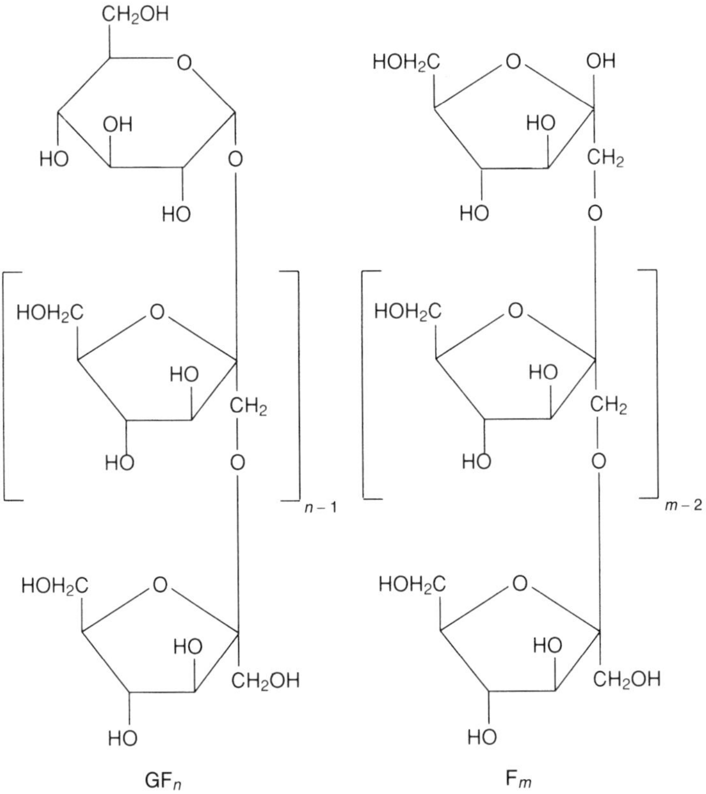

Fig. 10.4 Structure of oligofructose: mixture of gluco-fructans (GF_n and fructans F_m).

inulin (40%) and submitting this mixture to a very intense shearing action under pressure. A network of inulin particles is formed and this traps a large quantity of water molecules.

Using this technology a product can be developed which behaves both as a dietary fibre and a substitute fat, and can therefore be incorporated without the risk of phase separation in numerous food products: ice-creams, fermented milks, chocolate mousses and biscuits.

10.5.2 Oligofructose

Numerous commercial preparations contain oligofructose, which is a mixture of fructose-oligosaccharides consisting of two or more fructose units, bound by a β-(1→2) bond (Fig. 10.4).

A molecule of glucose is often found at the end of the chain. Oligofructose is obtained from inulin which is the reserve glycan in numerous foods. One of the most widely used raw materials is chicory root. Inulin is extracted from the root by means of a process that resembles the method used to extract sucrose from sugar beet (diffusion into water). This biopolymer is then hydrolysed and purified using techniques similar to those used in starch technology. However, other commercial preparations of fructo-oligosaccharides are obtained by means of the enzymatic grafting of one, two or three molecules of fructose onto one molecule of sucrose.

Short chain oligofructose solutions have a viscosity similar to those of sucrose solutions and they develop in a similar way. Consequently these oligofructoses are used to improve the cohesion and texture of sorbets and other food products such as ice-creams, pastry fillings and coatings for cakes, mousses and pastries.

In addition, these carbohydrates are remarkably stable, although this depends on the pH/temperature combination. At low pHs or high temperatures these sweeteners are hydrolysed into glucose and fructose. At neutral pHs, these gluco-fructans are stable at high temperatures in the order of 140–150 °C. At low pHs (below 3) stability is only obtained at temperatures of 70 °C and below.

In addition, short chain oligofructoses have a higher water-holding capacity than sucrose and a lower one than sorbitol. Also, the freezing point of these products is very close to that of sucrose.

This strong similarity with the properties of sucrose is repeated from an organoleptic point of view. The sweetening power is, however, reduced in relation to this disaccharide.

Oligofructoses whose DE (dextrose equivalent; see Section 11.4) is generally between 20 and 25, also possess attractive functional properties (good solubility, moderate sweetening properties, bulking agent, wetting agent), which explain their use in numerous sectors (confectionery, biscuit-making, dairy industry, pharmaceutical preparations, etc.).

10.6 Polyols (Sugar alcohols)

10.6.1 Alditols

It must be borne in mind that these are not monosaccharides, as they do not have any carbonyl groups. Nevertheless they are very profitable as they are derived from monoses resulting from hydrogenation and are found in fruits and various parts of plants.

It is easy to reduce the CO group chemically (sodium borohydride, hydrogen with catalyser, sodium amalgam, etc.). However, the biological method is more profitable. We know about the enzymes of oxido-reduction which cause the reaction in both directions (oxidation of the polyol into a monose). Certain alditols, derived from aldoses, widely found within the plant kingdom, probably come from the corresponding aldoses via reduction by means of the NAD or NADP enzymes.

However, these substances, unlike the monoses, are rarely combined;

although mannitol glucoside is found in certain seaweeds. The scientific name is that of aldohexose with the suffix 'itol', but, in the case of those well known for some time, a commonplace name is still used which generally designates the plant in which it was initially discovered.

The interest shown recently in these substances has been based on their special physico-chemical (water-binding, etc.) properties, in particular their non-dependence on insulin and their non-cariogenic effect, which explains their use in health foods.

Depending on the various metabolic methods used to break down the polyols, these compounds can be used in dietary treatments for obesity (non-caloric substance), for diabetes (non-insulinogenic by-product, delayed liberation of glucose) and cardiovascular disease.

The metabolism of these hydrogenated carbohydrates takes place using very well-known biochemical pathways: Embden–Meyerhof pathway, Horecker and Racker pentose phosphates cycle, Touster cycle and Krebs cycle.

10.6.1.1 D-*Sorbitol*

This, together with its isomer, mannitol, is the most widespread:

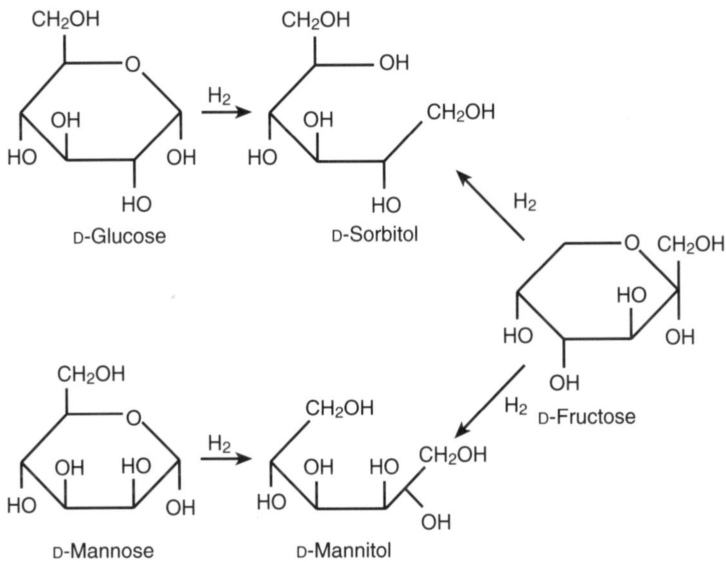

D-Sorbitol is an important constituent of numerous natural foods, in particular common edible fruits, whereas it is rare in mammal tissues. It is metabolised in humans, and used more efficiently than glucose by people with diabetes, producing the same quantity of energy (approximately

Table 10.2 Properties of the principal polyols

	Sucrose	Maltitol	Lactitol	Sorbitol	Mannitol	Xylitol	Isomalt*
Sweetening power (solution at 10%)	1.0	0.9	0.3	0.5	0.5	1.0	0.5
Hygroscopicity		+	–	+	–	+	–
Solubility (g in 100 ml water at 25 °C)	210		150	235	22	185	33
Dissolving heat (J/g)	–17	–80	–50	–112	–121	–155	–38
Energy value (kJ/g)	17	12	8.5	17	17	17	8.5

* Isomalt: glucose – sorbitol + glucose – mannitol.

17 kJ/g); consequently sorbitol is used in the manufacture of dietary foods for those with diabetes. Being not fermented by yeasts is another property that reinforces this use.

From a technological point of view, other properties have made sorbitol an important auxiliary substance in the food industry, and in particular in confectionery products where it often replaces invert sugar (Table 10.2). These properties can be summarised as follows:

- High water-binding capacity. Water is jointly responsible for its honey-like consistency; it can only evaporate very slowly in the presence of an adequate quantity of sorbitol.
- Resistance to heating (with the invertase process it is not possible to heat to over 70 °C).
- Retarding effect on the crystallisation of sucrose and glucose; the crystals formed remain small and undetectable in the mouth.
- Very little sweetening effect (approximately half that of sucrose); people can eat more without feeling uncomfortable.
- Relatively low viscosity of syrups.
- Ability to complex heavy metals, which helps to improve storage stability of fat products.

The sorbitol used in the food industry is not a product of extraction: it is manufactured by means of glucose hydrogenation. It is generally found in the form of a 70% concentrated syrup, which corresponds to its maximum solubility in water at 20 °C. Crystallised sorbitol has a very high dissolution enthalpy in water (112 kJ/kg) which gives the confectionery in which it has been incorporated a highly valued refreshing taste.

Sorbitol is involved in the metabolism of higher organisms through fructose. The sorbitol NAD dehydrogenase is found in the liver, the prostate and the kidneys. Sorbitol passes through the intestinal barrier by means of the simple phenomenon of diffusion (10–20 g per hour). However, the liver can convert between 30 g and 40 g of sorbitol into fructose in one hour. The oxidation of sorbitol and the resulting oxidative conversion of fructose are hepatic reactions which allow these compounds to rejoin the glycogenolytic ways by means of non-insulin-dependent mechanisms. The enzymes responsible – sorbitol-dehydrogenase, phosphofructokinase and hepatic aldolase – are not actually controlled by insulin. For people with diabetes, sorbitol presents the advantage of not producing a significant rise in blood glucose after it is taken by mouth. Glucose can be produced while the sorbitol is being metabolised.

10.6.1.2 Mannitol

This is also found in abundance in plants. Its content is higher than that of monoses in some fungi. Mannitol is metabolised in almost the same way as sorbitol. After diffusion through the intestinal wall, mannitol is oxidised into fructose by the mannitol-dehydrogenase. Mannitol is 50% excreted in the faeces and the urine, and the rest is oxidised in the liver. Mannitol is generally considered to have a caloric value of 17 kJ/g (Table 10.2). In addition to being poorly assimilated, mannitol encourages diuresis.

Mannitol, which is less hygroscopic than the other polyols, is suitable for bubble gums.

10.6.1.3 Isomalt

Isomalt is manufactured in two stages:

- Enzymatic transglucosidation converts the sucrose into isomaltulose (α-D-glucopyranosyl-1→6-fructose);
- Catalytic hydrogenation of the isomaltulose leads to the formation of the equimolar mixture of α-D-glucopyranosyl-1→ 6-sorbitol (GPS) and α-D-glucopyranosyl-1→6-D-mannitol (GPM).

Isomalt has a flavour of pure sugar, similar to that of sucrose, and does not leave any aftertaste. Isomalt has a synergistic action when it is combined with other alditols. So, by adding 10% of these polyols, a sweetening power comparable to that of sucrose can be obtained. Similar synergistic effects are obtained when isomalt is used in combination with intense sweeteners such as aspartame, saccharin and acesulphame-K. Finally, isomalt reinforces the flavouring effect in foods and does not generate any refreshing effect, unlike the other polyols.

Isomalt, which is considered to be non-hygroscopic, has sufficient solubility in water (30 g/l at 20 °C) to allow numerous applications. GPS and GPM, which are very stable chemically, can be hydrolysed by the α-D-glucosidases. However, as the speed of hydrolysis is very slow, isomalt does not constitute the ideal substrate for the micro-organisms present in food products.

More specifically, the slow speed of hydrolysis of the disaccharide bond from isomalt explains why, after ingestion by mouth, only approximately 50% of this carbohydrate is converted into energy.

The physical, chemical and biological properties of isomalt are the reasons behind the development of its use as a replacement for traditional sweeteners in the chocolate, dairy, confectionery and cake-making industries.

10.6.1.4 Xylitol

This is found in low concentrations in many fruits and plant products. It is extracted from wood hemicellulose. It is a sweetening agent, which has the same energy value, appearance and approximately the same sweetening power as sucrose (Table 10.2) but, like a number of polyols, it is recommended as a 'non-cariogenic sugar' within food industries.

In higher organisms, the metabolism of exogenous xylitol calls upon two methods fairly closely linked to one another: the Touster cycle and the cycle of pentose phosphates. In the Touster cycle, xylitol is the intermediary between L-xylulose and D-xylulose. Exogenous xylitol must be converted into D-xylulose by the cytoplasmic xylitol-dehydrogenase in the liver in order to be able to enter these cycles. At this level xylitol is 80% metabolised, but can be converted into glucose. This rate can vary between 20 and 80%. Conversion takes place in the liver depending on the organism's requirements in glucose. Production is slow with a delayed effect, which makes the use of xylitol advantageous for people with diabetes. It is well-tolerated in the human body (up to 100 g/day).

10.6.2 Cyclitols

The natural penta- or hexahydroxylated by-products of cyclohexane are numerous, especially in the plant kingdom, but also in the animal kingdom. As they do not correspond to monoses they are normally classified alongside alditols.

Hexahydroxylated cyclitols are interesting: they form nine stereoisomers (seven inactive, one dextro-rotatory and one laevo-rotatory). Among the inactive isomers is the sugary substance extracted from the muscle juice, which has been given several names: inosite, i-inositol, myo-inositol, meso-inositol.

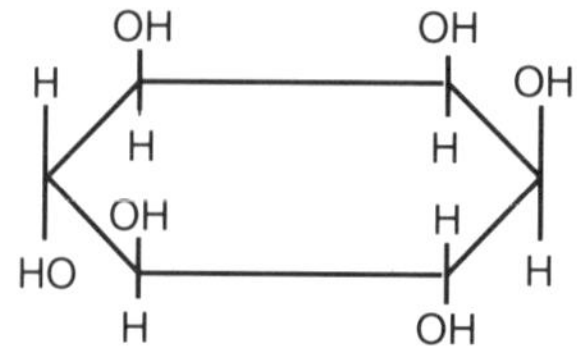

It is most commonly found in animals, in a free or combined state. It is present in various tissues: liver, muscle, blood, urine and especially in the sperm. Formerly it was considered to be a vitamin (B_7). However, no spontaneous deficiency has been observed in humans; however, in the rat, deficiency leads to skin problems, hair loss (alopecia) and asthaenia.

In the combined state, myo-inositol is associated with monoses (with galactose in sugar-beet galactitol) and especially in phosphoric acid. The mono-, di- or triphosphoric esters form inositophospholipids. Hexaphosphoric ester is phytic acid, which carries as many P atoms as C atoms (P = 28% and C = 11%); it is the most strongly phosphorylated plant constituent, and in food grains it can represent between 80 and 90% of the total phosphorus. Phytic acid is an exception in the living world, where phosphorylated polyesters are rare. In addition *in vitro* studies have revealed the inequality of charges; there are certainly six strong acid functions (pK = 1.8), but only two weak ones (pK = 6.3); the other four functions only appear at pH = 9.7, probably because of the hydrogen bonds between neighbouring radicals. Phytic acid, like oxalic acid, slows down the absorption of calcium in humans, following the formation of insoluble salts. Phytin is a complex salt of calcium and magnesium which is found in numerous plants; it is mainly extracted from *Corn steep liquor* (the soluble part of maize grains).

10.7 Intense sweeteners

These synthetic and semi-synthetic substances, and substances of plant origin, share the same glucophore structure (Fig. 10.5). For this reason all types of sweeteners will be discussed in this section, whatever their origin and method of preparation.

Current legislation relating to these intense sweeteners still varies enormously. Standardisation procedures have been started up in the countries within the European Union and are moving towards deregulating the use of these substances.

French law permits the incorporation of aspartame, saccharin and acesulphame-K in foods. Cyclamates are sold exclusively by chemists and their incorporation in foods is still prohibited: thaumatin (talin) has been authorised at the test stage for a period of three years.

10.7.1 Aspartame

Aspartame is a dipeptide of L-aspartic acid and L-phenylalanine in the form of its methyl ester. Its sweetening power is approximately 200 times that of sucrose.

Once absorbed, aspartame splits up into its constituent parts which are then metabolised in the same way as the other amino acids contained in

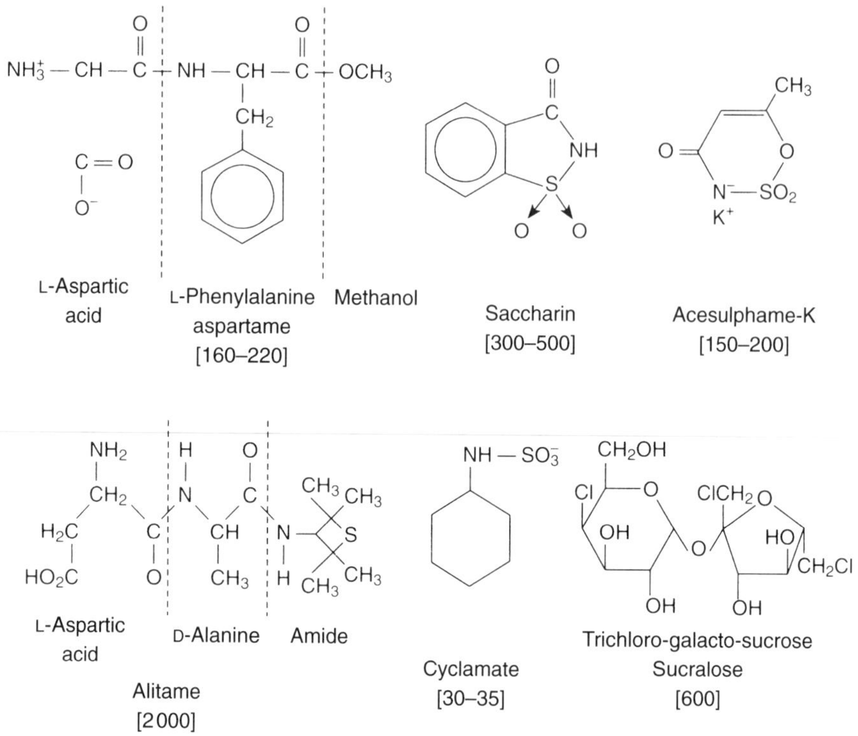

Fig. 10.5 Structure and sweetening power of intense sweeteners (the sweetening power is indicated between brackets; sucrose = 1).

our diet. This sweetener supplies the same quantity of energy as proteins (17 kJ/g).

Aspartame's solubility depends on pH and temperature. It is lowest at the isoelectric point (pH = 5.2). In solution, the ester bond is likely to be hydrolysed and this results in a reduction in sweetening power. When dry, below 8% moisture content and at ambient temperature, stability is excellent for several years.

Aspartame is now permitted in numerous countries (almost 100). The recommended daily allowance (RDA) is between 40 and 50 mg/kg. Extending the range of products containing aspartame increases choice for people with diabetes or those who are obese, while facilitating the prevention of caries and weight gain.

10.7.2 Saccharin

Saccharin is 1,2-benzisothiazolin-3-one-1,1-dioxide. Saccharin in aqueous solution is relatively unstable to heat; it can hydrolyse into *o*-sulphamoylbenzoate above 100 °C. The pH of a saturated aqueous solution

(0.35% w/v) is approximately 2.0. Saccharin produces salts, the best known of which are sodium salts (anhydrous or dehydrated) or ammonium, potassium or calcium salts.

When pure, its sweetening power is estimated at approximately 700 times that of sucrose. Commercial saccharins have a sweetening power generally estimated to be 500 times that of sucrose. A certain bitterness and metallic aftertaste have been attributed to the presence of impurities; additives, such as, for example, glucono-delta-lactone, are used to mask this aftertaste.

In spite of its instability to heat, provided that the heat processing periods are accounted for, saccharin can be useful in food or medical preparations that require the use of heat within an aqueous medium.

As the potential impurities of saccharin have been involved in, or held responsible for, variations in the results obtained in toxicity tests, the contaminating substances in saccharins have reduced considerably over recent years.

Although all studies have shown that commercial saccharins did not have any mutagenic effect, US recommendations nevertheless suggest using this sweetener with care for children and pregnant women.

10.7.3 Acesulphame-K

Acesulphame is 3,4-dihydro-6-methyl-1,2,3-oxo-thiazine-4-one-2,2-dioxide. Only acesulphame-K, i.e. the salt of potassium, is commercialised. Acesulphame is prepared from by-products of acetoacetic acid which are involved either as starting products or as intermediate products during manufacture.

The sweetening power of acesulphame-K is almost 200 times as high as that of sucrose. Mixed with other substances with a sweet taste, acesulphame-K presents effects of synergy which are particularly marked when it is mixed with aspartame or cyclamate. In addition, a very pleasant sweet taste is obtained by mixing acesulphame-K with sucrose or fructose. Likewise, the insipid sweet taste of the polyols is pleasantly increased by this intense sweetener. However, when used as the sole sweetener in order to obtain a very pronounced sweet taste, a certain degree of bitterness becomes apparent.

As a result of its high solubility in water (270 g in 1 l of water at 20 °C), its excellent stability when stored and when subjected to heat processes, the areas in which acesulphame-K can be used are extensive and varied.

Within the field of drinks, acesulphame-K is very easy to use in the form of sweetener mixtures. For example, as acesulphame-K and aspartame have a different time/intensity profile, by varying the mixture level it is possible to balance the sweetening flavour curve depending on the different raw materials used in manufacturing the drink. So, for citrus fruit juices, a mixture of two parts acesulphame-K and one part aspartame is particularly highly appreciated.

In the case of dairy products containing fruit preparations, acesulphame-

K can be used as the only sweetener. In low-calorie confectionery, jams and cakes, a small quantity of acesulphame-K is sufficient to give body to these products and to supply an initial impression of a sweet taste.

This sweetener is rapidly absorbed by the organism, but it is also rapidly eliminated without having been modified. Bearing in mind the favourable results obtained when studies on the innocuity of acesulphame-K in relation to health have been carried out, some countries have authorised its use with a limit value (RDA is 15 mg/kg in the USA), while others have set no limit.

10.7.4 Other sweeteners

This section will briefly consider those sweeteners whose use is either prohibited at the moment, or very strictly regulated. As this is a category of substances with very different chemical structures, we will refer to them according to their biochemical family.

10.7.4.1 Sweeteners of a protein or peptide type

Thaumatin is a protein extracted from the fruit of the *Thaumatococcus danielli*. Three sweetening proteins, called thaumatins 0, I and II can be extracted. Their molecular mass is approximately 22 000 Da. The mixture of thaumatins I and II is commercialised in the form of **Talin**. These proteins, which consist of 207 residues of amino acids and eight disulphide bridges, can withstand a temperature of 100 °C at a pH of 5.5.

The sweetening power of Talin is very high: 1500–2500 times that of sucrose. However, this very intense sweetening power is accompanied by a rather persistent sweet flavour (10–20 min) which is also contaminated by a liquorice-type flavour. So, Talin is used mainly in the food industry, as a flavour enhancer and very seldom as a sweetener. In Japan, Talin is used extensively, but mainly in combination with liquorice, amino acids, citric acid, succinic acid and lactose.

Monellin has been isolated from the berry of Nigeria (*Dioscoreophyllum cumensii*). This protein (MM 10 000) which has two names (sometimes referred to as monellin, and on other occasions, unileverin) has a very strong sweetening power. It loses its taste at 60 °C at a pH of under 2. In addition the berries must be stored at –20 °C if they are to avoid losing their sweet taste. This extreme fragility no doubt accounts for the limited development of this sweetener.

Miraculin is a glycoprotein (MM 40 000) contained in the red berries of a tropical plant *Synsepalun dulcificum*. This sweetener is very fragile: it is destroyed in particular by heat and trypsin. The glycan part contains between 6 and 7% arabinose and xylose; the aglucone part contains 16 amino acid residues but no tryptophan (sweet amino acid).

Alitame is a dipeptide from the L-alpha-aspartyl-D-alanine-amide series of compounds. This molecule, which resembles the aspartame molecule, does however have a sweetening power which is 10 times higher and more stable. Its intense sweet taste is due to the terminal amide (2,2,4,4-tetramethylthietanylamine).

The sweet taste of alitame, of good quality and without any aftertaste, and its good stability at pHs of between 2 and 4, seem to indicate a successful future for this dipeptide in IFPs, particularly in the field of fizzy drinks. In addition, this sweetener has a clear advantage over aspartame for people suffering from phenylketonuria.

10.7.4.2 Sweeteners of a carbohydrate type

Stevioside is a diterpenic type glycoside extracted from the leaves of *Stevia rebaudiana*. New varieties and hybrids of *Stevia* have been developed. They contain a higher proportion of rebaudiosides (anomers of stevioside) whose technological properties (solubility, stability) and organoleptic properties are better than those of pure stevioside.

The sweetening power of 80% pure stevioside is 300 but the sweet flavour is associated with another type of bitterness, consisting of an astringent sensation on the tongue and a liquorice aftertaste. The flavour profile of the rebaudiosides is more satisfactory, but associations with other sweeteners or other substances (histidine, hydrocolloids) have been tested to attempt to improve it.

Japanese cuisine uses this sweetener to reduce the intensely salty flavour of numerous traditional dishes, but without interfering with colour. It is also widely used in the manufacture of chewing-gums and in the tobacco industry.

However, because of the structure of its steviol core, this sweetener may have anti-androgenous properties. Its use is therefore prohibited in numerous countries (EU, USA).

The dihydrochalcones of some natural flavones have an extraordinarily high sweetening power. These flavones are prunin (extracted from the wood of the plum tree), naringin (principal flavonoid in grapefruit) and neohesperidin (present in certain varieties of orange (see Fig. 15.3)).

These are semi-synthetic sweeteners since their extraction from natural flavonoid glycosides is followed by catalytic hydrogenation. Dihydrochalcones are relatively stable at a neutral pH. The combined effect of heat and lowered pH accelerates breakdown. So, at pH 2, breakdown occurs after 24 hours at 75 °C and after 60 minutes at 100 °C.

The sweetening power of the various dihydrochalcones differs widely:

- That of dihydrochalcone prunin is located at around 0.4.
- That of dihydrochalcone naringin at around 110.
- That of dihydrochalcone neohesperidin (NeoDHC) at around 1000.

Non-glycoside neohesperidin by-products have been synthesised; they have an intense sweet taste (sweetening power of between 500 and 700) and their solubility in water is excellent.

The sweet taste of the dihydrochalcones is associated with a fresh taste, fairly similar to that of mint, and, in addition, it has a 'lingering' effect which is often found in sweeteners with a very intense sweet taste. These two characteristics (interfering menthol flavour and 'lingering' effect) pose problems of incompatibility between this flavour and numerous food products. So, combinations with other sweeteners have been developed; for example the NeoDHC/saccharin/cyclamate (25/64/11%) mixture generates a very pleasant sweet flavour.

Neohesperidin dihydrochalcone is the most useful by-product out of the whole group; at the moment it is marketed only in Belgium, where it is used in the beer industry, in drinks and in certain types of confectionery.

Sucralose is manufactured from sucrose using a process which involves the selective substitution of chlorine for the hydroxyl groups located in positions 4,1′ and 6′. Its official name is: 1,6-dichloro-1,6-dideoxy-β-D-fructofuranosyl-4-chloro-4-deoxy-α-D-galactopyranoside.

This sweetener with a sweetening power of 600, is highly soluble in water (28.2 g/100 ml at 20 °C) and very stable in the acid aqueous phase. So, sucralose appears to be one of the most stable sweeteners at a pH which corresponds to that of non-alcoholic drinks. In addition, because it has no reactive chemical function, sucralose is remarkably inert in comparison with other ingredients present in food products. This sweetener also behaves well during cooking.

Because of the taste qualities of sucralose which are very similar to those of sucrose, its high solubility and stability during storage and cooking, this sweetener would seem to have a promising future.

10.7.4.3 Sweeteners having different structures

Cyclamates, which are sodium and calcium salts of cyclamic acid, are very soluble in water and very stable (they are not altered by pHs from between 2 and 10 and they can endure temperatures of up to 500 °C).

The sweetening power of cyclamates is 30–35 times greater than that of sucrose. Often used in combination with saccharin (in the ratio of 10/1) – which allows the sweetening power to be doubled – cyclamates were used in drinks, biscuits, cakes, ice-creams and drugs in order to mask bitterness.

However, their incorporation in foodstuffs is prohibited in many countries. Problems of toxicology associated with these sweeteners are largely associated with the fact that they can be metabolised within the organism to produce cyclohexylamine, cyclohexanol and cyclohexanone.

Glycyrrhizine, which is extracted from the rhizome of *Glycyrrhiza glabra*, is a relatively powerful sweetening substance. This saponin, which results from the association of glycyrrhetic acid and glycuronic acid, has a sweetening power 50 times stronger than that of sucrose. This substance is a good flavour extender and has the particular ability of reinforcing sweet flavours. However, its use in the pure state as a sweetener is limited because the sweet taste is hampered by a strong and persistent aftertaste of liquorice. Combining it with products such as lactose and sorbitol allow this liquorice flavour to be somewhat reduced.

Since glycyrrhetic acid, whose structure is similar to that of deoxycorticosterone, can cause hypokalaemia, hypernatraemia and metabolic alcalose, it is likely that the authorised use of glycyrrhizine will be limited to that of an extender with a sweet taste.

Phyllodulcin, whose taste is fairly close to that of the dihydrochalcones, is extracted from the leaves of *Hydrangea thunbergii*. The sweetening power of phyllodulcin is approximately 400. The sweet taste of this sweetener is not pure and it is associated with a liquorice-type aftertaste which does not appear immediately but persists in the mouth ('lingering' effect).

Many other substances have a sweet flavour:

- The aromatic by-products of urea and thiourea, dulcin (PS 200) and suosan (PS 350).
- Perillartime from the group of oximes (PS 450).
- Finally, one of the most powerful synthetic sweeteners: *n*-propoxy-1-amino-2-nitro-4-benzene (PS 4000).

The toxicity of these substances has either been investigated or is currently being studied.

10.8 Uses of sweetening substances in confectionery and chocolate-making

This section considers the possibilities of using these substances in confectionery, jam-making and the chocolate industry, in which they play a dominant role.

10.8.1 Confectionery and jam-making

The main technological operations within the confectionery industry are shown in Fig. 10.6. Jam-making is the industry that converts fruits (in the form of purée, pulp or juice, depending on the name of the end-product) into a mixture having the appropriate gel consistency.

Originally sucrose and honey were the main sweetening substances involved in the formulations of confectionery and jam products. First of all, glucose or fructose, alone or in the form of invert sugar, then glucose syrups were introduced to extend the range of raw materials in the sugar products industry. Finally, this range of sweetening substances was enriched by loaded sweetening substances and intense sweeteners.

The traditional process for manufacturing **cooked sugars** consists of dissolving sucrose in water in the presence of starch hydrolysates (glucose syrup), raising the concentration in dissolved dry materials generally to above 98% by cooking (water evaporation), in order to obtain a mass which has a vitreous structure after rapid cooling.

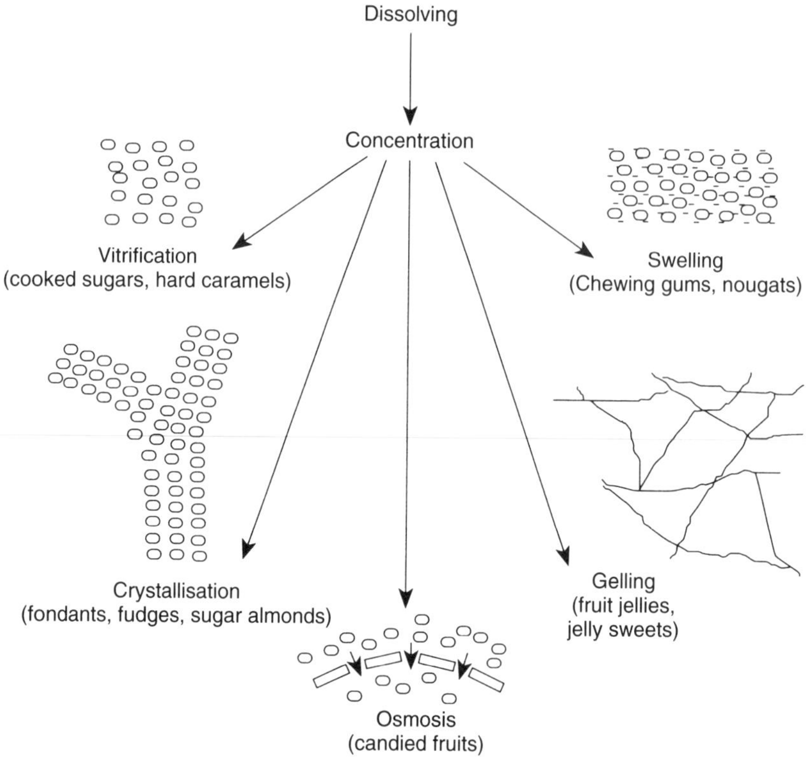

Fig. 10.6 Diagrammatic representation of the principal technological operations used in confectionery.

The transparency of this product is linked to the absence of crystals, and this crystallisation is inhibited by the low conversion rate (DE 36–39) of glucose syrups in quantities that vary from 40 to 60% of the total dry matter.

Hard **toffees** are similar in nature as they are not crystallised; their composition includes, in addition to sucrose and the dry materials, glucose syrup, milk proteins, lactose and fats. As a result of the Maillard reaction the milk ingredients develop a characteristic flavour and a brown colour.

The texture of toffees depends on their moisture content: a hard toffee is cooked at between 128 and 131 °C in order to obtain a moisture content of between 5 and 6%, whereas soft toffee, cooked at 118–120 °C has a moisture content of between 9 and 10%.

Substituting sucrose and glucose syrup with other sweetening substances is limited in certain cases by the keeping quality of the end-product. Thus, for example, incorporating large quantities of honey in cooked sugar increases the hygroscopicity of the end-product (the two main carbohydrates in honey – fructose and glucose – are 1.3 times more hygroscopic than sucrose). On the other hand, the hygroscopicity of cooked sugars and hard toffees based on isomalt or hydrogenated glucose syrup is lower than for similar traditional products.

Fondants and **fudges** are characterised by their partially crystallised structure, in which some of the sucrose – approximately 50% – is in a crystalline form in equilibrium with an aqueous sucrose-saturated liquid phase, containing carbohydrate molecules supplied by the glucose syrup (glucose, maltose, maltotriose, etc.). Fudges are caramels, part of whose sucrose has been deliberately crystallised.

Other sweetening substances (polyols, glucose syrup rich in fructose – high fructose sucrose content, HFSC) can be used, but must be adapted to the manufacturing technology.

The manufacture of **sugar-coated products** consists of coating a hard core with successive layers of sucrose microcrystals. This operation is carried out in turbines which alternate between spraying sugar syrup and injecting hot air.

Sugar coating requires syrups with high viscosity (maximum 200 cP) but whose supersaturation must nevertheless be considerable in order to encourage rapid crystallisation in the form of microcrystals. As a consequence, the viscosity variation curve of saturated solutions of sweetening substances (sucrose, sorbitol, xylitol, etc.) will define the upper or lower limits of the sugar-coating temperature. For example, the conditions for sugar-coating using maltitol are between 45 and 65 °C.

The manufacture of **crystallised fruits** is based on osmosis: water passes through the cell walls and the 'sugars' from the syrup penetrate the fruit. The speed at which the sugar from the syrup penetrates the fruit must be faster than the speed at which the water exits, to prevent the fruit from losing its shape under the effect of osmotic pressure. In addition, the difference in speed must be as low as possible. This means that there must be very little difference in concentration in soluble dry materials between the syrup and the fruit (in the order of 5%), and a constant and regular increase in the concentration in syrup sugars. The DM level of the crystallised fruit is approximately 75%: the resulting equilibrium relative humidity (ERH) is very low, thus allowing the end product to be stored.

The syrup is composed of a mixture of sucrose and glucose syrup, but the latter can be replaced by invert sugar, strongly converted glucose syrup.

The following observations can be made:

- The osmotic pressure of the glucose syrups is proportional to their DE.
- Sorbitol syrups have lower viscosity for the same DM.
- The anti-crystallising properties of sorbitol are very important in the technology of crystallised fruits.

Jam is a traditional method of conserving fruits by reducing their a_w thus creating a sugared gel rich in soluble dry matter. Pectins are the hydrocolloids most frequently used in jam-making.

The mechanism for forming the pectin network is well established (see

Chapter 12). With strongly methylated pectins, the role of sucrose is to bind the molecules of water from the hydration sphere of the pectin molecule. It is not possible to lower the sucrose content to less than 30% with poorly methylated pectins alone; beyond this limit other gelling agents such as carrageenans become essential.

Sucrose can be replaced by other sugars or polyols. For example, glucose syrups with 60–65 DE can replace up to 50% of sucrose in jams and lower the ERH of jams more effectively than the same mass of sucrose. In addition, this substitution inhibits the re-crystallisation of this disaccharide. With polyols (sorbitol, xylitol) it is possible to obtain either products whose texture is very close to traditional jams provided that their pectin content is increased, or products with different organoleptic characteristics.

Apart from pectin, the main gelling agents used in jam-making are gums and gelatin. Unlike pectin, the gels in gelatin are heat-reversible (see Chapter 9) and the role of sugars and polyols in the formation and stabilisation of gelatin gels is associated with the stabilisation of the triple collagen helix.

Manufacturing expanded confectionery products requires the preparation of a concentrated solution of sucrose and glucose syrup (90–95% DM). The expanded structure is obtained by mechanical whipping in the presence of an aerating agent (egg white, whey proteins, etc.). Bubble gums also have a low proportion of fats in their formulation.

These expanded products have a hygroscopic nature (ERH of between 45 and 60%) which is particularly marked since the product is not only in contact with the steam via the external surface but also via the air bubbles due to expansion. In some cases, beating is not only involved in obtaining the expanded structure but can also bring about sucrose crystallisation, thus affecting texture, and increasing firmness and conservation times by reducing the hygroscopic nature of the product (ERH increases).

In these products the sucrose and glucose syrup can be entirely substituted with mannitol and hydrogenated glucose syrup; this results in bubble gums whose texture characteristics depend on the mannitol level and the percentage of residual water.

10.8.2 The chocolate industry

Chocolate is a product obtained from cocoa paste and sucrose, with or without the addition of cocoa butter. It must contain at least 35% of dry cocoa products, at least 14% of dry non-fat cocoa and 18% butter.

Cooking chocolate can contain lower quantities of cocoa, whereas coating chocolate must be richer in cocoa butter (at least 31%). Milk chocolate must also contain not only dry milk products (at least 14%) but also a maximum of 55% sucrose, a minimum of 25% dry cocoa products and 25% total fats. White chocolate can only consist of cocoa butter (at least 20%), sucrose (maximum 55%) and dry milk substances (at least 14%). Chocolate can contain other ingredients such as hazelnuts, puffed rice, dried fruits, etc.

From a physical point of view, chocolate can be defined as a virtually anhydrous dispersion of very fine particles of a non-lipid nature (sucrose, lactose, proteins, minerals, etc.) in a solidified lipid phase, essentially consisting of triglycerides. In dark chocolate, these neutral lipids come from the cocoa alone, but in the case of milk chocolate or white chocolate, they can also come from milk. Lecithin is added to the chocolate to coat the 'sweet particles' and to emulsify the residual water in order to give the chocolate good pouring properties.

In chocolate, sucrose is found in a dispersed state in a continuous lipid phase, in the form of microcrystals which are sufficiently fine (average diameter 20 µm) not to be noticed when the product is eaten. This carbohydrate not only intervenes as a sweetener but is also involved in developing the flavour notes of chocolate during manufacture. In fact, following partial hydrolysis, and in the presence of cocoa proteins, it gives rise to Maillard reactions which are responsible for several flavour precursors. This is true for all types of chocolate, and particularly so for milk chocolate.

Replacing sucrose with other sugars or with polyols can reduce the caloric value of the chocolate, and, depending on the substitute, can allow slower or faster metabolism. In this way, the industry has extended its range of products by manufacturing chocolates using fructose and glucose, polyols, glucosaccharides and fructooligosaccharides. These products sometimes present very typical rheological and organoleptic characteristics (for example, spicy and burnt aftertaste).

A full description of the uses of sweetening substances would also have to include industrial applications which make use of sweet flavours: refreshing drinks, ice-creams, cakes, desserts, dairy products, etc.

In addition to their sweetening power, sweeteners are valuable in these products because of their flavour-enhancing properties, their osmotic and texture-creating capacity and their viscosity.

11

Starch products

11.1 Introduction

In spite of losing a little of its popularity, starch remains the most economical and the most versatile of all the nutrients. Maize, wheat and potatoes which contain 71, 76 and 74% starch (of the dry matter) respectively, are the agricultural raw materials most frequently used in order to extract this glycan. Rice and manioc, which contain more starch (approximately 90% of the dry matter), are two more sources but the volumes processed are lower. In fact, special starch characteristics correspond to each raw material and these determine its value as a texturing agent. The manufacturer takes into account numerous technical and economical criteria apart from the starch content; and this is why maize is the most widely used source in the world.

Starch must also be considered to be a source of carbohydrates capable of producing, by means of thermal, chemical or enzymatic processes, a full range of intermediate food products used in almost all the sectors of the agro-food industry, as well as numerous non-food applications. This large range of products can be classified into three large groups (Fig. 11.1):

- Starches.
- Modified starch.
- Starch hydrolysates.

The enzymatic processes in the digestive system are capable of converting all the complex molecules from the starch into glucose. The energy value of 17 kJ/g is therefore valid for all types of starch and their products of hydrolysis. However, the speed of assimilation of the various carbohydrates depends on their molecular size, with glucose, of course, being assimilated the most rapidly.

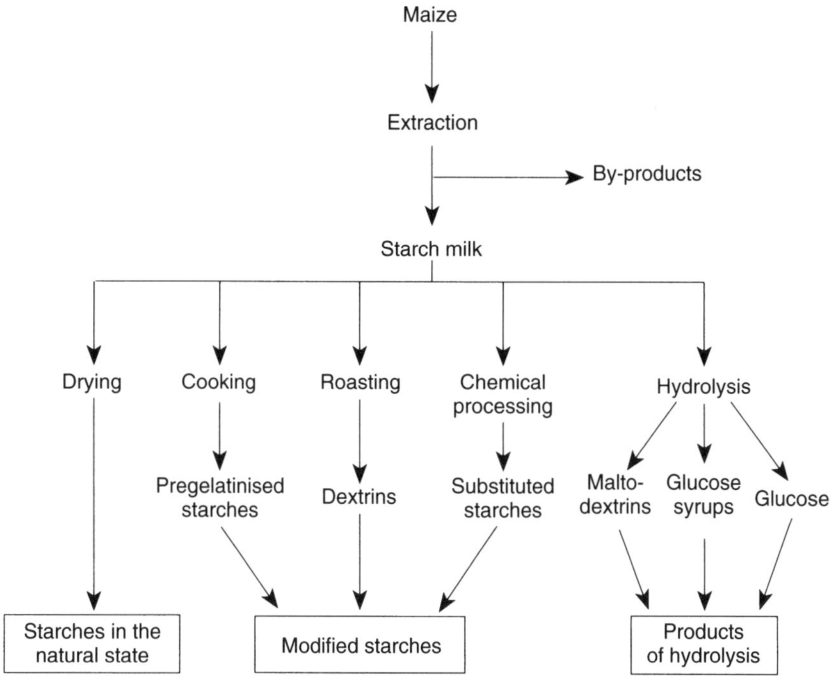

Fig. 11.1 The manufacture of starchy products.

11.2 Starches in the natural state

Starch is extracted from maize seed using a process of wet grinding. The process is based on a grinding, grading and a centrifuging operation aimed at separating the starch from the cellulose, lipids and proteins which are closely bound to it (see Chapter 3).

The wet grinding process starts with softening the seed by immersing it into a solution of dilute acid. Coarse grinding splits the grain so that the germ which contains lipid can be removed. The fine-grading operation allows the cellulose to be eliminated from the endosperm. Finally, centrifuging separates the less dense protein fraction from the denser starch. The starch is then washed and dried or left in a tank to be processed later. This aqueous suspension is known as 'starch milk'.

11.2.1 Basic structure

All types of starch consist of α-D-glucopyranose in linear chains with the bond $\alpha(1\rightarrow4)$, in amylose, or in branched chains by the $\alpha(1\rightarrow6)$ bond on $\alpha(1\rightarrow4)$ chains in amylopectin.

Table 11.1 Ratio between amylose/amylopectin and swelling rates of different starches

	Size (μm)	Amylose (%)	Amylopectin (%)	Swelling power	Solubility (%)	X-ray spectrum
Maize	5–25 15	24	76	24	25	A
Waxy maize	5–25 15	<1	>99	64	23	A
Amylon	5–25 15	70	30	6	12	B
Wheat	2–38 20	25	75	21	41	A
Potato flour	15–100 33	20	80	1000	82	B
Manioc	5–36 20	17	83	71	48	A

Amylose is the least abundant constituent and its molecule is the simplest and the smallest (MM of between 150000 and 600000). The $\alpha(1\rightarrow4)$ glucose chain is linear, but not straight; the monose residues are arranged in a helical structure, with up to six residues per turn. All these constituents form a sort of loose tube. Recent studies have indicated the presence of limited branching in certain amylose molecules. In an attempt to simplify the explanation and in order to understand the properties of amylose more clearly, our presentation will ignore these discoveries.

Amylopectin is much more abundant than amylose; the types of starch known as 'waxy' or 'glutinous' such as rice, maize or sorghum contain up to 95–97% of amylopectin.

Amylopectin has the same structure as amylose, except for the fact that the molecule has a bushy appearance because of the branchings which are connected to one —CH_2OH out of every 25 residues (approximately), and the molecular mass is much higher (several million). The side chains possess the same structure as the principal chains.

To sum up, all types of starch consist of either one of these molecules or both and the ratio between them varies according to the origins of the starch (Table 11.1).

11.2.2 Properties of starches

As the plant produces molecules of starch it deposits them in successive layers around a central hilum in order to form a compact grain. In this way, the nearby molecules of amylose and the external branching of amylopectin can combine by means of hydrogen bonds, in parallel, to constitute radially arranged oriented crystalline bundles, known as 'micelles'. These micelles maintain the structure of the grain and allow it to swell in hot water

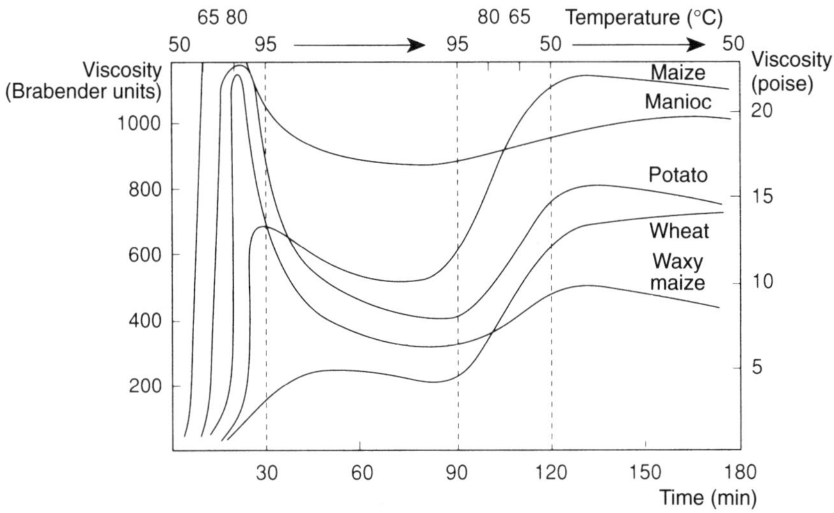

Fig. 11.2 Gelling curves – retrogradation of various starches (measurements taken using suspensions with 8% (p/p) starch).

without rupturing completely and without the individual starch molecules dissolving.

The crystalline structure of the micelles influences the way in which light penetrates the starch grain; non-hydrated (or non-gelatinised) starch grains make the plane of a polarised light deflect. This phenomenon is known as **double refraction**. When the radial arrangement of the mackle is disturbed, the double refraction disappears.

This reminder of the structure of starch within the grain allows us to understand why the hydrogen bonds that are established between the chains influence both the physical resistance and the solubility of the molecules via the hydroxylated groups: they allow fairly compact masses which have a certain degree of crystallinity, i.e. regularity in their spatial structure, to form. These bonds can be broken using an appropriate reagent or by heating. In this way solubility will be increased and crystallinity reduced.

The starch in grains is in fact almost insoluble in cold water. When the mixture is heated, a dispersion known as 'slurry' is produced, and viscosity is increased. There is a critical temperature which varies according to the structure of the starch; above this temperature the dispersion is irreversible and below this temperature it can be reversed. This temperature ranges from between 65 °C and 80 °C depending on the type of starch.

Starch **retrogradation** is an important phenomenon. It concerns the formation of interchain links between aligned molecules. This results in the phenomenon of 'syneresis', possibly with exudation of liquid and drop in viscosity (Fig. 11.2). Consequently there are also difficulties in enzymatic hydrolysis. Starch retrogradation has consequences in various fields: cakes

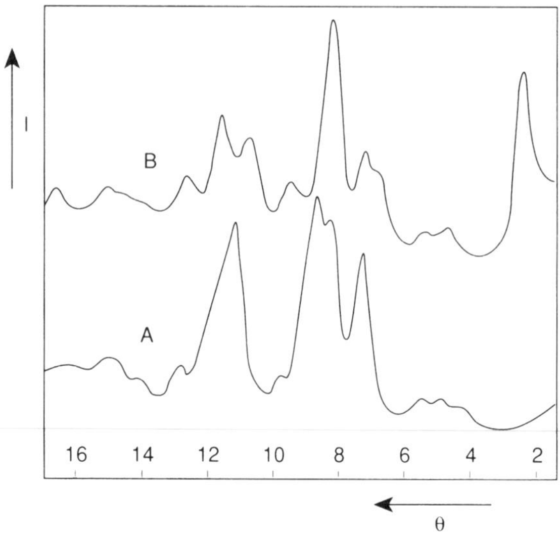

Fig. 11.3 Diffraction diagrams for (A) cereal starches and (B) tubers.

that do not rise, cream that has 'turned', bread that has gone hard (stale) without drying, liquefaction of slurry and glues, etc.

The speed of retrogradation increases as the proportion of linear amylose increases. This amylose is more accessible to reagents than amylopectin. For example, epichlorhydrin makes slurry less viscous and reduces retrogradation, which may not be able to take place, even in an autoclave.

Because of this, we believe that in the course of this starching–dissolving–retrogradation process the various types of starch behave differently according to their botanical origin. For example, the loss of double refraction occurs at a lower temperature for wheat starch than for potato or maize starch, with rice starch being the most heat resistant. On the other hand, potato and manioc starch have the highest hydration capacities and swell at the lowest temperatures.

Finally, starch that retrogrades strongly is slowly digested, so 'encapsulations' can be produced by using types of starch that are rich in amylose. Gels from starch of this type will retain a given shape and will retain it after retrogradation, with amylose gel constituting a peripheral gangue, which is slow to hydrolyse and, in particular, heat resistant. Chinese noodles are an example of this as they retain their shape in soups and resist lengthy boiling.

Depending on its plant origin, starch is found in two different **crystalline forms**, A for cereal starches and B for tuber starches, respectively (Fig. 11.3). We can easily check which of the two types of starch is involved by means of X-ray diffraction and then predict some of their physico-chemical properties. Likewise the study, by diffractometry, into the changes in structure

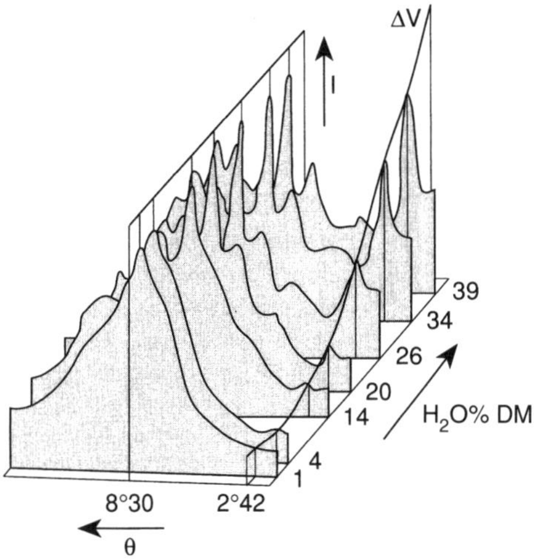

Fig. 11.4 Influence of hydration on the course of the diffraction diagram for potato starch.

due to hydration has allowed us to understand more clearly both the way the grain is organised and its swelling possibilities.

An awareness of these structural properties and the use of X-ray diffraction can partly allow us to explain the phenomenon of bread going stale. Type A starch in the original flour is 'gelatinised' when the dough is cooked subject to the joint action of water and temperature. It then loses its initial structure and its X-ray diffraction pattern contains no characteristic peak (Fig. 11.4). Then, during the process of staling, the starch recrystallises into the B form (retrogradation) with peaks appearing in the diffraction trace whose intensity increases with time. It is therefore possible to study how the process of staling changes with time, as well as the influence of certain factors such as concentration, temperature or the use of 'anti-staling' agents on the latter (see Section 11.5.2).

11.3 Modified starch

For the reasons given in the previous section, starch manufacturers have developed techniques to allow them to manufacture modified starches. Their properties are adapted to the needs of the industry.

11.3.1 Heat treatment

By **drying** starches in the native state at high temperatures until they are completely dry, with or without the addition of acid catalysers, **dextrins** are obtained. Their viscosity is lower and their solubility increases or decreases depending on the intensity of the heat treatment. Their film-forming capacity is higher than that of starch, which makes them useful in coating processes.

Recently, a large US pharmaceutical group registered a patent for a biodegradable plastic composed of 95% maize starch. This product is obtained by heating the starch under pressure so that the water remains in the starch instead of being evaporated. Water, acting as a plasticiser, allows the so-called vitreous transition temperature of the starch grain as it changes from a crystalline structure into an amorphous structure to be lowered. Transition can then take place at a temperature lower than that of starch grain depolymerisation. The mixture can be processed by extrusion at temperatures between 140 and 170 °C only. In order to obtain a product with relatively stable plastic properties, organic plasticisers are added (glycerol, sorbitol) and texturing agents (titanium dioxide, silica).

Cooking starch milk – for example over a heated roller – at a temperature higher than its gelling point results in **pre-gelatinised starches**. This cooking is immediately followed by drying to avoid retrogradation of the dispersed starch molecules. Another technique, **extrusion**, also allows a wide variety of pre-gelatinised starches to be produced. These processes can be applied equally well to starches in the native state and to modified starches, thus broadening the range of products.

In this way **cooking-extrusion**, which consists of submitting the raw material to the joint action of temperature (up to 250 °C) and pressure (up to 200 bar) for a relatively short period of time (between 10 and 60 seconds) and at a relatively low water content (10–40%) gives us a very wide range of products: crackers, snacks, soluble breakfast foods in granule, powder and flake form, flours, soups, instant drinks, etc. This increasingly widespread technique destroys the structure of the maize grain wholly or partially, resulting in high solubility when cold and excellent digestibility.

However, this disorganisation depends on both the composition of the types of starch (ratio between amylose/amylopectin and associated lipids) and the extrusion parameters (temperature, hydration, speed of the extruder screws resulting in a particular shearing rate). For example, with cereal starches, an amylose–lipid complex appears which is resistant to α-amylase *in vitro* whereas the starch in potatoes, which is lipid-free, is completely broken down into malto-dextrins.

To sum up, pre-cooking or pre-gelling using a traditional or modern technique produces a starch that is soluble when cold and swells in cold water. It is therefore easier to incorporate this in foods at low temperatures. This product is recommended for various instant preparations: sauces, cream fillings, mousses, spreads, etc.

11.3.2 Chemical processing

Grafting radicals by means of chemical processing allows products of varying viscosity to be obtained. The aims of these treatments are numerous and even occasionally conflicting. In the main, they involve increasing the stability of the starches when hot, when stored, when subjected to thermomechanical shearing, or on giving them cationic, anionic or hydrophobic natures.

However, starches that have been chemically modified must comply with the regulations specified by legislation so no more than 5% can be incorporated in foodstuffs.

Three types of modifications are summarised in Table 11.2: depolymerisation, cross-linking and stabilisation.

Table 11.2 Starches modified using chemical processes

Reticulated starches

Chloroepoxide by-products

Phosphate by-products

Di-esters

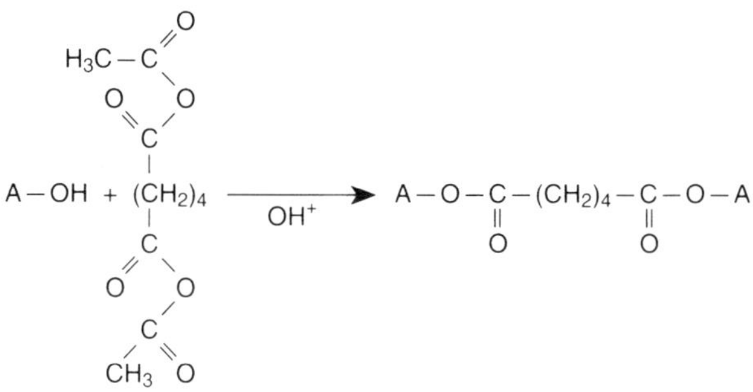

Aldehyde by-products

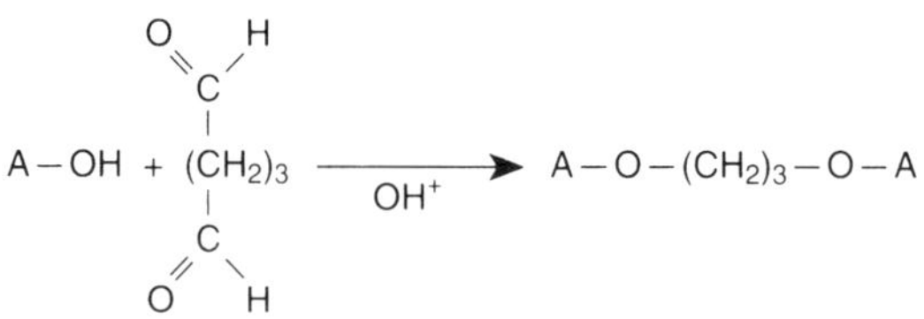

Stabilised starches

Starch esters

Hydroxyalkyl starches

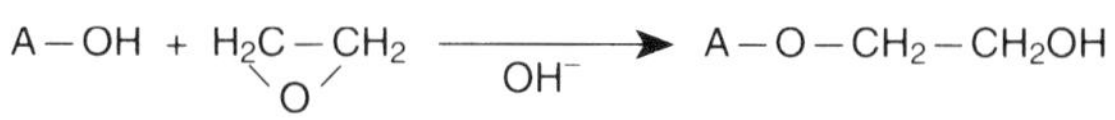

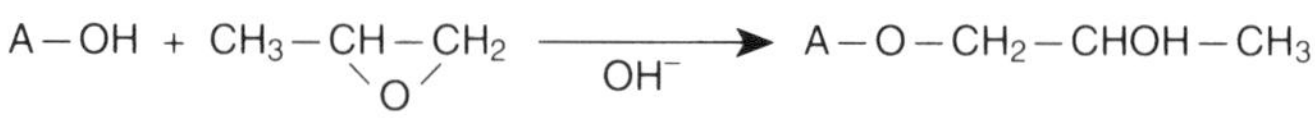

Table 11.2 *Continued*

Anionic starches

Alkaline oxidation

Maleation

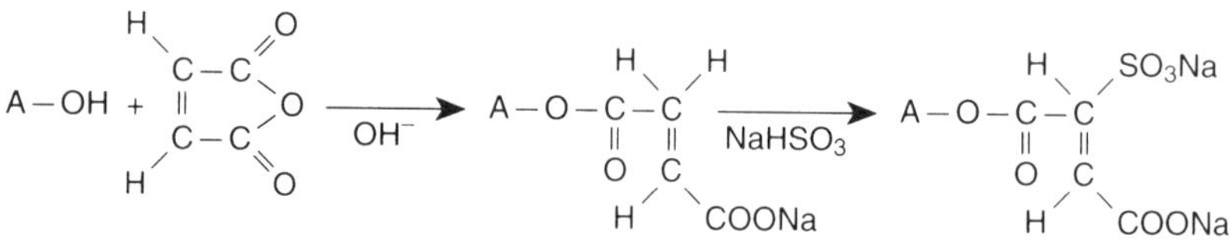

Cyanoethylation

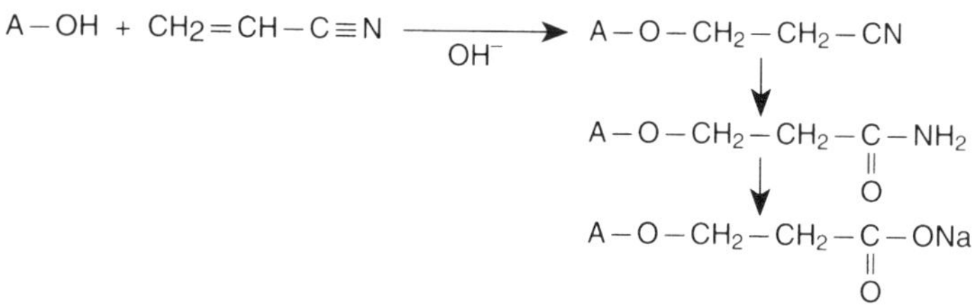

Carboxymethylation

Cationic starches

Aminotertiary by-products

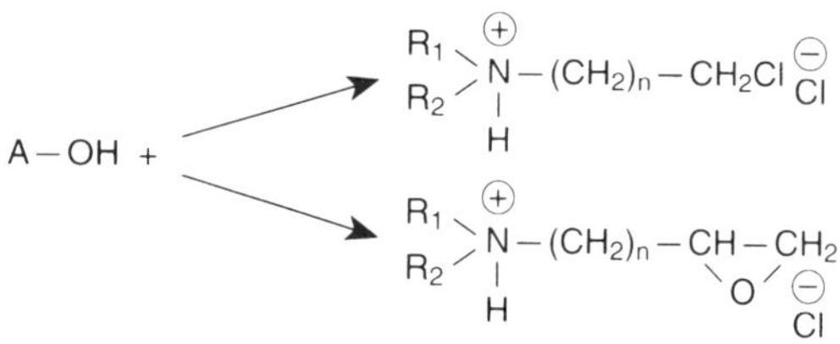

Quaternary ammonium by-products

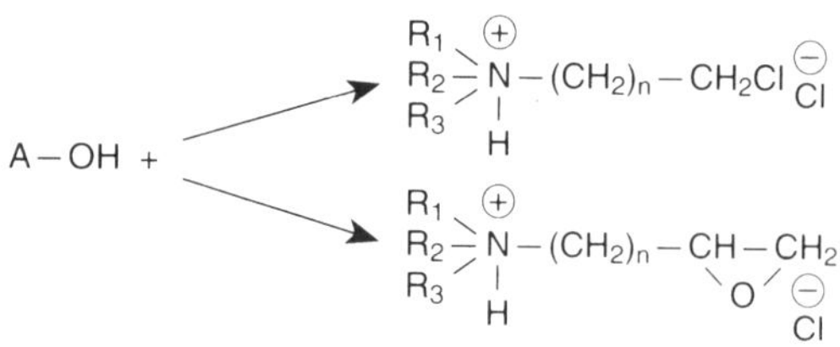

11.3.2.1 Cross-linked starches

Cross-linking allows the molecular chains to be restructured by joining them to one another. A reinforced molecular network is obtained which thus allows us to influence the rheological profile of the starch: overall, by taking advantage of the water retention capacity of a given granule, cross-linking establishes the viscosity at the desired value.

This process therefore allows the resistance of the starch under heat and in an acid medium to be increased. Only a very small number of these bridges are sufficient to modify the starch extensively. In most cases, one bridge for between 500 and 1000 D-glucose units is sufficient to obtain good stability without any modification in nutritional value.

Which are the bridging agents? Bi-functional reagents capable of reacting on two hydroxyls such as:

- chloroepoxide by-products – epichlorhydrin;
- phosphorylated by-products – phosphorus oxychloride, sodium trimetaphosphate;
- dianhydrides of acids – mixed adipic anhydride, acetic or citric-acetic;
- aldehyde by-products – formol.

These types of starch are used as thickeners (sauces, soups) or as temporary binders to hold together ingredients in a homogeneous suspension (ready-prepared meals, foodstuffs for children, salad dressings).

11.3.2.2 Stabilised starches

Introducing ester or ether groups into the starch molecule allows viscosity to be stabilised, especially at low temperatures.

The class of esters is dominated by the phosphorus acetates and esters. The ether class is dominated by hydroxyethylated and hydroxypropylated starches. These by-products have good water-holding properties at low temperatures which makes them particularly suitable for use within the field of deep-frozen or fast-frozen products.

11.3.2.3 Specific starches

These starches, which may or may not be charged, are manufactured for very specific uses. The most important of these are as follows:

- **Anionic starches.** Comparable to their opposite numbers within cellulose (the CMCs), carboxymethyl starches are extensively used in the fields in which a thickening quality is required: wallpaper paste, printing pastes and textiles.
- **Cationic starches.** The amino tertiary and ammonium quaternary by-product classes are the most common classes, and most commercial cationic starches belong to these two groups.

  These starches are widely used in the paper industry for manufacturing paper in which they create a network with the electronegative fibres of the cellulose which allows them to retain charged minerals more effectively. Properties connected with physical characteristics are added to these retention properties.

- **Hydrophobic starches.** Introducing a long chain alkyl group (by means of a reaction with an acid anhydride, for example) gives the starch hydrophobic qualities so that it cannot be moistened.
- **Bipolar starches.** Introducing a lipophilic group of the sodium succinate actenyl type gives the starch a bipolar character which allows it to act as an emulsion stabilising agent, in particular by reducing granular crystals of fats in foodstuffs.

11.3.3 Controlling the composition of starches and genetic improvements

The study of the functional properties of starches has shown that amylose encourages gelling on cooling and that amylopectin produces thick liquids that do not gel when cooled.

As has been stated previously, maize is currently the favourite substrate with starch manufacturers. Selectors have identified innumerable varieties of maize, from very waxy to very gelatinous, rich in either amylopectin or amylose. In addition, maize lends itself very well to gluten and starch separation.

In wheat the work of the selectors has mainly involved the quality of the gluten and its suitability for bread-making. However, at the moment, geneticists are working on creating types of wheat that can be used in fractionation with contrasted amylose and amylopectin contents. Starch from potatoes is the type that allows the widest technological applications. This explains why research is under way to reduce the level of amylose in potatoes. The concept consists of reducing the expression of a coding gene for an enzyme involved in the biosynthesis of amylose using an RNA anti-directive.

Fluidified starches are the result of a very gentle hydrolysis of starch, whose effect is to reduce the length of the molecular chains. The viscosity of this type of starch when hot is reduced without altering the formation of the gel which appears during cooling. The preparation of these starches which are highly sought-after in the manufacture of gelled confectionery leads on to a discussion about the family of the products of hydrolysis of this glucan.

11.4 Starch hydrolysates

Hydrolysis can be carried out using the specific glycosyl-hydrolases of the $\alpha(1\rightarrow4)$ bond, the $\alpha(1\rightarrow6)$ bond, or both types of bond at the same time.

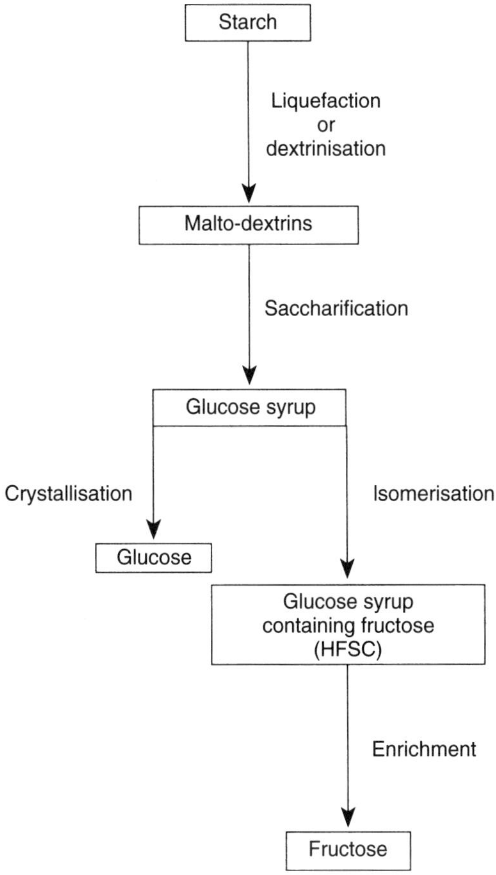

Fig. 11.5 Principal types of products of starch hydrolysis.

Acid hydrolysis has the advantage of being rapid and complete, but disadvantages are problems with colour and taste and salt concentration following any neutralisation. Depending on the method used, the composition of the hydrolysates will be very different (Fig. 11.5).

Malto-dextrins, glucose syrups and hydrolysates are commonly characterised by their 'dextrose equivalent', DE. The DE is defined as the percentage of reducing carbohydrates present in the syrup, in comparison with the total quantity of oligo-saccharides. So, for example, the DE of the malto-dextrins is lower than 20, that of glucose syrups between 20 and 97 and that of hydrolysates higher than 97 (Table 11.3). This definition does, however, have limitations in that, depending on the technological process used, two products of starch hydrolysis can have very different compositions while still having the same overall DE.

Table 11.3 Carbohydrate composition of malto-dextrins and glucose syrups

DE	12	17	28	38	61
Glucose	<1	<1	1	18	25
Maltose	3	5	16	11	42
Malto-triose	7	9	23	13	14
Other saccharides	90	86	60	58	19

11.4.1 Malto-dextrins

Isolating thermostable α-amylase has allowed acid dextrination to be progressively replaced by the enzymatic method which takes place at high temperatures (around 80 °C and at pHs close to neutrality). Hydrolysis by the α-amylases produces malto-dextrins whose DE varies between 3 and 20. This actually involves a starch-liquefying operation which aims to produce products that are easily digestible, have a low content in saccharides with low molecular mass and low osmotic pressure.

Because of this, malto-dextrins provide ingredients which are very useful in the formulation of infant foods, health foods and drip-feeds for the very sick, but they also have many other applications as intermediate food products.

Malto-dextrins are mainly used as texturising agents. These compounds allow thickening to take place and give certain products a smoothness (sauces, sterilised or frozen soups). They act as a binder in the production of cooked meats and allow good development of the bacterial flora which converts nitrates into nitrites. In confectionery, malto-dextrins replace arabic gum in 'gums' and hard liquorice. They increase the 'chewability' of chewing-gums while reducing their sweet flavour. In the same way the superficial crystallisation of sucrose (white spot) in sorbets is eliminated by adding these IFPs.

Certain malto-dextrins can be used as substitutes for fats (see Section 13.6.2.2), in products such as pastry fillings, low-calorie butters and margarines, spreads and salad dressings. These dextrins with a low degree of starch hydrolysis contain less than 5% glucose, maltose and maltotriose. So, at a concentration of between 20 and 25% they are converted after a few hours into a thermo-reversible gel which possesses a texture comparable to that of a fat. From a concentration of 20% upwards, the hardness of the gels, measured using penetrometry, reaches a useful value and the texture obtained resembles that of a fat. However, when malto-dextrin is incorporated in foodstuffs it is not necessary to resort to such high concentrations, because the additional effects of other ingredients that are soluble in water, such as proteins, starch or other stabilising agents, must be allowed for.

11.4.2 Syrups and glucose hydrolysates

Hydrolysis involving a mixture of α- and β-amylase, isoamylase (or pullulanase) and glucoamylase produces glucose syrups whose DE is high and which can reach 97; this represents a 97% glucose solution.

Acid hydrolysis allows glucose syrups with a DE of between 20 and 65 to be obtained. However, above 50 DE the products of hydrolysis break down to form coloured compounds with a bitter flavour. This is why mixed acid and enzymatic hydrolysis is used to produce glucose syrups with DE > 50.

Starch hydrolysates can be used as they are, or can then undergo conversion treatments of varying degrees. In the first case hydrolysates with low DEs are obtained and, in the second case, glucose or maltose solutions are still used.

Hydrolysates with 97 DE allow glucose to be prepared for injectable solutions after crystallisation and purification. Likewise, passing these syrups through an enzymatic reactor with immobilised isomerase glucose allows a glucose–fructose mixture to be obtained whose sweetening power is close to that of invert sugar. These syrups with a high fructose content manufactured in liquid form, can be used in all sweetened foodstuffs in which they are economically advantageous in comparison with sucrose.

Industrial uses of glucose syrups are mainly associated with their sweetening, crystallising and viscosity properties and with their water-holding ability.

As sweetening power is mainly associated with glucose and maltose the concentration increases with the DE, but parameters such as pH and temperature can influence its value (Table 11.4). The poor sweetening power

Table 11.4 Detection threshold and relative value of sweetening power of glucose syrups

	Detection threshold per 100 (m/v)	Relative value of sweetening power
Sucrose	0.46	100
Glucose syrups DE:		
15	3.25	14
18	2.50	18
25	2.20	21
37	1.68	27
42	1.21	38
43	1.54	30
52	1.07	43
64	0.80	58
78	0.73	63
Syrup with high fructose content	0.45	102

of glucose syrups with low DEs improves the palatability of these solutions, thus allowing a larger quantity of carbohydrates to be ingested. Products with a high glucose and maltose content can be crystallised easily. This property can sometimes pose problems, for example in coatings, and adding a small quantity of malto-dextrin can prevent crystallisation. Likewise, the development of a microcrystal phase by means of whipping is a solution adopted in other products (fillings for melt-in-the-mouth biscuits).

Products with low DEs are able to prevent water uptake. This property is of course made use of when storing food products that would be affected in various ways by water absorption.

Products with high DEs retain a high osmotic pressure in complex media and this ability is used to stabilise the water in a foodstuff and/or to limit the development of undesirable micro-organisms.

The viscosity of the products of starch hydrolysis depends on temperature, the carbohydrate concentration and especially their DE. Since its ability to increase viscosity decreases when the DE increases, products with low DEs are used as thickening agents whereas products with high DEs find applications on account of their plasticising properties.

Finally, the fermentability of the products of starch hydrolysis is used in numerous industries (pharmaceutical, agro-alimentary, chemical) where it is the starting point for numerous bioconversions.

11.4.3 Cyclodextrins

Cyclodextrin glucanotransferases (CGTases) [1,4-α-glucan-4-α-D-(1,4-α-D-glucan)-transferase] (EC 2.4.1.19) are enzymes that are able to convert the linear chains of starch into cyclic molecules known as Schardinger cyclodextrins. The action of this enzyme is complex as it appears to catalyse at least three reactions which involve the phenomena of cyclisation, bonding and hydrolysis.

CGTase is therefore able to convert the linear starch chains into cyclic molecules which are mainly α, β and γ cyclodextrins with six, seven and eight units of glucose respectively, linked in α(1–4) form. The long amylose chains are not the best substrates for this reaction. The maximum cyclohexaamylose content is obtained when the amylose is broken down into polymerisation chains (DP below 100), and contaminated with amylopectin, which suggests that the reducing extremities of the external chains of amylopectin act as amylose hydrolysis acceptors. The linear oligosides of starch are also good substrates and the branched carbohydrates can be incorporated in the cyclodextrins to produce connected cyclic molecules.

The advantage of the cyclodextrins is that they offer a hydrophobic cavity of average size (1.5 nm × 0.7 nm × 0.8 nm) whereas the molecule is hydrophilic on the outside (Fig. 11.6). This toric structure allows stable inclusion complexes to form, with a wide diversity of organic substances and also with salts and halogens. So hydrophobic substances can be dis-

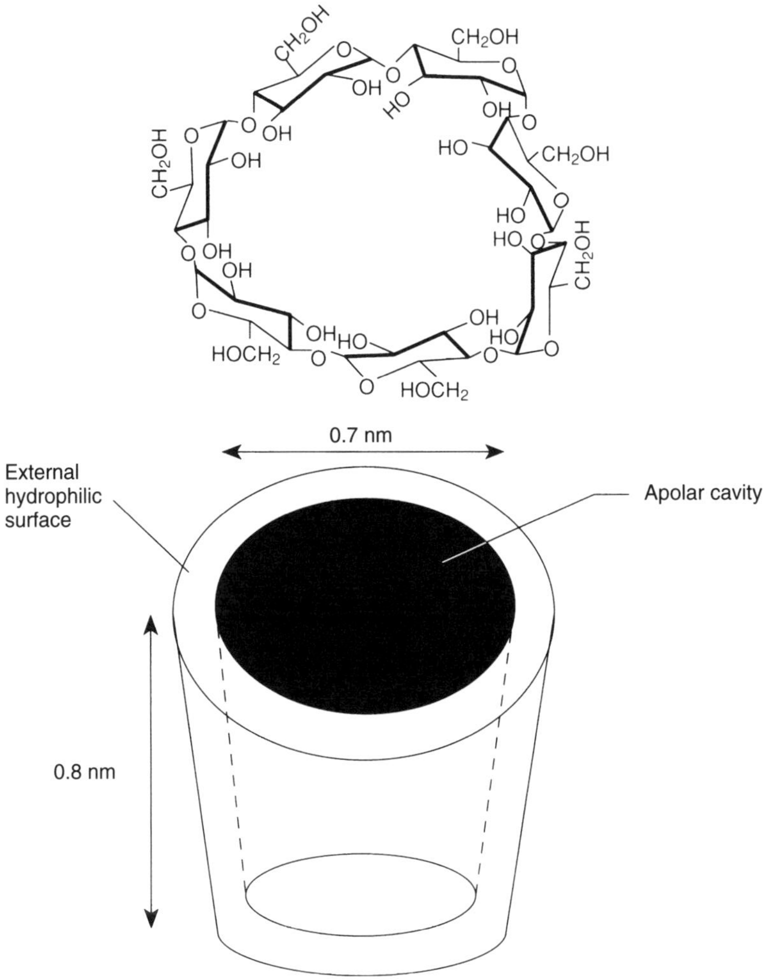

Fig. 11.6 Diagrammatic representation of β-cyclodextrin.

solved. Depending on their respective size, the 'guest' molecule is encapsulated fully or partially, with cyclodextrin acting as the 'host' molecule or receptor. In addition, the complex improves the stability of the 'guest' molecule not only in water but also in air in the case of dry products, as well as in relation to heat, oxidation and hydrolysis.

Some of the main industrial applications are as follows:

- Stabilisation of volatile substances, emulsions, aromatic compounds, spices, etc.
- Elimination of undesirable molecules (bitterness of medicines, cholesterol in foodstuffs).

- Modification of the chemical activity of a molecule by protecting some of its functional groups.
- Increased solubility.
- Protection in terms of oxidation.

Added to this is the fact that the method of preparing these complexes is very simple: water can be eliminated simply by stirring the cyclodextrin solution with the solution of the molecules to be encapsulated. The interactions between the 'host' and the 'guest' molecule are by nature hydrophobic, so dissociation is simple and gentle.

The principal industrial future for these cyclodextrins is currently within the pharmaceutical sector. This boom in pharmaceutical applications has its origins in the increased bio-availability conferred on the 'guest' molecules by the cyclodextrin complex, which involves increasing solubility and absorption. In other words, the same quantity of drug produces a greater biological effect when it is administered by a cyclodextrin complex. So the quantity of drug can be reduced while still maintaining the same therapeutic effect, which is a considerable advantage. We can even see a reduction in undesirable side-effects such as stomach irritation. This encapsulation could constitute tomorrow's galenic form.

Cyclodextrins can also allow the inclusion of other products in addition to pharmaceutical ones, mainly those of a phytosanitary type. An unexpected application has been the discovery that cyclodextrins increase the yield of cereals. Germination in seeds treated with β-cyclodextrin is delayed for several days, yet the final returns are between 20 and 40% higher than those for control seeds.

As far as the agro-food sector is concerned, β-cyclodextrin is used to remove cholesterol from fat in milk and to carry flavours.

In order to extend the applications of these molecules, several processes for manufacturing second generation cyclodextrins, which can carry ramifications to increase their solubility, have been developed. All the protocols use pullulanase in a reverse way to bind the chains of monosaccharides onto the cyclodextrins.

Cyclodextrins also have the ability to mime the role of biocatalysts. The cavity of cyclodextrins is in fact lined with hydroxyl groups which, depending on the pH of the solution, can be either in acid form (OH) or in base form (O^-). The resulting acid–base catalysis potential is similar to that of enzymes such as ribonuclease and the proteases. These properties of catalysis and binding within a cavity of molecules (depending on their size and their hydrophobicity) has led researchers to produce an artificial enzyme by synthesis which mimics the transaminases. By covalent means they have attached the phosphate coenzyme of pyridoxamine to a cyclodextrin. Indolpyruvic acid is then converted by this modified cyclodextrin into chiral tryptophan in a ratio of 3/1 for the L and D enantiomers respectively. So it would seem that the glucose units of cyclodextrin, optically active, are capable of binding substrates by regio-selection. If the speed of catalysis of these artificial enzymes is not as high as those of natural enzymes, it is, on the other hand, higher than that of non-catalysed reactions.

The regionally specific binding properties of the cyclodextrins may have more specific applications. Certain drugs often taking the form of racemic mixtures cannot be used for therapeutic purposes because of the undesirable side-effects of one of the enantiomers. However, using affinity chromatography, β-cyclodextrin has allowed a number of racemic products of interest in treatment to be dissolved. These are β-blockers, sedatives, antihistamines, anticonvulsants, diuretics, cardio-active alcaloids and opiate substances.

11.5 Interactions with other biochemical constituents

The existence of the synergies between starch and other biochemical constituents whether or not of a carbohydrate nature, has been well known for some time. This phenomenon is still poorly understood and its potential has still to be supported by experimental work.

11.5.1 Hydrocolloid starches

The viscosity of starch slurry can be greatly increased by the presence of water-soluble polysaccharides; for example in the presence of guar gum viscosity is multiplied by a factor of 10. Another consequence concerns the level of stability of the system when cold when retrogradation is slowed down.

The swelling of starch is hardly disturbed by the presence of hydrosoluble gum. However, this is localised in the continuous suspension phase whose volume decreases as the grains of starch increase. The consequence is a clear increase in gum concentration in this phase, resulting in an extremely spectacular increase in the viscosity of the continuous medium.

A practical consequence of this type of phenomenon is that it is possible to concentrate a solution of water-soluble gum 'artificially' well beyond the level possible by means of direct dispersion in water. Depending on the thickening and gelling nature of the polysaccharide, the type of starch and the respective concentrations, it is possible to prepare mixtures with very diverse rheological properties.

11.5.2 Amylose–lipids

Most glycans cannot establish special interactions with lipids. Only amylose can form complexes having very special structures with lipids whose chemical structures are well defined.

Three categories of lipids are known to form complexes with amylose. These are monoacylated lipids:

- free fatty acids;
- lysophospholipids;
- monoglycerides.

Free fatty acids and lysophospholipids (mainly lysophosphatidylcholine) are the two large classes of lipids included in cereal starches. The lipid content of these starches is actually in the order of 1%. Other types of starch contain hardly any lipids. These are in the main monoacylated lipids: approximately 80% of lysophospholipids and 20% of free fatty acids in the case of wheat starch; these percentages are reversed for maize starches.

On the other hand monoglycerides are manufactured on an industrial scale and are frequently used as food additives. They have similar properties to those of free fatty acids and lysophospholipids (see Chapter 13) in relation to starch.

We know that amylose has the capacity to form inclusion complexes with numerous hydrophobic compounds: iodine, certain types of alcohol, etc. The same is true for monoacylated lipids. This complexing capacity seems linked to the possibility of amylose presenting a helical conformation. The example most often put forward is that of the inclusion complex. The amylose helix whose structure consists of six glucose residues per turn has an internal diameter of 0.45 nm. The configuration of this helix is such that the interior is 'lined' with C—H, and therefore lipophilic groups, whereas the hydrophilic C—OH groups are directed towards the outside. The lipophilic agents whose diameter is compatible with the internal diameter of the amylose helix can therefore constitute stable inclusion complexes. For monoglycerides and lysophospholipids, only the fatty chain can penetrate the inside of the helix as the hydrophilic group is too voluminous to be included (Fig. 11.7).

It has been shown that it is beyond a weight monoglyceride/amylose ratio of approximately 1/20 that the complex formed precipitates but it has also proved that real amylose saturation only occurs at a ratio of 1/5 (which corresponds, assuming a molecular mass of 150000 for amylose, to a molar ratio of 100/1).

The essential characteristic of the amylose–lipid complex is therefore its insolubility in an aqueous medium. It has been shown by scanning calorimetry that this complex 'melts' at temperatures in the order of 95–110 °C depending on the nature of the lipids and the length of the carbonated chains and that the complex forms again when cooled.

The presence of lipids included in cereal starches allows their behaviour during the starching process to be explained. The second stage of the process at approximately 90 °C, characterised by a second swelling stage and significant amylose solubilisation, is probably linked, at least in part, to

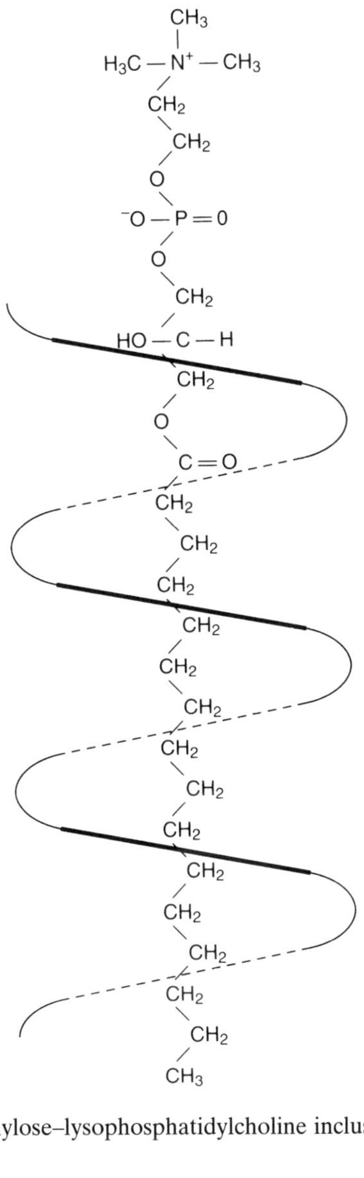

Fig. 11.7 Amylose–lysophosphatidylcholine inclusion complex.

the fusion of the amylose–lipid complex. The fact that this complex can form again on cooling definitely brings about the process of retrogradation during storage.

Therefore one way of affecting the behaviour of starches during the starching and retrogradation process will be to add the complexing lipids of amylose. This is currently carried out at an industrial level using distilled hydrogenated monoglycerides and stearyl lactylates (see Section 13.3.5). It has actually been established that non-hydrogenated monoglycerides are much less effective because they contain a large proportion of unsaturated fatty acids. The presence of double bonds seems to limit the inclusion possibilities of the fat chain inside the helix. The use of these possibilities of complexing amylose with monoglycerides is widespread in terms of cooked products.

Incorporating saturated monoglycerides in the products of industrial bread-making has the effect of limiting the speed at which the inside of the bread firms up. There is, however, a good correlation between the complexing power of the amylose and the anti-staling effect. Not enough is known about this to be able to propose an overall scheme for the mechanisms involved. It is believed that these emulsifiers intervene at different stages in the bread-making process. During kneading, the monoglycerides are distributed throughout the dough and can be adsorbed on the surface of the starch grains. This adsorption has the effect of limiting the swelling of the starch grains during cooking. On cooling, the amylose–monoglyceride inclusion complexes form, and the amylose complexed in this way precipitates, 'coagulating' the inside of the starch grain. This makes reorganisation of the amylopectin chains more difficult. Since amylopectin crystallises as the bread firms up, we can understand the retarding effect these amylose–monoglyceride complexes have while this phenomenon is occurring.

11.6 Uses of food starches

The increase in the food uses of starch products is fairly consistent and is based essentially on two factors, namely advances within the cooked meals industry and the consumer's search for a more balanced diet.

Starch products are used in food preparations in order to effect certain characteristics, such as texture, appearance, moisture content, consistency and stability during storage. They allow more costly ingredients to be replaced and facilitate manufacture. They are used to thicken or make more fluid, to clarify or to make opaque, to attract water or to repel it, in order to produce short or long, smooth or pulpy textures, soft or crispy coatings. Starches and their by-products are used in types of manufacture as diverse as jams, chilled or frozen products, drinks, baked products, dehydrated or extruded products. The main properties of starches in various food products are listed in Table 11.5.

As is the case for any intermediate food product the selection of a starch product for a formulation is effected in accordance with the nutritional and functional properties sought in the end-product. This is not always a simple choice. Starch actually fulfils several functional roles. For example in bakery and sweet pastry products, starch can do the following:

Table 11.5 Properties of starches in various food products

Adhesion	Breadcrumbing, coatings
Anti-adhesion	Bread, gums
Aid to fluidity	Fats, baking powder
Anti-staling	Baking products
Moulding	Confectionery
Encapsulation	Essential oils and aromas
Thickener	Sauces, soups
Shaping	Meat, pet products
Gelling	Confectionery
Icing	Confectionery, cake-making
Binding	Meat-balls, potatoes, extruded products
Foam reinforcement	Marshmallows, drinks
Water holding	Meat, baking products
Stabilising	Drinks, emulsified salad dressings

- Dilute gluten in the dough.
- Absorb a quantity of water whose amount depends on its state of damage, its physical state (pre-gelled) or chemical (modified).
- Provide a surface area allowing interactions of other constituents in the formulation with gluten.
- Liberate, via enzymatic hydrolysis, fermentable carbohydrates.
- Make the dough flexible by means of partial gelling during cooking.
- Absorb water from gluten hydration, thus bringing about an increase in rigidity of the protein film and the loss of good expansion capacity.

From a nutritional point of view, the uses of starch products allows digestibility and the caloric value of the foodstuff to be adjusted.

As far as functional properties are concerned, four qualities predominate in the use of starch and its by-products, namely sweet taste, texture, preservation and ease of use. In comparison with sucrose, each sweetening substance possesses a relative **sweet flavour** which depends on numerous parameters such as concentration, acidity and the temperature and nature of the carbohydrates present. In addition it is very often necessary to take into consideration the effects of synergy which result from the association of various sweetening substances. Adding glucose syrup whose DE is between 40 and 95 allows the relative sweetening taste to increase in strength by altering the concentration of solutions.

Even if this is closely associated with the percentages of dry matter and temperature, the **viscosity** of a glucose syrup depends in the first instance on its molecular composition and, as a consequence, on its DE.

Hygroscopicity is an important property since it justifies the use of glucose syrups in numerous applications. In fact, the range of products derived from starch contains not only products that, in solution, have a greater water activity than sucrose, which is the case with glucose syrups having a low DE, but also products that have a water activity lower than that of sucrose. These are glucose syrups with high DEs or syrups containing fructose, whose hygroscopicity increases as the DE increases.

Controlling the crystallisation of sweet solutions containing sucrose or lactose can be effected by adding glucose or glucose syrup. The inhibition of crystallisation depends on the presence of molecules of high molecular mass and so it is more pronounced with glucose syrups of low DE.

On the other hand, carbohydrates of low molecular mass have the greatest effect on lowering the freezing point as well as on raising the boiling point and osmotic pressure.

In conclusion, starch products, whose range continues to grow, are opening up new uses within the field of food processing. In particular they improve the quality of end-products in terms of **visual appearance**, **texture**, and **sensation in the mouth**. They can also offer economic advantages by replacing costly ingredients and by improving returns.

12

Hydrocolloids and dietary fibres

12.1 Definitions and classification

Among the numerous macromolecules of natural origin some possess the property of being able to disperse easily and thus increase viscosity quite significantly, sometimes producing a gelling effect. These food-thickening and gelling agents are known as water-soluble gums or **hydrocolloids**.

Depending on their origins, we can differentiate between:

- gums of plant origin which are mainly of a carbohydrate nature (carrageenans, alginates, pectins, starches, xanthan, cellulose by-products, etc.);
- gums of animal origin which are protein by nature, essentially caseinates and gelatin.

This chapter will consider the principal thickening and gelling agents used as additives. Gelatin and starches, considered to be nutrients, are covered in Chapters 9 and 11 respectively.

In addition to their thickening and/or gelling capacity, these biopolymers can also be used for their other properties: stabilisation of suspensions and emulsions, water-holding capacity, binding properties, formation of complexes with proteins. This is why macromolecules are among the most important ingredients in the 'dietary fibre' complex (Fig. 12.1). This term is used to designate the various plant ingredients which form part of the diet but are not broken down by enzymes within the human body. In practice, the nature of the fibres varies, as will become clear, depending on the point of view under consideration:

- Physiological point of view (non-digestibility by human digestive enzymes).

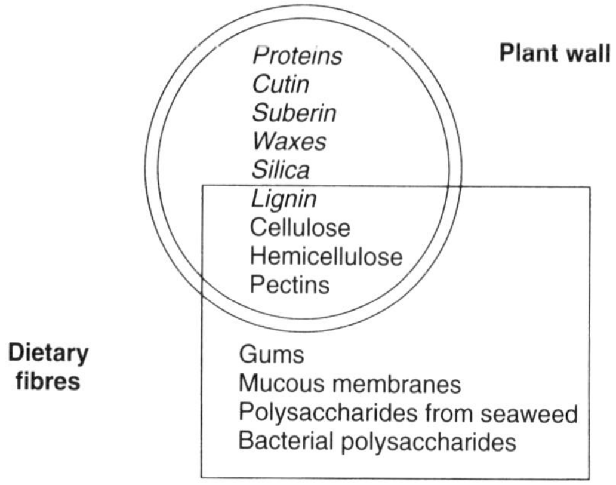

Fig. 12.1 The principal constituents of the 'dietary fibre' complex and its relationship with the plant wall.

- Analytical and biochemical point of view (depending on methods of quantitative analysis).
- 'Raw materials' point of view (origin of fibres and process).

12.2 Parietal plant polymers

The cell walls of most plants are mainly composed of glycans and lignin. Diagrammatically, in the primary wall polysaccharides constitute an amorphous phase steeped in a loose network of cellulose fibres; the secondary wall is thickened by a deposit of cellulose in layers arranged to form a very compact fibrillar framework offering high mechanical resistance with possible impregnation of lignin.

12.2.1 Glycans in the wall

12.2.1.1 Cellulose and by-products

This biopolymer is a result of the exclusively linear condensation of D-pyranoglucose units, in $4C_1$ conformation, joined to each other via β-1$\rightarrow$4 bonds, whose degree of polymerisation can reach 10000. All the hydroxyl groups are in an equatorial position in comparison with the glucose cycle layout, which makes the polymer extremely stable. The molecules are stabilised by other hydrogen bonds. As the basic structure takes the form of a very rigid chain, these bonds can join together in parallel in a microfibrillar structure. These microfibrils, which are 3.5 nm in diameter, appear in a form whose degree of crystallisation depends on their location. They come together to form fibres of 15–20 nm in diameter. Cellulose is insoluble in aqueous solvents and very resistant to chemical degradation; it can be extracted in hot diluted alkaline and acid solvents.

Cellulose can be dissolved by means of conversion into **esters** (cellulose nitrate, cellulose acetate, etc.) or **ethers** (methylcellulose, carboxymethylcellulose, etc.). These cellulose by-products are soluble in water and their solubility is determined by two parameters, viscosity and substitution (Table 12.1).

Carboxymethylcellulose (CMC) can be considered to be a weak acid which is not soluble in either a solution of strong acid or in organic solvents. Its anionic nature explains its reaction with amphoteric proteins at pH_i which prevents it from precipitating during heat processing or storage. During the CMC/protein reaction a soluble complex is formed, thus allowing numerous drinks and sauces to be prepared. In dairy products such as ice-creams and chilled desserts, CMC produces a smooth texture since it regulates the growth of ice crystals in frozen products and prevents syneresis in gelled desserts. This cellulose by-product is also used to improve body in products with a low sugar content.

Non-ionic cellulose ethers such as hydroxypropylcellulose (HPC) and hydroxylpropylmethylcellulose (HPMC) are soluble in water only at certain temperatures. HPC is soluble in cold water, but insoluble in hot water. It is also soluble in several polar organic solvents: the more polar the solvent, the clearer the solution. The solubility of methylcellulose (MC) depends on the degree of substitution (DS). For a DS of between 1.4 and 2.0 it is soluble in water, and it becomes soluble in organic solvents from a DS of 2.4.

All the cellulose ethers form solutions with varying degrees of viscosity in water. The gelling phenomenon depends on the level of viscosity, which itself depends on the solubility of the product, the temperature, its concentration, its degree of polymerisation, factors connected with the medium (pH, other components of the medium, such as starches, salt, etc.). So MC and MHPC which are soluble in cold water gell at a water temperature of between 50 and 90 °C depending on the degree of substitution of the cellulose chain. This hot gelling effect can be useful in certain applications (manufacturing doughnuts, reconstituted breadcrumbs-coated products, etc.).

Cellulose by-products act as thickening, binding, texturising or water-binding agents. Combinations of these can produce the desired effects of consistency, appearance, body and palatability in various products. For example, MC, MHPC and CMC can be used in bakery products; they allow the consistency of doughs to be adjusted, water holding to be improved and the shelf-life of certain cakes to be extended by reducing the speed of staling.

12.2.1.2 Hemicelluloses

Experts include the non-cellulose and non-pectin glycans of the plant cell wall in this category. These are mixed polymers containing neutral

Table 12.1 Principal cellulose by-products

Carboxymethylcellulose (CMC)

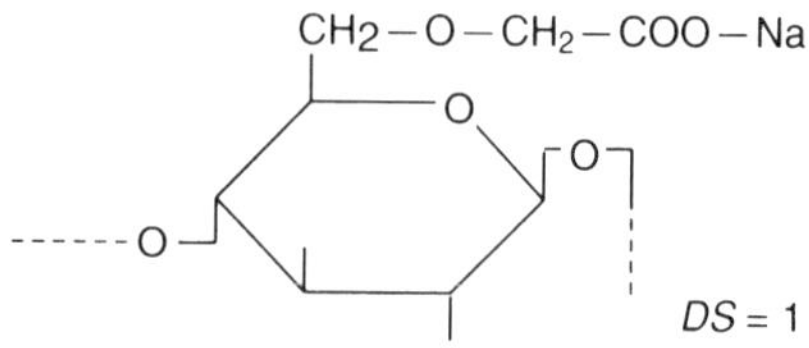

Hydroxypropylcellulose (HPC)

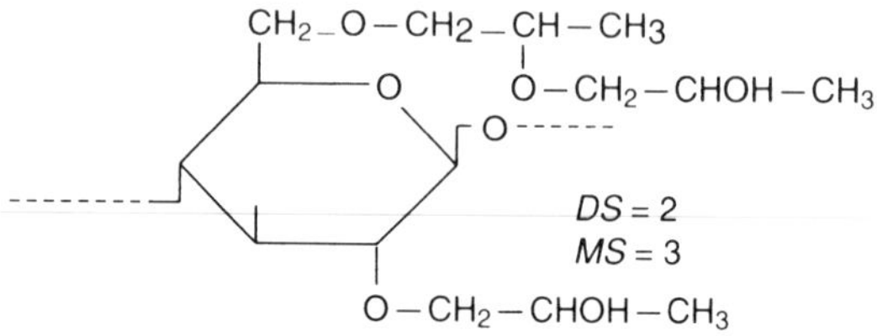

Methylcellulose (MC)

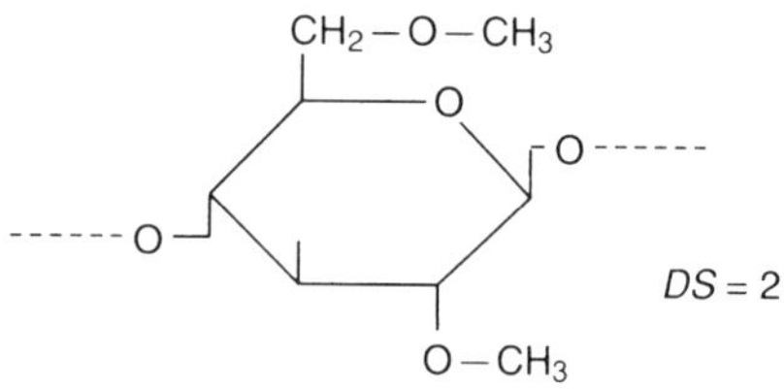

Methylethylcellulose (MEC)

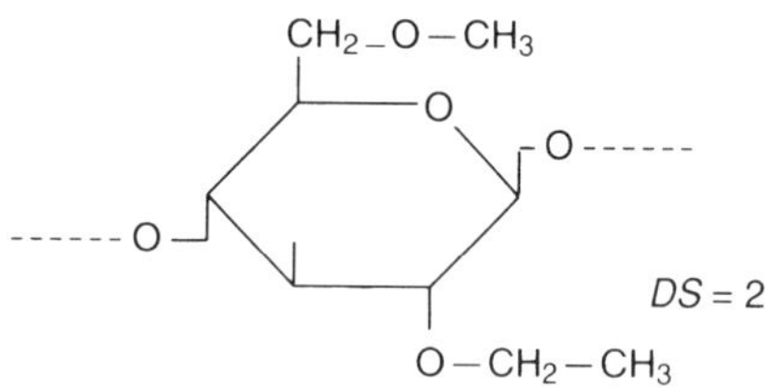

Hydroxypropylmethylcellulose (HPMC)

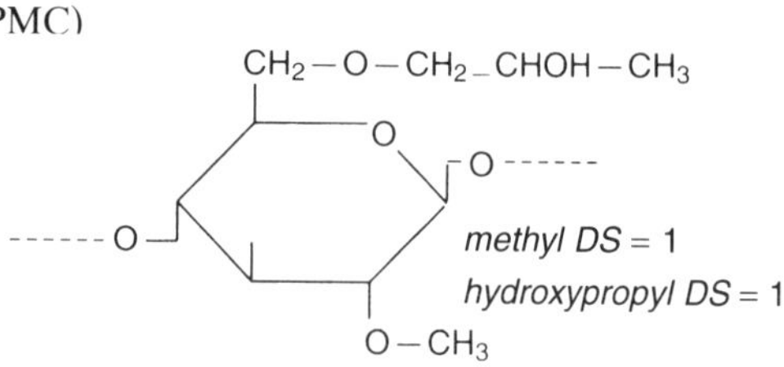

DS: average number of hydroxyls per glucose unit which have reacted.
MS: average number of substitution molecules combined with cellulose per glucose unit.

monosaccharides (xyloses, arabinose, mannose, galactose, glucose) and acid ones (essentially glucuronic acid and its methyl-4 by-product).

Some of the main types of hemicellulose are as follows:

- **Xyloglucans**, which consist of a cellulose-type framework (D-glucopyranosyl units bound in β-1→4) to which D-xylose units or short chains formed of xylose, galactose and fructose are connected on the C_6 of glucose.
- **Xylans** are non-cellulose polysaccharides comprising a principal chain, as their name implies, of D-xylose units bound into β-1→4 with different types of substitutes such as glucuronic acid or arabinose on the C_2 and/or the C_3 of the xylose, as well as side chains which contain varying amounts of xylose and arabinose. Xylans are abundant in wheat bran (up to 40%) and their structure, connected to varying degrees, explains their structural relations with lignin (Chapter 10).
- **Mannans**, consisting of D-mannose units bound in β-1→4 with D-galactose cycles on the C_6, distributed at random throughout the chain. The galactomannans of guar and carob are used as thickeners. Hemicelluloses are therefore miscellaneous glycans, acid or neutral, which can be extracted by dilute alkaline solutions.

12.2.1.3 Pectins

Typically found in the middle lamella, pectins consist of D-galacturonic acids of $4C_1$ conformation, bound in α-1→4 (Table 12.2). L-Rhamnose units bound by α-1→2 bonds cause a deflection of approximately 90° from the axis of the linear chain. The acid groups can be esterified by methanol. The degree of methylation is variable and determines the potential technological uses of the pectins extracted. The alcohol functional groups carried by the C_2 and C_3s of the galacturonic acids can be acetylated. There are also side chains, and neutral monosaccharides (arabinose, galactose) mainly connected to the rhamnose units.

Alongside these dominant rhamnogalacturonans are some grains or tubers of galactans, arabinans and arabinogalactans.

In pectins extracted within an acid medium, galacturonic acids are found mainly in the form of methyl esters. Certain types of pectins are then demethylated by acid or alkaline means. When demethylation is effected using ammonia, the ester group can be converted into amides.

Pectins are characterised by their degree of esterification which represents the percentage of galacturonic acids found in the form of methyl esters. A degree of amidation (DA) is also used.

It is possible to differentiate between:

- highly methylated pectins: HM pectins whose degree of esterification is higher than 50 (degree of esterification >50);
- low methylated pectins: LM pectins whose degree of esterification is lower than 50 (degree of esterification <50).

Table 12.2 Structure and properties of pectins

Origin	Structure	Solubility	Effect	Properties	Behaviour: Rheology	Behaviour: Temperature	Behaviour: Mechanical	Behaviour: Freezing/thawing	Behaviour: Storage environment	Applications
Pressed apple pulp, citrus, orange skin etc.	Polygalacturonic acid esterified to varying degrees by methanol.			Gel in acid and sweet medium >60%. Gel not heat reversible. With 'rapid sets' higher temperature and setting time shorter than with 'short sets'.	'Soft' elastic gel.	Swells and moves into solution when cold in absence of H^+ or Ca^{2+}.	Broken gel.	Depends on sugar content.	Retrograda-tion possible.	Acid preparations such as jam, confectionery. Acid dairy products.
E 440	High methoxyl (HM) pectin degree of esterification >50%.	Soluble when hot in water or in milk.	Gelling agent.							
	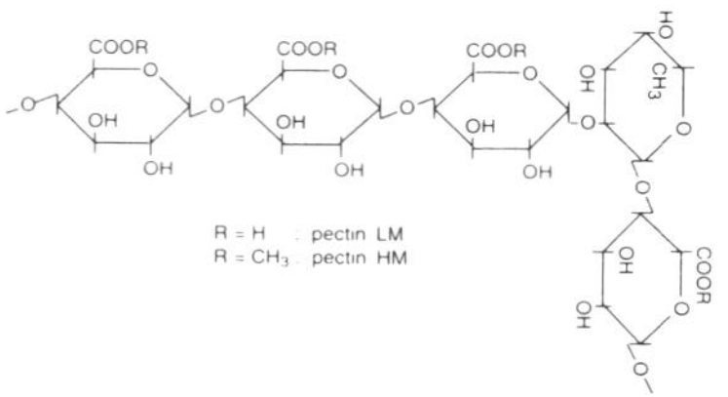									
	Low methoxyl (LM) pectin degree of esterification <50%. May or may not be amided.	Soluble when cold.	Gelling agent.	Gel in less acid medium but requires the presence of calcium. Gel heat reversible. Gel strength depends on pH and calcium content.	Cohesive-ness of gel depends on calcium content.					Milk desserts. Jam with low dry extract content. Pure fruit sugar.

Within the LM pectins are also the non-amide pectins (DA = 0).

The presence of side chains (neutral carbohydrates), and also methoxyl groups on the C_6 of the galacturonic acid, helps to spread the chains apart and thus facilitates pectin hydration. The higher the degree of esterification, the greater the solubility when cold. This is why the acid groups of low esterified pectins are partially neutralised in order to improve their solubility when cold, provided that the water is not particularly hard (approximately 10° hydrotimetric titre, HT).

To sum up, pectins are polysaccharides which are essentially acid in nature and can be extracted by water, calcium chelating agents, and dilute acids which are capable of forming gels. As a result of this ability and their polyelectrolytic properties, these glycans have a key role in the mechanisms of action of dietary fibres.

Finally, in cereals, glycans have a wall of non-cellulose β-glycans. These are linear chains that consist solely of D-glucose residues bound by β-1→4 and β-1→3 bonds. It is difficult for these polymers to aggregate and form viscous solutions.

12.2.2 Lignin

Apart from a few special cases such as wheat bran, plants which form part of the human diet contain low quantities of lignin. Lignin is a three-dimensional heteropolymer formed of monomers of the phenylpropane type. The three aromatic cyclic compounds found most frequently are the coniferylic, coumarylic and sinapylic alcohols. Monomer composition varies according to type and the bonds between these units are multiple and extremely diverse (ester, ether, etc.). The hydrophobicity of lignin can participate in certain types of adsorption.

The plant wall contains a number of molecules which cannot be broken down by endogenous monogastric enzymes. They are resistant to acid hydrolysis and determined by analysis in the lignin fraction.

12.3 Polysaccharides from seaweed and micro-organisms

12.3.1 Carrageenans

Carrageenans are sulphated polysaccharides extracted from red algae. They are prepared on an industrial scale using methods based on two of their properties:

- They are soluble in hot water.
- They are insoluble in polar organic solvents.

Three major fractions have been identified within this large family of glycans, and used for their technological capabilities. The κ and ι fractions produce gels, whereas the λ fraction is non-gelling. All these constituents contain galactose residues, sulphated to varying degrees and bound alternately in 1→3 and 1→4 (Table 12.3).

In **λ-carrageenan** the macromolecular chains can separate very easily from one another and as a result have no tendency to reassociate in order to produce gels. The reasons for this are, on the one hand, the way in which the sulphates are directed towards the interior or the exterior of the spiral: electrostatic repulsion and the steric spatial requirement prevent the monofibrillar chains coming together to form a network; on the other hand the irregular spiral structure makes formation of inter-macromolecular bonds difficult. λ-carrageenan can therefore act only as a thickening agent and its solutions behave in a pseudo-plastic way; that is to say viscosity decreases with shearing action.

Total hydration of the ι- and **κ-carrageenans** takes place by means of heating. On cooling, the molecules approach each other to create junction zones. The alternation of the $4C_1$ and $1C_4$ structure allows the macromolecule to be arranged in a regular spiral interrupted from time to time by bends which cause it to deviate. In these two fractions, the sulphate groups are located on the external face of the spiral alone; the internal faces of the chains can therefore come together and in this way form double spirals.

The network of ι-carrageenans formed by a succession of double spirals and bends produces a very transparent elastic gel. In addition it is thixotropic. The cation K^+ has a special action on the κ-carrageenan. Its small size in the hydrated state allows it to overlap in the helix and partly to neutralise the sulphated groups. The double helixes can then come together and form aggregates. This results in hardening of the gel, shrinkage of the structure and expulsion of water, producing syneresis and making the gel opaque. This phenomenon increases as the number of bends decreases. The hydrated Na^+ cation is too large to overlap into the helix and the Ca^{2+} cation is likely to form bridges between two sulphate groups of two different double helixes, thus forming inter-macromolecular bonds. A small quantity of Ca^{2+} increases the force of the gel but if the concentration is too high then the effect is detrimental.

12.3.2 Agar

Agar is also extracted from red algae. It is a mixture of several polymers formed of linear chains in which galactose units bound in β-1→4 and 3,6-anhydrogalactose units bound in α-1→3 alternate. Lesser quantities of methyl-ethers, sulphate esters and pyruvate cetals can also be found on these residues. Agar is used as a gelling agent in cake-making and in certain dairy preparations.

12.3.3 Alginates

Alginates are extracted from brown algae. They are linear co-polymers of D-mannuronic acid and L-guluronic acid. Each molecule contains homo-

Table 12.3 Structure and properties of carrageenans

Origin	Structure	Solubility	Effect	Properties	Behaviour					Applications
					Rheology	Temperature	Mechanical	Freezing/ thawing	Storage environment	
Red seaweed E 407	Sulphated galactose and 3–6-anhydrogalactose.			• Proteins of an amphoteric type precipitate with the carrageenans when the pH of the solution is lower than the isoelectric point. At higher pHs the protein/carrageenan reaction can be obtained. This reaction allows gel strengths which are higher in milk than in water to be obtained. • Gelling takes place during cooling within a precise temperature zone which is relatively high and depends on the concentration of the gelling agent used and the salts present. • The melting temperature of the gel is higher than the gelling temperature.						
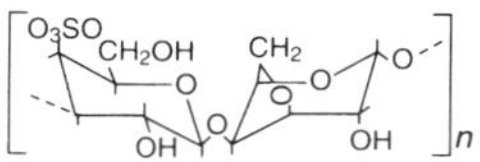		κ: soluble when hot (65–70 °C) in water and in milk	Gelling agent.	Reactivity with milk proteins. Heat-reversible gel. Synergy with carob.	Gel which breaks with syneresis.	Heating: slight swelling at average temperature, moves into solution at 60–70 °C. Cooling: no viscosity when hot, abrupt increase at gelling point.	Gel sensitive. Gel ground.	κ alone has tendency not to dissolve, retrogradation, syneresis.	κ alone: gel retrograded syneresis.	Milk desserts. Chocolate milks. Ice-creams and frozen creams. Meat preserves, pet food. Air deodorising gels. Immobilisation of enzymes via inclusion. *In vitro* meristem culture.

Table 12.3 *Continued*

Origin	Structure	Solubility	Effect	Properties	Behaviour: Rheology	Behaviour: Temperature	Behaviour: Mechanical	Behaviour: Freezing/thawing	Behaviour: Storage environment	Applications
	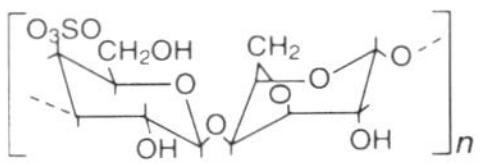	ι: soluble when hot (55 °C) in water and milk.	Gelling agent.	Reactivity with milk proteins. Heat-reversible gel.	Gel elastic without syneresis.	Heating: considerable swelling when cold, moves into solution around 50–60 °C. Cooling: little viscosity when hot, abrupt increase at gelling point, very good behaviour towards heat treatments.	Very good. Sensitive but picks up on resting, thixotropic.	Very good.	No change.	Desserts. Sauces. Salad dressings. Ices and ice-creams. Cosmetics.
	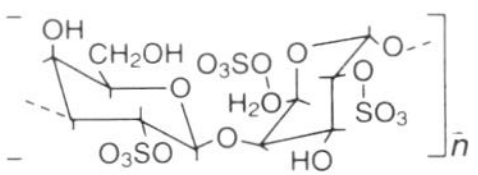	λ: soluble when cold in water and milk.	Thickening agent.	Reactivity with milk proteins.	Gel very weak in milk.	Very good in neutral medium; in acid medium depoly-merisation depending on H^+ temperature and time.	Not very sensitive.	Very good.	Very good.	Desserts. Sauces. Salad dressings. Toothpaste. Cosmetics.

mannuronic regions and homo-guluronic regions as well as regions in which the two types of residue alternate. The polyuronide produced when the two acids interlink is called alginic acid (Table 12.4).

The two acids are epimers of carbon 5, that is to say, the only difference is the position of the C_5–C_6 link above or below the average plane of the cycle. On the other hand the spatial requirement and the interactions due to the acid group of the C_6 force the conformation of the cycle to have this acid group in an equatorial position, which is why β-D-mannuronic acid is always $4C_1$ and guluronic acid $1C_4$.

These two monomers are not randomly distributed within the macromolecules, but distributed in segments of approximately twenty units. The way in which these segments assemble in variable proportions depends on the species, the part of the seaweed involved and, to a lesser degree, the age of the seaweed and the place in which it is gathered. Monomer distribution determines the gelling properties of the alginates; so, for example, the guluronic sequences have a conformation which is best suited to gelling by calcium. As with pectins, a molecular 'egg-box' network is formed through the association of polyuronate sequences by Ca^{2+} chelation.

Furthermore, all the operations involved in the manufacturing process are based on the differences in solubility:

- Alginates of alkaline metals are soluble in water.
- Alginic acid and its calcium by-product are virtually insoluble in water.

12.3.4 Microbial polysaccharides

The use of microbial polysaccharides is increasing enormously. Within the food industry, those used most frequently are xanthan, dextrans and curdlan.

Xanthan gum is a glycan with a high molecular mass produced by the aerobic fermentation of *Xanthomonas campestris* as a culture on glucose. The principal chain is made up of D-glucose units bound in β-1→4 and carrying side chains on every other residue on the C_3, consisting of a triholoside formed by a β-D-glucuronic acid surrounded by D-mannose units. On approximately half of the terminal mannose units a pyruvic acid is bound to the C_4 and C_6; non-terminal mannose carries an acetyl group on the C_6 (Table 12.5).

It must be specified that if the molecule is completely regular in terms of the presence of branchings in the case of one glucose out of two, there is nevertheless a certain degree of irregularity because all the side chains are not necessarily acetylated and pyruvated. Often only one side chain out of two is pyruvated but this proportion varies according to the source used and the culture conditions.

As xanthan is insoluble in organic solvents, it is recovered by

Table 12.4 Structure and properties of alginates

Origin	Structure	Solubility	Effect	Properties	Behaviour: Rheology	Behaviour: Temperature	Behaviour: Mechanical	Behaviour: Freezing/ thawing	Behaviour: Storage environment	Applications
Brown seaweed *Fucaceae* *Laminariaceae* *Macrocystis* E 401 to E 405 depending on the salt used including E 401: sodium alginate,	β D-mannuronic acid (M) 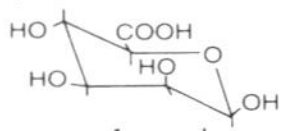 α L-guluronic acid (G) 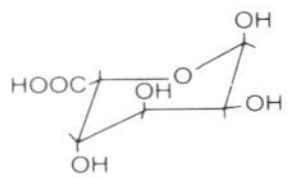 Possible sequences (links 1→4) –M–M–M–M –G–G–G–G –M–G–M–G	Alkaline by-products of alginate: • soluble in water when cold • soluble in milk in the presence of complexing salts of calcium • soluble in milk when hot	Thickening agent	• The progressive and limited addition of Ca^{2+} or H^+ ions to a solution of alkaline alginate can obtain non-heat-reversible gels which can be liquefied by complexing agents or solution of neutralisers. • The massive addition of Ca^{2+} or H^+ ions leads to precipitation. • Absorbing power: Possibility of retaining in the order of 25–30 g per gram of alginate.	Pseudoplastic.	Heating: swells and moves into solution when cold in the absence of Ca^{2+} and H^+. Loses part of its viscosity during high heat treatments.	Not very sensitive if thickening agent.	Thickened solutions. Good behaviour.	Thickening. No change.	Ices and ice-creams. Products for cake-making. Toothpaste. Cosmetics. Textile printing. Paper printing. Water flocking. Restructured products. Immobilisation of enzymes by inclusion. Absorbing products.
			Gelling in the presence of calcium and acid.		Elastic gel.	Preheating: viscosity varies with temperature.		Good.	Gel retrogradation of varying degrees.	

Table 12.5 Structure and properties of xanthan gum

Origin	Structure	Solubility	Effect	Properties	Behaviour					Applications
					Rheology	Temperature	Mechanical	Freezing/ thawing	Storage environment	
Biopolymer on hydrocarbonate substrate using *Xanthomonnas campestris* E 415 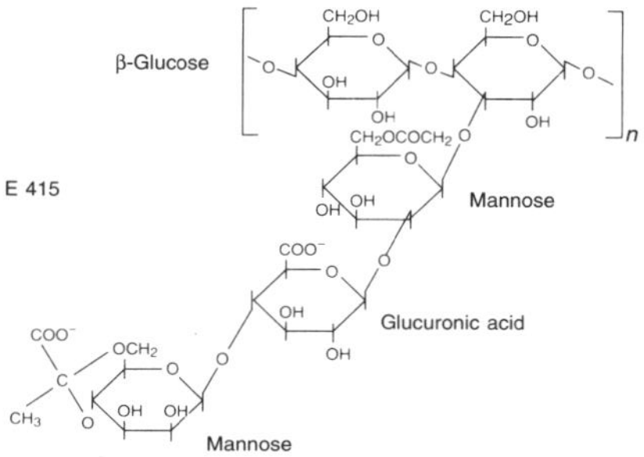	Glucose, mannose and glucuronic acid.	Soluble in water and milk when cold.	Thickening agent.	High suspensoid power. Remarkable stability in acid medium Stability to salts.	Plastic.	High viscosity when hot. Heating: swells when cold and moves rapidly into solution. Viscosity not very influenced by temperature; very good behaviour in acid medium.	Stability towards shearing.	Very good.	Few changes.	Sauces and salad dressings. Toothpaste. Cosmetics. Animal feed (milk suspension for veal). Agriculture. Cleaning, painting products.
				Synergy with carob.	Elastic and very cohesive gel.	Cooling: gels with carob.				

precipitation in isopropylic alcohol. The coagulum obtained is separated, washed, pressed and dried then ground to the desired particle size.

The presence of glucuronic acid and pyruvic acid gives xanthan a marked anionic character. In commercial products these functions are neutralised by the cations Na^+, K^+ or Ca^{2+}.

This conformation is confirmed by the fact that the molecules of xanthan adopt a single or double helix conformation which can join up with rigid chains which have no tendency to associate with each other. This rigid structure can be melted and can result in a less viscous disordered state. In the presence of small quantities of salt, this orderly state is stabilised and the transition temperature exceeds 100 °C. Since the useful technological properties, namely high viscosity which varies only slightly with temperature and pseudoplastic nature, are due to this structured form, it is advisable to use xanthan in the presence of electrolytes.

Dextrans are not so important from an industrial point of view. These are polymers that are synthesised by bacteria of the *Leuconostoc* type cultivated on sucrose. They consist of a chain of α-D-glucose structures bound in 1→6 with considerable connections with C_4. In addition, we can find varying proportions of 1→3, 1→4 and 1→2 bonds. Dextrans are generally soluble and used as emulsifying and stabilising agents.

In the same way, the industrial applications of **curdlan** are limited. This is an insoluble neutral polysaccharide which forms a resistant gel when heated in an aqueous suspension. It is synthesised by bacteria of the *Agrobacterium* type. It is a linear polymer of D-glucose bound in β-1→3, capable of establishing the long microfibrils which form the gel by auto-association. After curdlan is added, the volume of the product is not reduced by cooking.

Another relatively new gelling agent of microbial origin is **gellan**. This gum is the generic name for an extra-cellular polyoside produced by aerobic fermentation of *Pseudomonas elodea*. The conditions of fermentation (carbohydrate substrate, pH 7, temperature 30 °C) are fairly comparable to those required for producing xanthan gum.

This glycan can take two forms: a native form (acetylated) and a substituted form (de-acetylated). The chemical structure consists of linear macromolecules whose unit of repetition is a tetrasaccharide comprising two glucose units, one glucuronic acid and one rhamnose unit. Approximately 25% of the repetition units carry an acetyl group bound to C_6 of one of the β-D-glucose residues. Gelling takes place by means of dispersion in hot water, in the presence of monovalent or divalent ions. The gel is resistant to acid pH, heat and enzymes present in food. Gelling can take place at very low concentrations: from 0.05%. Gellan would therefore seem to have a promising future since we know that the lower the amount of gelling agent which needs to be added, the more the impact of the aroma and flavour of the product.

12.4 Other polysaccharides used as food additives

12.4.1 Gums

'Gums' is the general term used for glycan exudates of certain ligneous plants. Among the best known are arabic gum, karaya gum and tragacanth gum. They are generally used in food for their emulsifying and thickening properties.

Extracted from acacia, **arabic gum** has a linear principal chain of D-galactose units bound in 1→3 and carrying branched side chains connected to C_6 which contain galactose, arabinose and glucuronic acid. **Karaya gum** is an exudate of an Indian tree known as *Stercularia*. This is an acetylated polysaccharide characterised by its acid nature and by low solubility. It contains mainly galacturonic acid, galactose and rhamnose. **Tragacanth gum** is exuded by bushes of the *Astrogalus* type native to Asia Minor. Its structure is schematically made up of a linear chain of D-galacturonic acids bound in α-1→4 carrying linear side chains containing xylose, galactose and fucose on the C_3.

12.4.2 Galactomannans

Numerous leguminous grains contain galactomannans with a similar structure, but only guar and carob are used extensively (Table 12.6). The monomers of these glycans are β-D-mannose and α-D-galactose. α-D-galactose units are connected in 1→6 on the principal chain consisting of β-D-mannose units bound to each other by 1→4 bonds. Guar and carob are differentiated by the number of connected galactoses.

Statistically the ratio of these two monosaccharides is as follows:

- Guar: 1 galactose for 2 mannoses.
- Carob: 1 galactose for 4 mannoses.

In fact this distribution is merely a random one. Galactose substitutes, especially in the case of carob, are not distributed uniformly along the chain, but are assembled in blocks. This characteristic, as we will see, is very important in terms of properties. Although consisting exclusively of neutral monomers without any ionic substitutes, guar is soluble when cold. The presence of large quantities of galactoses connected to the principal chain of mannoses encourages the chains to move apart and allows the solvent, water, to penetrate. Carob, which possesses fewer galactoses and some zones described as 'smooth' regions, as they contain no galactoses at all, will combine in pseudo-crystalline aggregates which prevent water from entering. Heating is required to dissociate these aggregates.

These two biopolymers produce viscous solutions which behave in a pseudoplastic way. Viscosity increases with concentration and for a given solid concentration, viscosity and pseudoplasticity increase with the degree of polymerisation.

Table 12.6 Structure and properties of galactomannans

Origin	Structure	Solubility	Effect	Properties	Behaviour: Rheology	Behaviour: Temperature	Behaviour: Mechanical	Behaviour: Freezing/thawing	Behaviour: Storage environment	Applications
Guar Seeds of *Cyanopsis tetragonolobus (leguminous)* E 412	1 galactose 2 mannoses 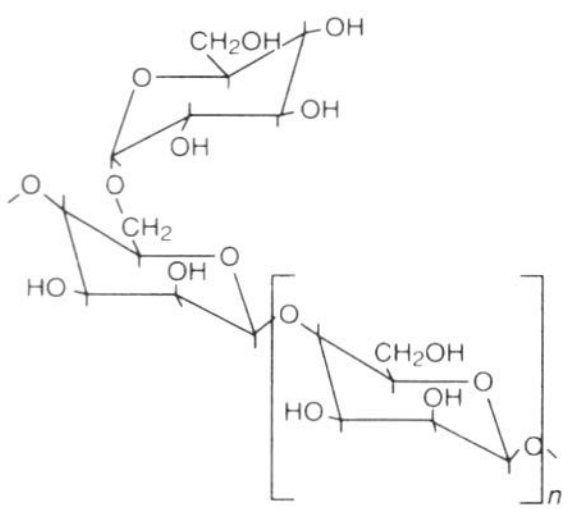 Monomer motif of the galactomannans Guar $n = 1$; Carob $n = 3$	Fully soluble when cold in water or milk.	Thickening agent.	Texture cut short by carrageenan.	Pseudo-plastic.	Heating: swells and dissolves when cold; loses part of its viscosity during severe heat treatments.	Not very sensitive.	Very good.	Very good.	Ices and ice-creams. Sauces. Pet foods. Drilling sludge. Explosives. Para-medical products.
				Reticulation to borax.	Thixotropic gel.					
Carob Seeds of *Ceratonia siliqua* (Mediterranean tree) E 410	1 galactose 4 mannoses	Total solubility at 80 °C.	Thickening agent.		Pseudo-plastic.	Cooling: viscosity depends on temperature.	Not very sensitive.	Good.	Good.	Ices and ice-creams. Pet foods. Deodorising atmosphere gels.
				Synergy with κ-carrageenan: elastic and cohesive gel.	Gel. Gel.					

In these two glycans viscosity drops significantly – although this is reversible – when temperature increases. Since the viscosity obtained with these two polysaccharides is more or less similar, the choice between them is guided by the properties which result from their differences in structure. So, for example, carob, with fewer substitutions, is not soluble when cold, but its regions without substitutes are able to associate with the polysaccharides which are found in helicoidal form in solution (xanthan, carrageenan).

12.4.3 Chitin

It may seem surprising to find chitin included in this chapter. In fact this glycan can come from plants such as fungi and yeasts, although the major source remains the shells of crustaceans.

Its structure comprises a linear chain of *N*-acetyl-D-glucosamine units (*N*-acetyl-2-amino-2-deoxy-D-glucose) bound by β-1→4 bonds. In its native form, chitin is insoluble in water and organic solvents and forms crystalline structures. It can be de-acetylated in an alkaline medium in order to produce a polymer – known as chitosan – which is still insoluble in water but whose salt form obtained by adding acid is a water-soluble cationic colloid. This polysaccharide can be used in the food industry as an emulsifier and thickening agent. Moreover, other by-products of chitin, similar to those prepared from cellulose, can be obtained, but do not have significant industrial applications at the moment.

12.5 Food utilisation of glycans

12.5.1 Thickening – Gelling

As we have just seen, the properties mainly sought after in these polymers are the viscosity or gelling power of their solutions. These properties are based on the behaviour of these macromolecules in an aqueous medium, in particular their conformation and hydrodynamic volume. From a diagrammatic point of view, three main types of conformation can be distinguished:

1 Random coil formation.
2 Rigid extended conformation.
3 Spiral conformation.

The various conformational possibilities of the principal hydrocolloids are shown in Table 12.7.

Random coil conformation corresponds to a statistical distribution within the space of the macromolecular chain. The hydrodynamic volume of the macromolecule depends on the nature of the glycosidic link or bond, the molecular mass and polymer–solvent interactions. Numerous

Table 12.7 Main types of hydrocolloid conformation

Conformation	Hydrocolloid	Conditions
Random coil	Galactomannans (guar and carob)	
	Pectins HM	
	Xanthan	High temperature
	λ-carrageenan	
	ι- and κ-carrageenan	High temperature
	Agar	High temperature
Rigid extended	Alginates	Presence of calcium
	Pectins LM	Presence of calcium
	Pectins HM	pH < 3, presence of sucrose
	Xanthan	Low temperature
Spiral	ι- and κ-carrageenan	Low temperature
	Agar	Low temperature

macromolecules can, under certain conditions – particularly temperature – adopt a random coil formation. When the flexibility of the chain is restricted, the rigid extended conformation is imposed. Finally, when stable intramolecular bonds can be established, the spiral conformation takes over.

12.5.1.1 Solubility of the hydrocolloids

This functional property is associated first of all with the properties of the grain: particle size, porosity, specific surface area, presence of impurities with varying degrees of solubility. Solubility also depends on the actual nature of the polymer. Generally speaking, glycans contain numerous hydroxyl groups which give them a marked hydrophilic nature.

However, since the characteristics of solubility can be explained by competition between solute–water and solute–solute interactions, it would be a good idea to make a distinction between:

- neutral linear molecules;
- neutral branched molecules;
- negatively charged molecules (polyelectrolytes).

Among the neutral linear biopolymers we need to be able to identify those that consist of bonds (1→4) such as cellulose and amylose which are very difficult to dissolve. The explanation of this phenomenon lies in the strong interactions between macromolecules which have very dense crystalline zones, not easily accessible by water. In dextrans, which are other types of linear glucans, the flexibility of the bond (1→6) makes them soluble in water.

Galactomannans, which are polymers with (1→4) bonds, are partially soluble in cold water because of the presence of side chains of galactose distributed all along the mannan chain. The more numerous these groups, the higher the solubility. So, for example, the solubility of guar gum is greater than that of carob in which the mannose zones without galactose substitutes can 'stack', forming pseudo-crystalline zones. These are stabilised by hydrogen bonds, thus preventing water from penetrating.

As far as the solubility of polyelectrolytes is concerned, this depends on their ionic state. In their salt form they dissolve easily whereas in their acid form they appear virtually insoluble.

To sum up, solubility can be improved by the presence of the following:

- Ionised groups (sulphates, carboxylates, etc.) which increase the hydrophilic nature.
- Branchings which move the principal chains apart and encourage hydration.

Solubility can be reduced by anything that encourages bonds between two glycan chains, such as, for example, regular regions without any branching and without any ionic substitutes, or the presence of multivalent cations such as Ca^{2+} in anionic molecules.

12.5.1.2 Rheological properties

The thickening or gelling properties of glucan solutions depend on the form adapted by the macromolecules, their rigidity and their possibilities of associating between each other. If no association takes place between the polymers, this is referred to as a pure **thickening** agent. A hydrocolloid of this type produces viscous solutions whose viscosity depends on molecular mass; at equivalent mass concentration a polymer of high mass produces a solution whose viscosity is higher than that of a solution of low mass.

In addition to molecular mass, the shape and rigidity of the macromolecule will determine the rheology of the solutions. We can identify two main types of behaviour:

1 **Very branched or globular macromolecules** occupy a very small volume and can be assimilated into spheres. As there is little to hinder the mobility of the solution, viscosity is low and the behaviour of the solutions is close to **Newtonian behaviour**. When the shearing rate (speed) of the solutions ($\dot{\gamma}$) is varied, the shearing stress (τ) is virtually proportional to $\dot{\gamma}$ and viscosity is constant whatever the shearing force. Other types of hydrocolloids can behave in this way when they are depolymerised.
2 **Unfolded macromolecules** can occupy a very large volume in comparison with that occupied by branched molecules. This is a considerable hindrance to the mobility of the solutions that contain them. Solutions of high viscosity can be obtained whose behaviour is of a **pseudoplastic** type. As the macromolecules direct themselves progressively in the direction of the shearing force, their viscosity decreases when the shearing rate (speed) increases. The ratio between τ and $\dot{\gamma}$ can be described by a power law of the following type:

$$\tau = \kappa \cdot \dot{\gamma}^n$$

where $\dot{\gamma}$ = shearing rate
τ = shearing stress
κ = consistency coefficient

This type of rheology is characteristic of pure shear-thickening agents, such as sodium alginate or the galactomannans.

In addition, unfolded macromolecules which are difficult to distort produce solutions which are even more pseudoplastic. These molecules can join up with straight chains which, beyond a critical concentration, can no longer turn when at rest. They find a position of equilibrium, stabilised by weak interactions. This behaviour is characteristic of xanthan solutions.

The formation of a three-dimensional network results in the formation of a **gel**. The macromolecules are bound to each other on restricted zones known as junction zones. The energy of the bonds in these zones and their number are responsible for the rigidity of the gel and for whether or not it is reversible.

Junction zones are most frequently obtained by assembling regular portions of macromolecules – with or without the help of a cross-linking agent such as the divalent ions – in the form of spirals or folded ribbons, whereas the irregular portions produce free fibres which break the junction zones and allow a network to be formed.

In concluding this section, we can look at the case of alginates which can be used as thickening agents or as gelling agents depending on whether or not they are in the presence of calcium ions. The alginates with the most gelling power are of course those that are the richest in homogeneous blocks of guluronic acid. The cross-linking cation must not be present at too high a concentration otherwise the colloid will go past the gelling stage and will precipitate. On the other hand, if the cation concentration is insufficient, then the solution will thicken and behave in an increasingly pseudoplastic way.

12.5.1.3 Phenomenon of synergy

Solutions of glycans are formed from one single molecular type or can be made up of the association of several polysaccharides. Because of their different and complementary properties, hydrocolloids are often used in combinations. A few examples follow.

Molecules of xanthan in their rigid helical form can combine with galactomannans via the zones without connected galactoses. So, carob gum, whose very few galactose units are distributed in a very irregular way, produces gels with xanthan whereas with guar gum there is merely an increase in viscosity. Xanthan/carob gums are heat reversible, very cohesive and maximum gel strength is obtained at a xanthan/carob ratio of around 1.

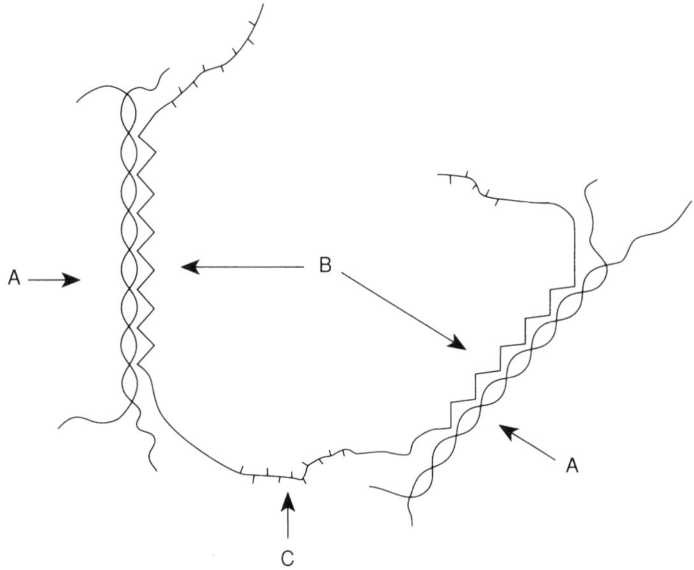

Possible interactions between the dual helices of carrageenans (A) and the 'smooth' regions of a molecule of carob (B), with the 'tufted' regions (C) acting as flexible joints.

Fig. 12.2 Mechanism for the interaction between the κ- or ι-carrageenans and the galactomannans.

Likewise, the parts of the carob macromolecule can move closer to double spirals of κ-carrageenan to form a network whose configuration is close to that of the network obtained with ι-carrageenan (Fig. 12.2). The gel obtained is elastic and transparent.

One application of this is in the stabilisation of ice-creams or sorbets by adding a mixture of carrageenan and guar gum. Likewise, in the meat processing industry, κ-carrageenan is always combined with several other hydrocolloids. This is what happens, for example, in the technique of vacuum massaging ham after injecting brine containing the gelling agent.

12.5.1.4 Interactions with proteins

By mixing solutions of proteins and glycans, it is possible to obtain three types of results:

1 A system of **two liquid phases** (water emulsion in water). This can be obtained when the macromolecular compounds are principally in different phases. This phenomenon is due to the limited thermodynamic compatibility of the proteins and glycans in an aqueous medium.
2 A **bi-phasic system** (complex by coacervation) can be obtained when the two macromolecular compounds are in the same concentrated

phase. This result is associated with the formation of an insoluble electrostatic complex between the protein and the anionic polysaccharide.

3 **Homogeneous and stable solutions** can be obtained in a bi-phasic system in which the two macromolecular ingredients do not interact, or if they do then they exist in the form of soluble compounds.

We can quote several examples. Carrageenans have a marked anionic character and are therefore likely to react with cationic polyelectrolytes such as proteins which can form a precipitate with these glycans when the pH of the solution is lower than the isoelectric point. So, for example, if we use an alkaline type gelatin, carrageenan/gelatin mixtures at a pH of less than 5.2 can be obtained.

In the case of casein, protein/carrageenan interactions take place at pHs that are higher than the isoelectric point. Gelling of carrageenan in milk is due to the interaction between carrageenan and casein and to the carrageenan molecules combining with each other. As there is an interval between the structuring temperatures of carrageenan/casein on the one hand, and carrageenan/carrageenan on the other hand, it is a good idea to move very rapidly through the temperature zone between the gelling temperatures of these two mixtures in order to avoid defects in the end-product.

Thanks to this synergy between casein and carrageenans, the quantities needed for gelling in dairy products are approximately five times lower than for gels in water (0.2% instead of approximately 1%). In milk intended for drinking, they are in the order of 0.03% instead of 0.2% in an aqueous drink.

In order to obtain simple stabilisation in dairy products such as whipped cream, λ-carrageenan can be used on its own.

12.5.2 Function of fibres in the diet

The dietary fibres described in this chapter represent a physiological unit. The mixture of these different polymers must be considered to be a whole limit which has a clearly defined action within the organism.

Understanding the dietary role of fibres entails determining their composition and their structure as well as their physico-chemical properties. In order to describe their behaviour it is essential to make a distinction between **soluble fibres** (pectins, galactomannans, etc.) and **insoluble fibres** (biopolymers from plant walls). Research into the relationship between structure, physico-chemical properties and the nutritional effects of water-soluble fibres has made a great deal of progress. On the other hand, the problem of insoluble fibres is much more difficult to tackle because of their anatomical complexity (several tissues), their chemical complexity (different types of polymers, essentially polysaccharides) and their physical complexity (several phases). In addition, a plant wall is a complex structure capable of undergoing considerable variations depending on the degree of maturity, the botanical origin and the environmental conditions, degradation, etc.

12.5.2.1 Physico-chemical properties

In the presence of water, **soluble fibres** dissolve completely. Aggregates or microgels often remain in these solutions and these interfere with their properties.

As stated previously, the solubility of glycans in water depends on the hydroxyl functions which are capable of interacting on an intramolecular level or with the water molecules. The linear chains of the regular structure assemble easily via strong intermolecular bonds whereas the presence of branches is a factor that will always increase solubility as it hinders interchain association.

The glycans that make up the ionic sites distributed over the principal chain dissociate within an aqueous phase into polyions and into counterions. This polyelectrolytic property explains why glycans are easy to dissolve when in a salt form as a result of electrostatic repulsion, and are virtually insoluble in an acid form.

Among the important properties of this type of macromolecule we can pick out the considerable increase in the viscosity of their solutions in comparison with the pure solvent, even at very low concentrations.

Insoluble fibres are characterised by their capacity for ionic exchange. This property is associated with the presence of uronic acids and, in the case of wheat bran, with phytic acids. They then behave as resins which are weak cation exchangers and, generally, monofunctional. The total capacity, expressed in milliequivalents (meq) per gram of dry product, depends on their origin, their maturity and the processes employed when extracting them or incorporating them in foodstuffs. This capacity varies between 0.3 and 3.1 meq/g. It is higher for vegetables than for cereals. It is important to know this value for clinical application of fibres, for example, in order to control their metal sequestering power. However, relations between the ion exchange capacity and the absorption of electrolytes are difficult to establish because this capacity must be measured during transit in order to take into account degradation and fermentation.

The fibres, whether or not they are water-soluble, can absorb and hold water. When a fibre swells in water this involves the water held in the interstitial structures and the water bound by hydrogen bonds or dipolar interactions with the chemical groups carried by the fibres. Their water-holding capacity closely depends on their composition (hydrophilic or hydrophobic polymers), the physico-chemical conditions of the medium (pH, ionic strength, temperature) and the method of preparation (hydrothermal processing, method of drying). This hydration capacity is extremely important in terms of nutritional physiology since it is involved in gastric elimination, the weight of stools formed and fermentation.

This quick reminder of the principal physico-chemical properties involved in the nutritional effects of fibres shows the importance of macroscopic aspects of these ingredients. The two essential characteristics are its particle size and porosity. Consequently, numerous tests for modifying the properties of fibres – essentially by means of mechanical and enzymatic processes – have been carried out. So, for example, enzymatic processes can result in partial solubilisation, but also in a certain amount of dietary fibre loss. Although these processes generally reduce physico-chemical properties they give rise to changes in the texture of the fibres which improve their sensory properties.

12.5.2.2 Role in human nutrition

Although at the moment there is no clearly established proof, fibres play a part in protecting against certain diseases (colonic diverticulitis, cancer of the colon and rectum, abdominal and hiatus hernia, haemorrhoids and varicose veins, coronary and metabolic disease).

Currently the mechanisms of action are not established, but it seems highly likely that it is the physico-chemical properties of fibres that are important rather than their chemical composition. Under certain conditions, the fibres act on intestinal transit and are involved in the carbohydrate and lipid metabolism. However, in all cases, the sites of direct action are digestive ones, because these molecules are not metabolised by the enzymes in the human digestive tract. Some are partly digested by the enzymes of the bacteria from the microflora in the colon.

As a result of their viscous properties, water-soluble fibres have an effect on carbohydrate metabolism by reducing the absorption capacity of the carbohydrates ingested. The metabolic effect of insoluble fibres is clear after ingestion over a sufficiently long period. They appear to modify the secretions of gastro-intestinal hormones and the mechanisms that regulate the carbohydrate cellular metabolism.

In the same way, quantitative and qualitative modifications in dietary lipid assimilation may explain the principal effects of these ingredients on lipid metabolism. For example, numerous fibres have the effect of reducing cholesterolaemia. For this reason, the study of fibre adsorption properties has mainly concerned biliary acids.

13

Lipid chemistry – fat substitutes

The raw materials used in lipid chemistry are animal fats and vegetable oils. The intermediate products involved are the fatty acids, esters, fatty alcohols and fatty amines.

There is very little point in making distinctions between solid fats and liquid oils according to their physical state at 20 °C. The term 'fat' tends to be used to cover the whole range. An 'edible' or 'nutrient' fat is usually a mixture of animal and/or vegetable fats. Nutrient fats are essentially composed of triglycerides, which easily represent the dominant fraction. These apolar lipids are accompanied by phospholipids and sterols. The composition in fatty acids varies considerably according to the origin of the fats used and the modification processes they have undergone.

The fields of application covered by lipid chemistry are enormous. Some examples are as follows:

- Detergents, which enhance the value of soaps, sulphates and sulphonates of fatty alcohol, and amines.
- Cosmetics, which use by-products of fatty acids as emulsifiers, and surfactants.
- Paints, which contain products such as alkaline resins based on glycerol and fatty acids.
- Plasticisers which also use the properties of unsaturated fatty acids and epoxidised oils.

The scope and volume of this work allow us to consider only the dietary applications of these products. So, after a brief reference to fat crystallisation, we will describe the structures, properties and utilisation of fatty acids, glycerides and phospholipids, as well as their by-products. Given their importance from an industrial point of view, special attention will be given to the physico-chemical and functional properties of the amphiphilic molecules. At the end of the chapter, the different types of fat substitutes will be examined.

13.1 Lipid crystallisation

The temperature of lipid phase crystallisation depends on lipid composition. It increases with the length of the chains and the saturation level of the fatty acids. The length of the chains increases the van der Waals bonds, and therefore the stability of the crystals. In addition, the carbon chain is bent at the unsaturation points of the fatty acids, resulting in the inhibition of molecules. This reduces the intermolecular bonds and, as a consequence, the melting point. What is more, the triglycerides that form the dominant fraction in almost all fats used in foodstuffs have a high spatial requirement and can take on several ordered arrangements, and therefore several crystallisation forms (Fig. 13.1).

An initial form known as α which is easy and quick to obtain is crystallisation according to an aligned-face hexagonal system, that is to say as a straight prism with a hexagonal base, which leaves the carbon chains with a poorly defined layout, and wide amplitude of movement.

These crystals develop spontaneously into a more organised form, with a stronger intermolecular bond. They are therefore denser and have a higher melting point. This form, known as β′, contains *ortho*-rhombic crystals, i.e. straight-sided prisms with a lozenge-shaped base.

Subsequently, the molecular rearrangements continue very slowly, enlarging the few crystals in β form that had initially formed. These crystals correspond to the densest possible organisation for this type of molecule. Their system is tri-clinical with an oblique prism having a base in the shape of a parallelogram. The increase in melting point which accompanies these transcrystallisations is considerable: for a triglyceride constituted by fatty acids (FA) in C_{10}, it rises from −10 to +17 and then to +32 °C for the α, β′ and β forms respectively.

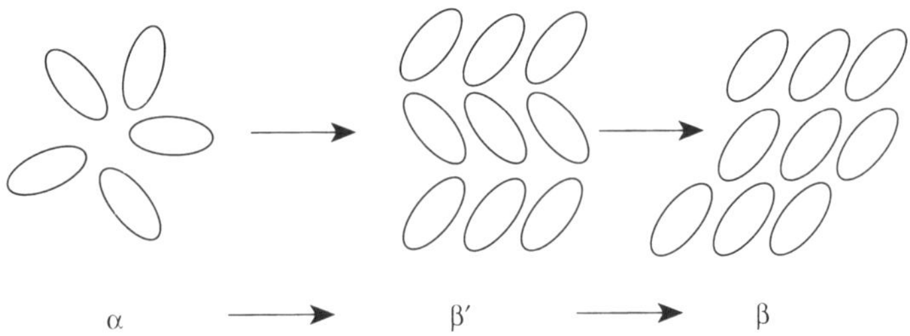

Fig. 13.1 Shapes of lipid crystallisation.

The phenomenon of retraction which accompanies the increase in density of the crystals takes the form of a modification in the structure of the fat which becomes granular over time. In order to avoid this type of texture, the fats are tempered. This involves mechanical action to multiply the crystal nuclei. Tempering consists of reheating the fat beyond the melting point of the $\beta^{1\prime}$ form – but below that of the β form – and then cooling it to increase the quantity crystallised in the β form.

As most natural fats are mixtures of triglycerides with widely differing melting points, the phenomena we have just described become complicated for two main reasons. Firstly, eutectic crystals form after the intercrystallisation of several glycerides with different melting points, and secondly, phenomena of absorption of the remaining liquid phase are observed on the surface of the crystals.

13.2 Fatty acids

As the main constituents of glycerides in terms of mass, the known fatty acids are extremely numerous, especially within the plant kingdom.

13.2.1 Structure

Fatty acids all have the COOH group at the end of:

- a non-polar side chain:
  - saturated with a linear, branched or cyclic structure;
  - unsaturated: monoenes, polyenes, both conjugated and unconjugated;
- a polar side chain: alcohol acids, ketonic acids.

13.2.2 Predominance and proportions

The major fatty acids are extremely widespread and, generally speaking, present in large quantities. These fatty acids are not very numerous but alone represent almost 95% of the fatty acids present in the oils and fats used in foodstuffs and industry.

Some of the most abundant are:

- palmitic acid, saturated in C_{16};
  $CH_3—(CH_2)_{14}—COOH$
- stearic acid, saturated in C_{18};
  $CH_3—(CH_2)_{16}—COOH$
- oleic acid, monoene in C_{18} (*cis* form);
  $CH_3—(CH_2)_7—CH{=}CH—(CH_2)_7—COOH$
- linoleic acid, diene in C_{18} (*cis* form)

$$CH_3—(CH_2)_4—CH=CH—CH_2—CH=CH—(CH_2)_7—COOH$$

The proportion of fatty acids present in fats in foods can vary considerably. Table 4.7 shows the average composition of the principal plant fats. We can see, on one hand, that there are ranges of the same order with regard to all the saturated and unsaturated acids, and secondly that the oils are very rich in unsaturated acids (in *cis* form) whereas fats are richer in saturated acids.

13.2.3 Physical properties

The **melting point** is a fundamental physical characteristic. It is more or less the same for fatty acids and for the saturated homogeneous triglyceride (the same acid multiplied by three).

The important points are as follows. For a given form, the melting point of the fatty acid rises as the length of the hydrocarbon chain increases. In the series of saturated acids, for example, the rise varies by between 6.5 and 9.5 °C with every two atoms of carbon:

lauric acid (C_{12}):	44.3 °C
myristic acid (C_{14}):	53.9 °C
palmitic acid (C_{16}):	63.1 °C
stearic acid (C_{18}):	69.6 °C
arachidic acid (C_{20}):	76.5 °C
lignoceric acid (C_{24}):	86.0 °C

For a given length of chain, the melting point drops with the number of double bonds: the drop is greater for the *cis* form than for the *trans* form (rare):

$C_{18:0}$ stearic acid	69.6 °C
$C_{18:1}$ vaccenic acid (*trans*)	44.0 °C
$C_{18:1}$ oleic acid (*cis*)	13.4 °C
$C_{18:2}$ linoleic acid (*cis*)	–5.0 °C
$C_{18:3}$ linolenic acid (*cis*)	–11.0 °C

The *trans* configuration retains more or less the same layout as the carbon chain whereas the *cis* configuration introduces a bend of approximately 30°.

As a result of the presence of the carboxylic group and a hydrophobic carbon chain, the **structure** is **amphiphilic**. Since hydrophobicity increases with a rise in the C number while COOH dissociation decreases, only the short chain (C_4 to C_{10}) fatty acids are known as volatile, that is to say, they are steam-distillable and only the first two (C_4 and C_6) are soluble in water. For the other fatty acids, the properties associated with hydrophobicity continue to increase, in particular insolubility in water.

13.2.4 Hydrogenation

Hydrogenation consists of saturating all or some of the double bonds of non-saturated fatty acids with hydrogen.

In the case of **selective hydrogenation**, the linolenic acid content of certain vegetable oils will be reduced in preference to the linoleic acid content and their stability to oxidation will be increased. This technique is used for oils that are particularly sensitive to oxidation and heat and that contain, as in the case of rape and soya, a relatively high percentage of linolenic acid. This process causes a degree of isomerisation of the unsaturated fatty acids, and, because of this, favours the formation of geometric (*trans*) and position isomers. Selecting good experimental conditions (catalyst, temperature, pressure and agitation conditions) can limit the production of the isomers:

linolenic acid →	linoleic (*cis*)	→	oleic acid (*cis*)	→	stearic acid
	isomers of linolenic and linoleic acid		monoene position isomers		
			elaïdic acid (*trans*)		

The purpose of **non-selective hydrogenation** is to prepare solid fats that can be used in the manufacture of margarine, for example. This type of hydrogenation aims to saturate a high proportion, or sometimes even all of the double bonds of unsaturated fatty acids. This type of reaction also encourages the formation of isomers, in particular *trans*-isomers, which helps to raise the melting point even higher.

In industrial practice, hydrogenation is carried out by passing very pure hydrogen in the presence of a catalyst (copper or nickel salt) over the fat brought to between 150 and 200 °C. Parameters such as temperature, pressure, speed of hydrogen injection, and the nature and concentration of the catalyst, vary according to the type of hydrogenation. As unsaturation is defined as the measurement of the iodine index which corresponds to the number of grams of iodine taken up by 100 g of fat, the result of hydrogenation can be expressed by the reduction (ΔI) in this index.

The reaction is heterogeneous since the system comprises three phases: gas (hydrogen), liquid (fat to be hydrogenated) and solid (catalyst). It is exothermic and releases something in the order of between 100 and 150 kJ per mole and per double bond. Consequently, a drop of one unit of iodine index will raise the temperature of the oil in the reactor by between 1.6 and 1.7 units, hence the reactor of a heating/cooling system needs to be equipped with a temperature regulator.

Although hydrogenation stabilises the fat against oxidation and modifies its behaviour during crystallisation, it does, however, cause a drop in nutritional value because of the conversion of some or all of the linoleic and linolenic acids.

13.3 Glycerides

Triglycerides predominate within the category of apolar lipids. Di- and monoglycerides are generally present as no more than 2% of the lipid total. In order to obtain mixtures of triglycerides with well-defined functional properties, trans-esterification or fractionation must be used.

The purpose of **trans-esterification** and **inter-esterification** is to modify the glyceridic structure of the fats by means of intra- and intermolecular rearrangement of the fatty acids on the glycerol. This molecular rearrangement can take place either on one single or on a mixture of two fats. As a result, numerous combinations of trans-esterification are possible. Hilditch has shown that heterogeneous triglycerides are the most numerous. He considers that a fatty acid that represents a proportion of less than a third of the total quantity of fatty acids will only be present once per triglyceride; when the proportion is between one-third and two-thirds it can be found twice in one triglyceride; beyond two-thirds it may be found three times, resulting in homogeneous triglycerides. This conformation thus explains the existence of a maximum of heterogeneous triglycerides in natural fats. The Hilditch laws are good approximations. There are others, such as, for example, Kartha's random distribution.

It is therefore possible to predict the composition, in terms of various triglycerides, of the final mixture at equilibrium. This is the result obtained in the case of **random trans-esterification**. So, this technique of **non-directed trans-esterification** allows us to prepare a margarine using fully hydrogenated sunflower oil or palm oil which has the advantage of not containing *trans* fatty acids.

In the case of directed trans-esterification, if the temperature is lowered, for example, the change of state of a triacylglycerol from the liquid phase into the solid phase is encouraged; in this way equilibrium can be altered in the liquid phase and gradually the desired mixture is produced.

In addition, the trans-esterification which takes place in the presence of catalysts such as sodium ethylate or sodium methylate at a temperature of between 100 and 160 °C can – after changing the position of fatty acids on the glycerol – modify the digestibility of the triglyceride.

In addition, as the lipase hydrolysis reaction is reversible, researchers have adjusted the displacement of the chemical equilibrium during enzymatic synthesis within a biphasic medium (aqueous phase/organic phase). Thus, if we incubate one or several triglycerides and one or several free fatty acids in the presence of a lipase, an exchange takes place between the free fatty acids until equilibrium is achieved, with the fatty acid or acids being distributed at random on the glycerol. However, with a specific (1–3) lipase only positions 1 and 3 are modified. It is also possible to incubate a mixture of different triglycerides in the presence of a lipase. The fatty acids exchange from one triglyceride to another until equilibrium is achieved. This is **inter-esterification**. Once again a specific lipase allows the desired triglycerides to be selected. In addition, the products associated with the reaction are much fewer than those produced during chemical trans-esterification.

Finally, triglycerides with medium length chains (MCTGs) present a different metabolic pathway in the organism from that of long chain triglycerides (LCTG) of 16 to 18C or more.

LCTGs are hydrolysed in the intestinal tract, then re-esterified in the intestinal cells. After passing into the lymph ducts, these triglycerides are found once again in the general blood circulation, including in the chylomicrons. These complexes, which also contain apoproteins, phospholipids and cholesterol, finally end up in the liver where the various constituents are metabolised.

The MCTGs cross the intestinal barrier in the form of triglycerides and/or free fatty acids and, on exiting, go directly to the liver via the special circuit of the portae venae where most of them are oxidised to produce energy. As a consequence, these triglycerides have very little involvement in adipose tissue deposits. The speed of MCTG absorption–oxidation is close to that of glucose but their energy value is approximately twice as high.

The MCTGs do not provide a sufficient or balanced lipid supply, as the essential fatty acids (EFAs) are missing. This led to the concept of inter-esterifying MCTGs and LCTGs that are rich in EFAs, in order to obtain 'structured' lipids. Currently manufactured and approved in the USA, structured lipids resulting from the inter-esterification of copra oil, a natural oil very rich in medium length chain fatty acids, and various fats rich in EFAs (soya oil, canola, fish) can be obtained. These products, which are in a liquid form, can act as the lipid phase in numerous preparations: dressings, mayonnaises, margarines, cakes, sweet pastries, etc. If these lipids are incorporated then foods retain the taste and texture to which consumers are accustomed. However, our knowledge of the long-term effects of using this type of lipid is still limited. Additional research is necessary.

13.3.1 Producing monoglycerides

Monoglycerides are an important category of emulsifying agents for use in foodstuffs. Various processes can be used to manufacture monoglycerides. The most popular one is based on a trans-esterification reaction between triglycerides and glycerol (Fig. 13.2). The raw materials most frequently used are lard, suet, cotton or sunflower oil. In general, these fats have been hydrogenated in advance in order to saturate the fatty acids. Depending on the molar ratios and the reaction temperature, the mixture obtained contains varying quantities of monoesters, which represent the fraction with the desired surfactant properties.

The purification process is based on distilling the monoglyceride at a temperature in the order of 200 °C under vacuum of lower than 0.1 mm Hg.

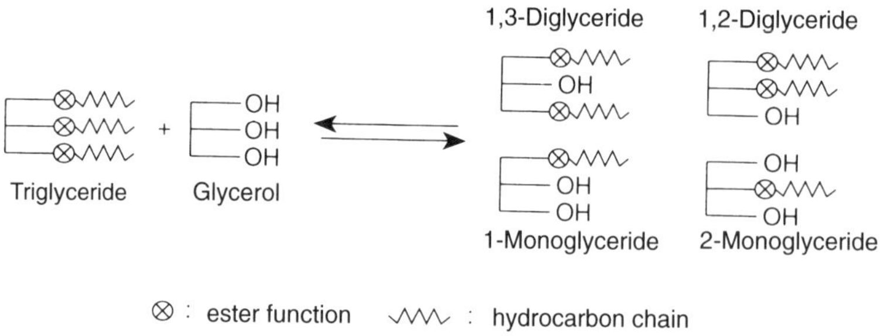

Fig. 13.2 Diagrammatic representation of the reaction of trans-esterification in the manufacture of mono-diglycerides.

The monoglycerides are more volatile than di- and triesters and are recovered after evaporation–condensation. The product obtained can contain up to 95% monoesters and, because of this purity, distilled monoglycerides are used increasingly.

13.3.2 Principal types of monoglycerides and by-products

It is possible to obtain acid by-products of monoglycerides from distilled monoglycerides. The most popular are the by-products obtained from various organic acids: citric, lactic, acetic, diacetyltartaric (DATEM) (Fig. 13.3). These monoglyceride by-products are used to improve the stability of emulsions.

DATEMs are known for their useful properties in products used in bread-making, where they act as dough-packing agents and as agents for controlling the texture of the end-product. Citroglycerides are used in the margarine industry as antioxidants. These citric by-products also allow the antioxidants in fats to dissolve and thus increase their anti-oxidising efficacy. Succinylated by-products are used in fats and as dough-packing agents in the cereal products used in cooking. Acetoglycerides are emulsifiers which are particularly useful in stabilising foams. When they are fully hydrogenated, the acetoglycerides have the special ability to form highly flexible plastic solids or transparent films. In this way they can be used as agents for coating various food products (fruits, meat products) so that these products can be protected from oxygen and from atmospheric moisture.

Sucroesters are esters of fatty acids obtained by the direct esterification of sucrose by methyl esters of fatty acids, whereas sucroglycerides are mixtures of monoglycerides and sucroesters obtained by trans-esterification of sucrose and triglycerides.

There is a wide range of sucroesters obtained by means of varying degrees of esterification, with the consequent range of emulsifying properties (see Chapter 10). Sucroesters are relatively hydrophilic mixtures so they are easily dispersible in water. Their taste is less pronounced than that

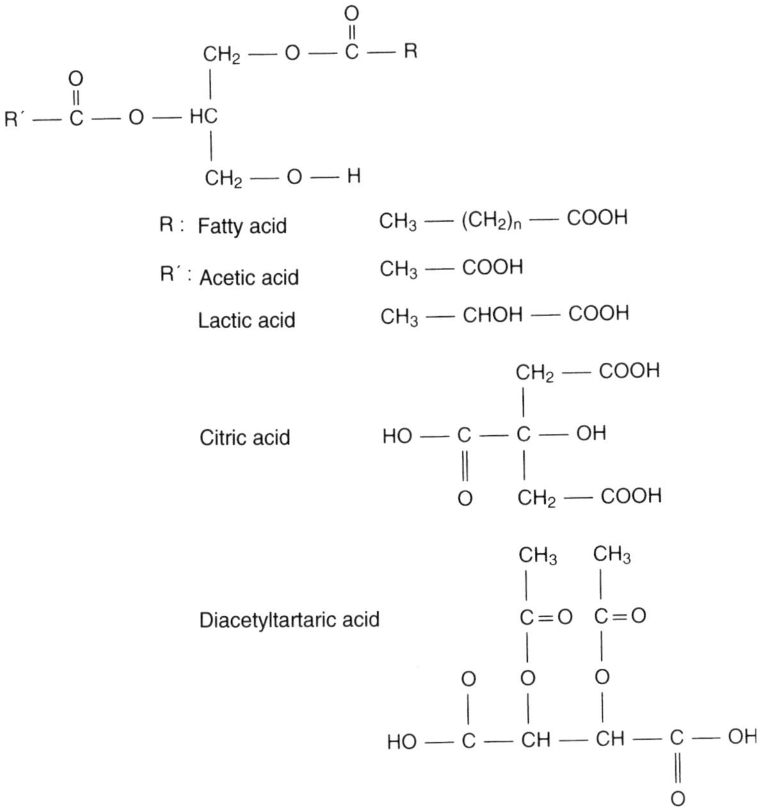

Fig. 13.3 Chemical structure of monoglycerides esterified by various organic acids.

of the monoglycerides and, in addition, they are easily hydrolysed by heat. Theoretical possibilities for use are numerous: ice-creams, cooking products, margarines, emulsified sauces.

Esters of propylene glycol are often used in mixtures with distilled monoglycerides and in particular as a swelling agent in the biscuit and cake-making industry. These esters consist of a mixture of mono- and diesters of propane-1,2-diol of fatty acids obtained from food oils and fats. As the chemical structure of the polyol has two alcohol functions instead of the three in glycerol, these molecules are less hydrophilic than the monoglycerides of fatty acids.

13.3.3 Polyglycerol esters of fatty acids

When hot and in the presence of a catalyst, glycerol dehydration under vacuum results in a mixture of polymerised glycerol. By fixing the reaction conditions, it is possible to produce a mixture of di-, tri- and tetraglycerol and, by acting on the nature of the fatty acids that esterify the alcohol functions, a whole range of products that are hydrophilic to a greater or lesser degree. These emulsifying agents can be used to prepare emulsions of water in oil, to modify the rheological characteristics of melted chocolate – together with the lecithin – or even to prepare products that can give margarine and desserts a 'glassy' quality.

13.3.4 Esters of sorbitans (Spans) and polysorbates (Tweens)

Sorbitan is a by-product of sorbitol obtained by forming an oxygen bridge between the carbons 3 and 6 (Fig. 13.4). The sorbitan esters are non-ionic surfactants, obtained from the esterification of sorbitol with fatty acids of natural origin (copra, olein, suet, stearin).

They can be hydrolysed enzymatically or chemically and show good tolerance with oral LD-50 (lethal dose) values of over 25 g/kg in rats and are considered to be non-toxic. Their RDA is fixed at 25 mg/kg. Their principal physico-chemical properties are shown in Table 13.1.

The use of sorbitan stearate can improve the swelling and firmness of foams by controlling syneresis. Sorbitan tristearate allows very good control of fat crystallisation in chocolates, margarines, ice-creams and cakes.

Polysorbates are non-ionic surfactants, obtained by the ethoxylation of sorbitan esters. They are hydrophilic emulsifiers with HLBs of between 12 and 18. Their RDA is fixed at 10 mg/kg and their physico-chemical characteristics are summarised in Table 13.1. The use of polysorbates is recommended for maintaining starch substances in suspension – thus avoiding starch retrogradation and slowing down the rate at which bread goes stale – and for dissolving aromas or distributing colourants in a homogeneous way.

Sorbitan esters and polysorbates are effective emulsifiers and stabilisers which are often combined to obtain optimum hydrophilic–lipophilic

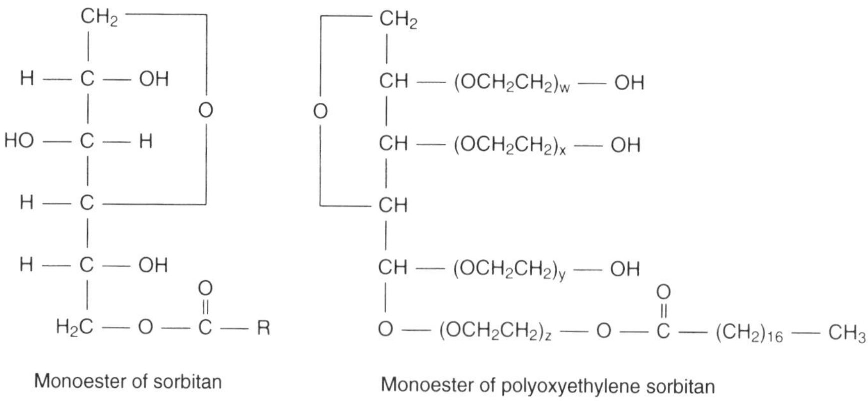

Fig. 13.4 Structure of the by-products of sorbitan.

Table 13.1 Principal physico-chemical properties of the esters of sorbitans and polysorbates

Chemical nature	Appearance at 20 °C	HLB	Maximum acid index	Saponification index	Hydroxyl index
Monostearate sorbitan	Scaly	4.7	10	147–157	235–260
Tristearate sorbitan	Solid	2.1	15	175–190	65–80
Mono-oleate sorbitan	Liquid	4.3	6	145–160	190–215
Polysorbate 20	Liquid	16.7	1	40–51	97–113
Polysorbate 40	Liquid	15.6	1	41–52	90–105
Polysorbate 60	Pasty	14.9	1	45–55	81–96
Polysorbate 65	Pasty	10.5	2	88–98	44–60
Polysorbate 80	Liquid	15.0	1	45–55	65–80

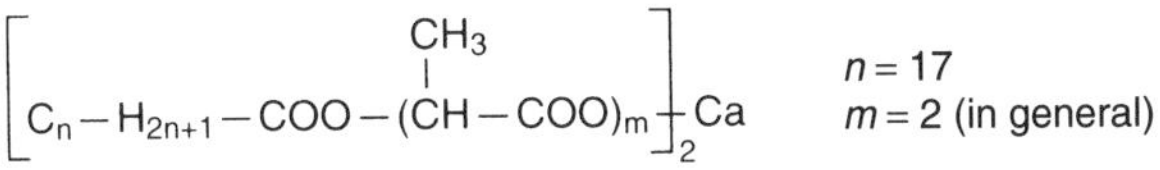

Fig. 13.5 Chemical structure of the stearyl-lactylates of calcium.

balance and to stabilise food ingredients in a dough or in a liquid lipid phase (creams, sauces). They are used as technological aids, as agents for dissolving lipo-soluble vitamins in water, or even as emulsifiers within the pharmaceutical, cosmetic and phytosanitary sectors.

These additives are authorised in numerous countries (including the USA and the UK), but not in France.

13.3.5 By-products of lactic acid

The by-products of lactic acid are the sodium or calcium stearyl-lactylates. They are obtained by esterification of the polymerised lactic acid, usually in a dimer form, by stearic acid (Fig. 13.5). These surfactant agents are mainly used in bread-making in order to improve the rheological characteristics of the dough, the texture of the crumbs, the volume of the bread and to extend its shelf life.

13.4 Phospholipids

Legislation and the trade refer to the mixtures of various polar lipids (essentially phospholipids and glycolipids), accompanied by varying quantities of triglycerides, as lecithin.

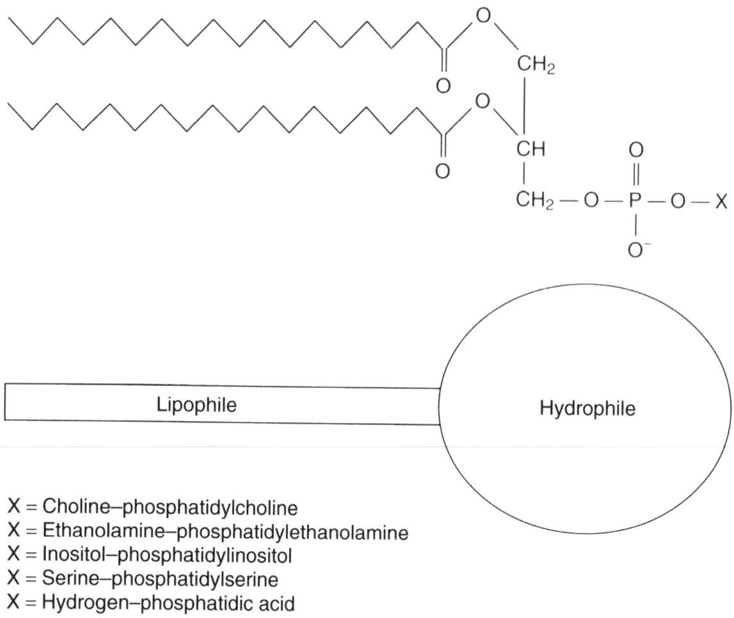

Fig. 13.6 Chemical structure of lecithin.

Phospholipids are diesters of fatty acids and *S-n*-glycerol-3-phosphate in which the phosphate group is also esterified by choline, ethanolamine, inositol or serine. The composition of the phospholipids depends on the origin of the lecithins. So, non-fractionated soya lecithin contains approximately 22% of phosphatidiyl-choline whereas lecithin from eggs contains approximately 70% of this molecule. The phospholipids from these two sources can also be differentiated by their essential fatty acid content. Phospholipids of soya have a much higher content than those from eggs (almost two-thirds of the total fatty acids as opposed to one-fifth).

13.4.1 Natural and synthetic lecithins

It is not very economical to extract lecithins from egg yolk. Commercial lecithins are usually extracted by solvent from soya oil which contains a fairly high proportion of this (between 2 and 3%). In fact, current commercial products are mixtures of phospholipids (approximately 50%), triglycerides (35%) and glycolipids (approximately 10%).

The principal active ingredients of lecithin are phospholipids, containing phosphatidylcholine (PC), phosphatidylethanolamine (PE) and phosphatidylinositol (PI) in virtually equal proportions (Fig. 13.6). PC has oil/water emulsion stabilising properties whereas phosphatidylethanolamine and, to a lesser degree, phosphatidylinositol have water/oil

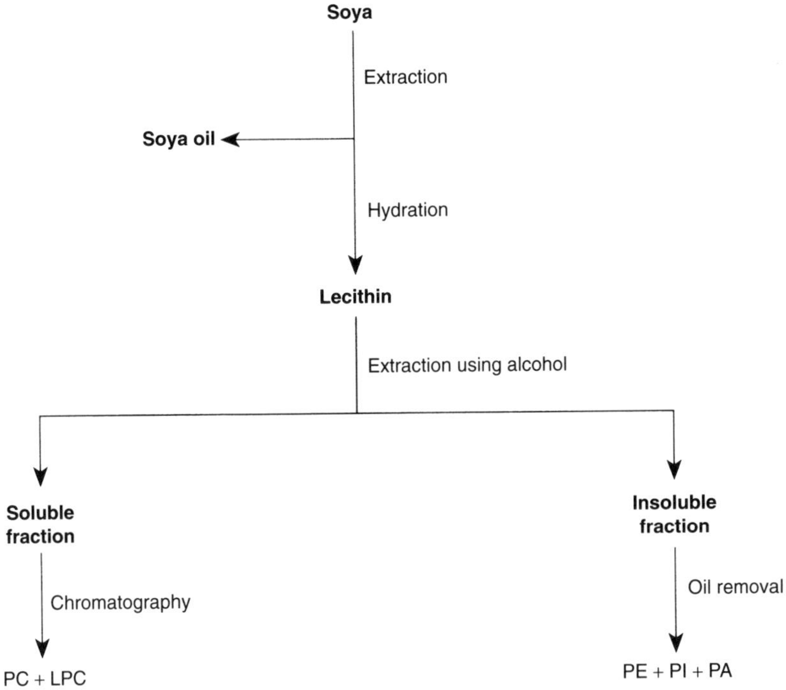

PC : Phosphatidylcholine
LPC : Lysophosphatidylcholine
PE : Phosphatidylethanolamine
PI : Phosphatidylinositol
PA : Phosphatidic acid

Fig. 13.7 Fractionation of soya phospholipids.

emulsifying properties. As a result of this antagonism, the basic mixture has relatively limited emulsifying properties.

Procedures for fractionating soya phospholipids have been developed (Fig. 13.7). Two fractions are separated, one which is soluble in the aqueous alcohol phase with predominance of phosphatidylcholine and lysophosphatidylcholine, and another which is insoluble, with predominance of phosphatidyl-ethanolamine or -inositol and phosphatidic acid.

Lysophospholipids are generated from phospholipids by means of the action of an A2 type phospholipase which releases the fatty acids from the position 2 of *S-n*-glycerol occupied by preference by polyunsaturated fatty acids.

From hydrogenated colza oil, after glycerolysis, phosphorylation and neutralisation of the acid obtained by ammonia gas, a mixture comprising triglycerides (40%), neutral phospholipids (15%), ammonia salts of mixed phosphatidic acids (40%) and ammonia salts of phosphoric acid (5%) is obtained. This mixture is used in numerous countries – as an emulsifying agent. Known as lecithin YN, the mixture has the advantage of not having an unpleasant taste, of being a standardised industrial product and of being more effective than soya lecithins.

In fact, the chemical modifications that can be used to improve the properties of the lecithins are numerous (hydrogenation, acetylation, succinylation, phosphorylation, etc.), but the commercial food products are principally hydroxylated and acetylated lecithins. Hydroxylation is carried out in the presence of hydrogen peroxide and increases the hydrophilic nature of the lecithins, while bleaching them. These lecithins have an emulsifying power which is higher than that of natural lecithins but they have lost all their nutritional value (oxidation of the carotenoids). Acetylation, which mainly involves phosphatidylethanolamine, enriches the lecithins with anionic and more hydrophilic lipids and increases their emulsifying properties. Acetylation is carried out while the oils are being refined by adding acetic anhydride. It has been proved that acylation of the amino function of the phosphatidylethanolamines with fatty acids of various chain lengths or various degrees of unsaturation allows molecules with different physical properties to be generated. Finally, the unsaturated fatty acids of lecithins can also be hydrogenated in the presence of hydrogen and a catalyst (nickel, palladium). These hydrogenated lecithins are apparently excellent stabilisers of fats in chocolate (anti-bloom agent).

Finally, certain manufacturers use buttermilk powders. The composition of these products from the dairy industry is very different from that of plant lecithins, firstly because of the presence of proteins (20–30%) and free carbohydrates (40–50%), and secondly because of their low phospholipid content (2–10%). In spite of this reduced content, buttermilk powders can have surfactant properties which are reinforced by the presence of proteins.

13.4.2 Uses for lecithins

Because of the bipolarity of its molecular structure (hydrophilic and lipophilic poles), lecithin is able to lower surface tension and to stabilise emulsions. There is no general rule about the use of lecithins because their method of action is complex. Depending on the desired aim, the right type of lecithin can be selected from the large range of products available on the market.

The food uses of phospholipids are numerous. In the margarine industry lecithin is generally used in combination with mono-diglycerides in order to maintain the water/oil emulsion (20/80) and to act as an anti-splashing agent. In the chocolate industry it is well known that adding lecithin can modify the rheological characteristics of the melted chocolate, and thus economies can be made with other costly fats such as cocoa butter. It is interesting to note that beyond a certain level of incorporation, the lipid additive causes a reversal of the viscosity curve. The current trend is to use it in association with ammonium phosphatides which do not trigger this phenomenon and are more neutral from an organoleptic point of view.

In the bread and biscuit-making industries the action of lecithin, often in combination with other emulsifiers, is involved in several reactions:

- Increase in the water-holding capacity of the dough and the volume of the end-product.
- Stabilisation of gluten proteins.
- Improvement of distribution of fats.
- Solving problems of sticking and breaking.
- Elimination or slowing down of the phenomenon of starch retrogradation.

One of the other applications is the use of lecithins for coating dehydrated products rich in lipids. In this way, the tendency to form lumps is inhibited and the products will rapidly absorb water and thus sink and disperse easily in the liquid in which they are to be reconstituted.

Finally lecithins have a nutritional advantage in the human and animal diet since they supply essential fatty acids, phosphatidylcholine and improve digestibility. Phospholipids are also extremely important in the molecular structure and biochemical function of all the biological membranes. Because of its action on the metabolism of cholesterol and the biosynthesis of acetylcholine, phosphatidylcholine is an active constituent in the pharmaceutical industry. However, phospholipids are also widely used as aids in the pharmaceutical industry as emulsifiers for perfusion solutions and in the manufacture of liposomes (Fig. 13.8).

13.5 Characteristics and functions of emulsifiers

Emulsifying agents are amphiphilic compounds whose chemical structure has both hydrophilic and hydrophobic functions. This special chemical structure gives them their emulsifying properties since by locating themselves at the oil/water interface these molecules can help to stabilise a thermodynamically unstable system. It is clear that certain proteins can also be considered to be emulsifying agents (see Chapter 2).

However, this functional definition is restrictive as, in fact, the role of emulsifying agents goes far beyond their capacity to stabilise emulsions. The amphiphilic structure gives these molecules unique physical properties which allow them to form complexes with biopolymers (starch, proteins, etc.), and to control the crystallisation of fats. These properties have important consequences for both the texture and preservation of foods and on the 'workability' of raw materials used in the composition of foods.

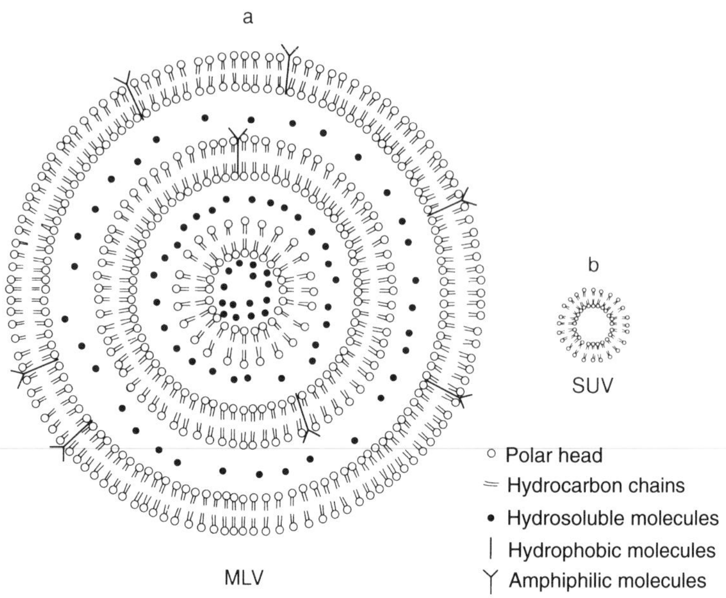

Fig. 13.8 Diagrammatic representation of multilamellar vesicle (MLV) and small unilamellar vesicle (SUV) liposomes.

13.5.1 Physico-chemical properties

13.5.1.1 Solubility

The effectiveness of an emulsifier is above all linked to its solubility in each of the two phases. The solubility of any surfactant agent is characterised by its hydrophilic lipophilic balance (HLB) (see Section 2.4.2.1).

The HLB values of emulsifiers fall within a range of 1 to 20. The lower the value, the more lipophilic the emulsifier and vice versa. Generally speaking, emulsifiers with an HLB of between 1 and 6 stabilise water/oil emulsions whereas those whose HLB is between 12 and 18 stabilise oil/water emulsions. We can also make an approximate estimation of the HLB of a mixture of emulsifiers by calculating the arithmetic average of the HLBs of each constituent.

Knowing this index allows an initial choice of emulsifier to be made, but it does not allow the effectiveness of a particular emulsifying mixture in stabilising a given emulsion to be predicted accurately. In fact, the HLB measurement does not take into consideration a number of parameters such as hydration of the polar part and interactions with the other compounds present in the food (ions, other polar lipids, proteins, etc.) which will frequently drastically alter the physico-chemical properties of the emulsifying agent. In addition, the HLB value will not be modified by the introduction of a double bond in the fat chain (oleate and stearate, for example) when their properties are actually very different. So, two emulsifying agents with the same HLB will not necessarily have the same functional properties.

Hydration of the polar head is also a factor rarely taken into consideration in terms of the use of emulsifying agents, whereas it does make an important contribution to the organisation and rheological properties of the interfacial films.

13.5.1.2 Physical state and mesomorphism

The effectiveness of an emulsifying agent in the formulation of a food product depends on the physical state in which it is found. We know that this physical state depends on chemical structure and on two essential parameters, temperature and water content, which determine the organisation of lipid chains in relation to one another. Thus, for each type of emulsifying agent, we can describe different characteristic structures, known as **mesophases**.

In their crystalline form, in the absence of water, amphiphilic molecules are crystallised in the 'a' form (Fig. 13.9a), the hydrocarbon chains are arranged in parallel with each other and the polar groups are located side by side. When in contact with water, and at a temperature in excess of the Krafft temperature (T_c), the hydrocarbon chains have better mobility and water can penetrate between the polar layers (Fig. 13.9b). This structure is a liquid-crystalline **mesophase** of a **lamellar** type and the properties of this mesophase largely determine the conditions for use of the emulsifiers.

This lamellar mesophase, also called the 'NEAT' phase, is characterised by the respective thicknesses of the lipid and aqueous phases, d_a and d_w, which are 3.8 and 1.6 nm respectively. On cooling, to a temperature below the Krafft temperature, a recrystallisation of the lipid chains takes place, although the layered structure is retained: this is the 'gel a' phase in which d_a is slightly increased (Fig. 13.9c). This 'gel a' phase is metastable and quickly moves into the more stable 'coagel with water' state. The lamellar mesophase and the crystalline 'gel a' state are generally the most useful from the point of view of the physical state of the emulsifier when being used.

In the case of a high water content, the 'NEAT' phase can be retained in two different ways. There is either unlimited swelling of the lamellar mesophase, or relatively limited swelling capacity, and a dispersion is formed. This dispersion consists of birefringent droplets whose internal construction is of the lamellar mesophase type. This phase is formed spontaneously by numerous phospholipids in suspension in water and the structures formed are sacs in which the phospholipids are organised into dual layers similar to those of cellular membranes. These structures, still called liposomes, have potentially important applications within the agrofood industry as molecule carriers (encapsulation of aromas) or enzyme carriers (Fig. 13.8).

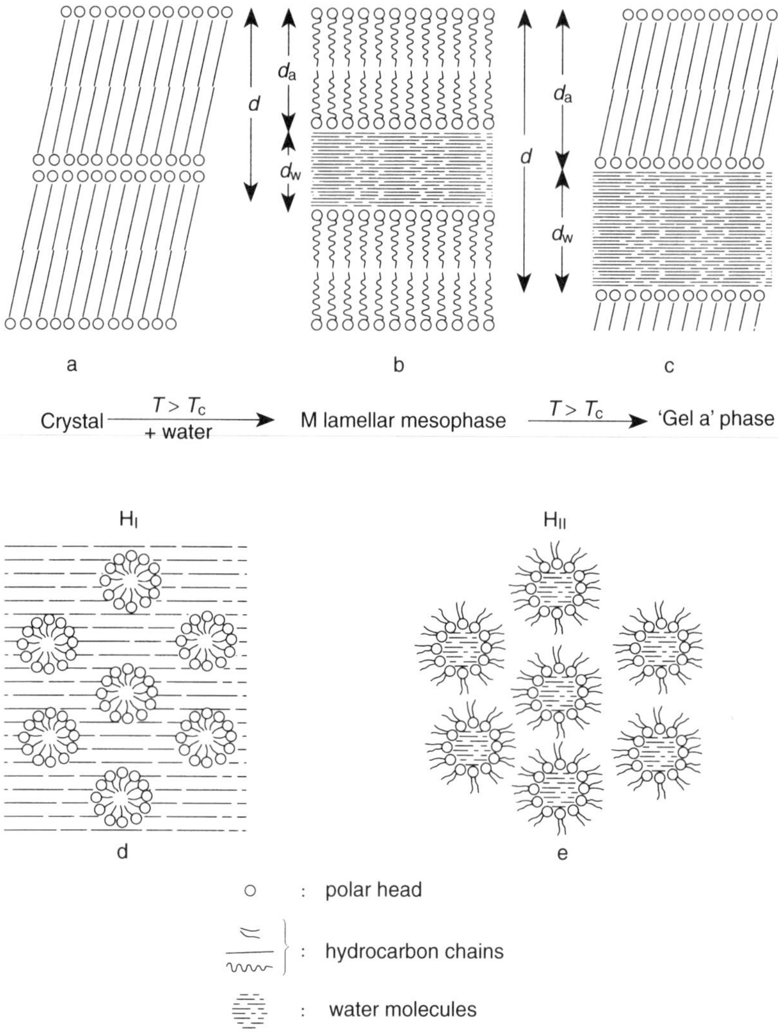

Fig. 13.9 Diagrammatic representation of various mesomorphic states.

There are two types of **hexagonal mesophase**. In the hexagonal I phase or the normal hexagonal phase, the lipids form infinite cylinders with hydrocarbon chains which constitute the internal part of the cylinder (Fig. 13.9d). The polar groups are directed towards the continuous aqueous phase: this kind of mesophase is observed with very hydrophilic surfactants. The hexagonal II phase, or inverse hexagonal phase, corresponds to a structure in which the water is imprisoned inside the cylinder whereas the hydrocarbon chains are directed towards the outside (Fig. 13.9e). This type of mesophase can be seen, for example, in the case of weakly polar emulsifiers or mixtures of emulsifiers.

The **cubic mesophase** type is sometimes observed at high temperatures. Currently six different types of cubic phase have been identified. This arrangement is the most complex of the lipid organisations.

Two major types of cubic structures can be distinguished: those in which the lipid and aqueous phases are continuous and those in which one of the phases is discontinuous, either the aqueous phase (for example micellar aggregates in water), or the lipid phase (for example inverse micellar aggregates in oil). There are considerable differences in viscosity between these different types of cubic structure. The phases consisting of micellar aggregates (continuous aqueous phase, for example) are much less viscous than the discontinuous phases. This latter type of structure, which can be seen with the monoglycerides, is generally to be avoided because of the very high viscosity of the mixture. Because of its very high viscosity this mesophase is also known as 'viscous isotrope'. However, these structures can be compared to microemulsions and have potentially important applications, especially in controlling the catalytic activity of certain enzymes or in the encapsulation of different hydrophobic or hydrophilic molecules. These cubic mesophases predominate in systems that contain unsaturated monoglycerides. They can also combine with phospholipids under certain conditions of hydration and temperature, or in the presence of other lipids. For example, I-palmitoyllysophosphatidylcholine forms cubic mesophases at 25 °C and at 45% water.

It is possible to link the geometry of the molecule to its mesomorphic behaviour. This simple relationship takes into consideration hydration, inter- and intramolecular interactions and the dynamics of the molecule (axial rotation in particular). When the spatial requirement of the polar part of the amphiphilic molecules is similar to that of the hydrophobic part, the general shape of the molecule is cylindrical and lamellar structures are formed. When there is an imbalance between the polar and the hydrophobic parts, the general shape of the molecule is a normal or inverted cone compatible with the formation of hexagonal II and micellar phases respectively (Table 13.2).

Numerous important parameters intervene in the mesomorphic behaviour of the amphiphilic molecules, in particular the length of the chains, the pH of the medium, the ionic strength and the presence of cations in the medium. So, for example, for phosphatidic acid the presence of calcium or a low pH results in the formation of the hexagonal phase (Table 13.2). In addition, certain parameters, such as temperature and the water content of the medium, modify the hydration of the polar part of the amphiphilic molecules, which cause changes in phase. So, dehydration of the polar head results in the formation of hexagonal II or cubic structures from the lamellar or hexagonal I phase.

Finally we know that the presence of proteins interacting with the amphiphilic lipids can have a considerable effect on the mesomorphism of emulsifying agents. In this way hydrophobic proteins can make phos-

Table 13.2 Relationship between the molecular geometry and the mesomorphism of the polar lipids

Lipids	Phase	Molecular geometry
Lysophospholipids Detergents	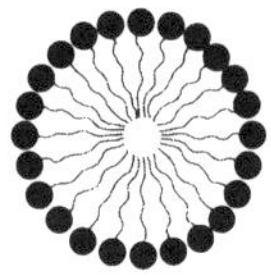 Micellar	 Inverse cone
Phosphatidylcholine Phosphatidylinositol Phosphatidylglycerol Phosphatidic acid Digalactosyldiglyceride	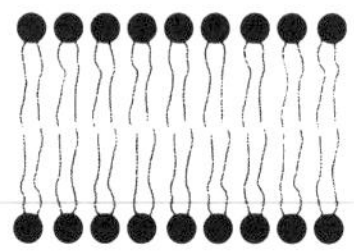 Lamellar	 Cylinder
Phosphatidylethanolamine Phosphatidic acid (Ca^{2+} and pH < 6 or pH < 3) Monogalactosyldiglyceride	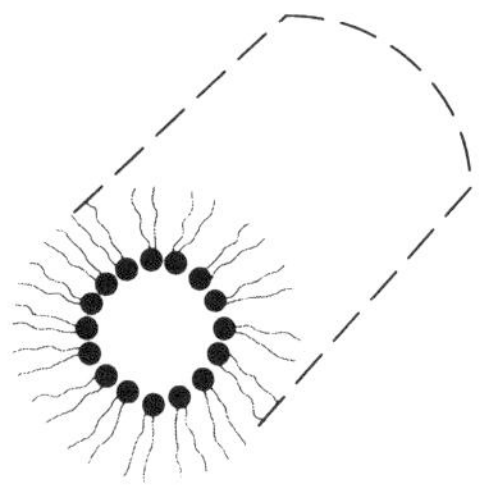 Hexagonal II	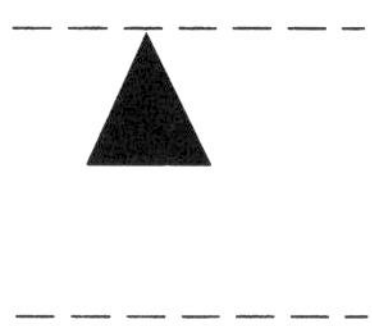 Cone

phatidylcholine convert from a lamellar structure to a hexagonal structure and lysophosphatidylcholine from a micellar structure to a lamellar structure. Certain basic globular proteins can behave in a way similar to calcium by interacting with anionic phospholipids.

13.5.2 Functional properties

The chemical structure of the amphiphilic compounds produces physical properties which can stabilise the oil/water and air/water interfaces.

13.5.2.1 Stabilisation of emulsions

The HLB index allows the effectiveness of an emulsifying agent to be predicted empirically. Optimum stability of water/oil emulsions is located in the HLB regions whose optima are located around 5–6. Outside these values, the speed of droplet coalescence is clearly increased and the emulsion is destabilised. In addition, at equivalent HLB, a mixture of emulsifiers is more effective than one emulsifier on its own.

Generally speaking, in an oil/water emulsion, a lamellar structure will form at the interface, whereas for a water/oil emulsion it is much more likely that the structure will be cubic or hexagonal II. The type of mesophase formed depends on temperature. For example, the lamellar mesophase tends to be converted into a cubic or hexagonal II mesophase under the effect of heat. The existence of these liquid crystalline structures can explain why saturated or unsaturated monoglycerides are capable of stabilising oil/water emulsions in spite of their low HLB. In fact, if the emulsifier concentration is sufficient, mesophase multilayers can form at the interface and stabilise this type of emulsion.

13.5.2.2 Capacity and stability of the aerated system

In 'foam' systems, the function of the surfactant agent is to contribute to producing a large volume and correct texture and to ensure stability against syneresis. The cohesion of the expanded emulsion is provided by the agglomeration of the fat globules which form a network inside the foam lamellae. These are emulsifiers with an 'a' tendency (Fig. 13.9) (monostearate of propylene glycol, acetylated or lactylated monoglycerides). These emulsifiers are not polymorphic and can only exist, below their melting point, in the crystalline 'a' form. They can easily be dissolved in fats and form a rigid crystalline film on the surface of fat globules.

Basically these emulsifiers operate as follows: in the absence of an emulsifier, proteins are adsorbed at the fat globule–water interface and prevent the flocculation of fat globules: the presence of an emulsifier with an 'a' tendency causes their desorption from the interface and the proteins dissolve once again in the aqueous medium, making it possible for the fat globules to coalesce. The consequence of this is a slight reduction in expansion but an increase in the rigidity of the foam.

A similar mechanism can explain the role of the monoglycerides or more polar emulsifiers in ice-cream. In this case, the destabilisation and agglomeration of fat globules takes place during the freezing process. The same is true in the case of cake dough containing fats in which expansion is generally obtained from egg proteins, but is impeded by the liquid fats which destabilise the structure of the foam. Incorporating emulsifiers with an 'a' tendency limits this effect by preventing the droplets of oil from coming into contact with the proteins of the aqueous phase.

In addition, in the case of cakes with a low fat content, the expansion of the dough can be improved by incorporating distilled saturated monoglycerides. It is essential for this to be added in the form of a liquid-crystalline dispersion or in the form corresponding to a 'crystalline a gel'. It is especially important to avoid any overheating at the time of incorporation which might result in a change of phase and in the viscous isotrope cubic mesophase. This incorporation results in a more uniform distribution of air in the dough, correct expansion and satisfactory cake volume. The mechanism of action of the monoglyceride is probably due to the formation of a monoglyceride–protein film on the air/water interface and this film produces a foam that is more stable against coalescence.

13.5.2.3 Interactions with the biopolymers

The formation of complexes between the surfactant agents and starch and/or proteins constitutes one of the essential foundations of the use of emulsifiers within the food industry, in particular as anti-staling agents in bread-making products. The capacity of starch to form inclusion complexes with lipids is dealt with in Section 11.5.2.

The hydrogenated distilled monoglycerides and the stearyllactylates are the most reactive towards starch. Non-hydrogenated monoglycerides are much less effective since they contain a high proportion of unsaturated fatty acids and the presence of the double bonds limits the possibilities for including the fat chain within the helix. Likewise, soya lecithin and the organic by-products of monoglycerides have very limited effectiveness. On the basis of the inclusion complex, although it is hardly surprising that emulsifying agents consisting of several aliphatic chains are difficult to complex by amylose, the same is not true of monoacylated emulsifiers such as sorbitan monostearate or propylene glycol. In fact, other factors such as the dispersibility of the lipid in the aqueous phase are also important. So, monoacylated lipids that form lamellar or micellar structures in water are better complexing agents of amylose than lipids forming hexagonal structures. So it seems that associations between lipid molecules can limit the complexing of monoacylated emulsifying agents. In dough, interactions with proteins could also obstruct the complexing of lipids with amylose.

Incorporating saturated monoglycerides in industrial bread-making products has the effect of limiting the speed of firming up the breadcrumbs. However, although there is good correlation between the complexing power of the amylose and the anti-staling effect, we still do not have sufficient knowledge to propose a scheme for all the mechanisms involved. It is believed that these emulsifiers intervene at different stages in the bread-making process. During kneading, the monoglycerides are distributed throughout the dough and can be adsorbed on the surface of the starch grains. This adsorption has the effect of limiting the swelling of the starch grains during cooking. When cooking begins, the amylose–monoglyceride inclusion complexes are formed by diffusion of the monoglyceride inside the starch particles. On cooling, the complexed amylose precipitates, 'congeals' the interior of the molecules and thus makes reorganisation of the chains of amylopectin more difficult.

In addition, incorporating ionic emulsifiers (stearyl-lactylates and diacetyltartric esters of monoglycerides) and a few neutral emulsifiers (sucro-glycerides, for example) results in an improvement in tolerance to kneading, water holding, gas holding and product texture. It is also possible to manufacture breads with a high protein content by incorporating protein concentrates, such as soya. The mechanisms involved have not yet been clearly explained. However, it does seem certain that the emulsifying agents are involved in a major way in gas holding during bread-making. Their interfacial properties and their capacity to disperse in gluten are therefore determining elements in the action of these lipids in cooked cereal products.

13.6 Fat substitutes

13.6.1 General information

The role of substitutes is to give the foods in which they are incorporated a rich, creamy texture, normally attributable to the presence of lipids. The qualities expected of a fat substitute can be summarised as follows:

- They must be non-toxic and not produce side effects that are disagreeable to the consumer.
- They must supply fewer metabolisable calories.
- They must possess physico-chemical properties close to the fats they are replacing in order to give the low-calorie foods rheological characteristics similar to those of the conventional food.
- They must confer organoleptic qualities identical to those of the fat they are replacing.

Technologists can distinguish two groups from among the fat substitutes. The first group includes the 'fat equivalents'. These possess all the characteristics of the lipid fraction, but their bio-availability has been lowered in order to reduce their caloric value. The second group is that of the 'fat replacements'. These are products that can replace fats with 'a thickening mixture high in water' – thus having a reduced energy contribution – but that give a sensation on the palate which is as close as possible to that of the fat.

13.6.2 Different types of substitute

In this section, substitutes will be discussed in connection with the three main classes of major nutrients.

13.6.2.1 Modified lipids

The principle that prevails in the design of this type of lipid is the reduction of molecule digestibility by means of structural modifications that reduce accessibility to the lipases responsible for lipid digestion (essentially triglycerides).

Numerous molecules have therefore been proposed (Fig. 13.10):

- Triglycerides of α-branched fatty acids: if a carbon-containing radical is substituted as an α of the carboxylic function of fatty acids in an

A. Triglycerides of α-branched fatty acids

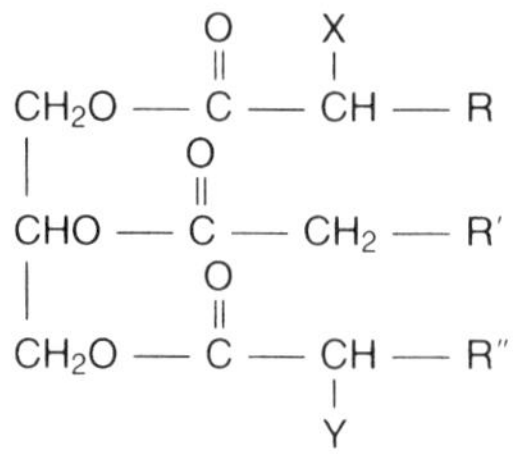

B. Glycerol diethers

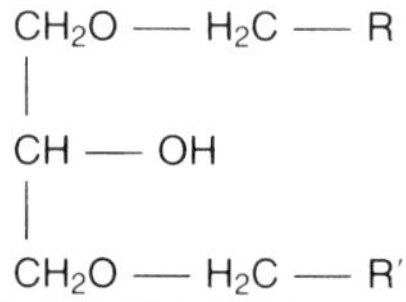

C. Polyesters of polycarboxylic acids

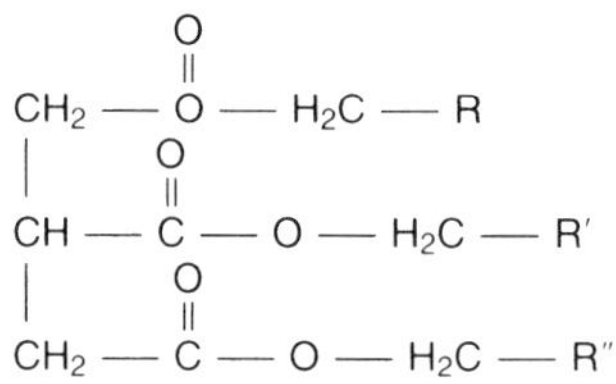

D. Polyesters of polyglycerols

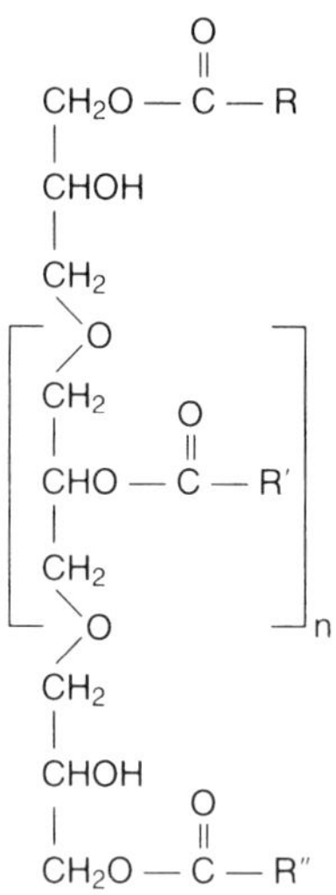

E. Esters of acids and fatty alcohols (waxes)

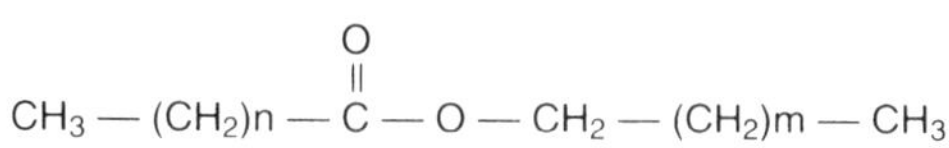

m, n = 18, 20

Fig. 13.10 Structures of fat substitutes of a lipid nature.

external position (as X or Y on Fig. 13.10A), hydrolysis will be weak or even zero.

- Glycerol diethers: fatty alcohols are substituted for fatty acids. The carbon-containing chain of these alcohols can contain between 8 and 30 atoms of carbon (Fig. 13.10B).
- Polyesters of polycarboxylic acids, such as citric acid, succinic acid, etc.
- Polyesters of polyglycerols (Fig. 13.10D) whose degree of hydrolysis depends on the number of glycerol chains. In addition the number of fatty acids esterifying the hydroxyl functions of the polyglycerolic chain will act on the properties of the molecule and its energy value.
- Esters of long chain fatty acids and alcohols (most frequently between 18 and 22C). The best-known of these waxes is jojoba oil which contains over 10% of erucic acid ($C_{22:1}$).
- Sucrose polyesters or sucro-esters which are a mixture of hexa–hepta- and octaesters of sucrose with long chain fatty acids (generally between 8 and 22 atoms of C) (see Chapter 10).

Triglycerides of α-branched fatty acids, glycerol ethers, polyesters of polycarboxylic acids and sucrose polyesters are not authorised for use anywhere in the world and do not seem likely to be used in the medium term. Polyglycerol polyesters are authorised as food additives. Waxes, in particular jojoba oil, seem to have no further future as fat substitutes.

13.6.2.2 Carbohydrate substitutes

Fat substitutes based on carbohydrates are obtained by increasing the viscosity of the aqueous phase and by using the gelling properties of glycans. Some are based on starches, such as partially gelatinised maize or manioc starch. These products, which possess significant water-holding properties, have a caloric value of approximately 16 kJ/g.

Other substitutes of a carbohydrate kind are glucose polymers obtained by means of thermal processing in the presence of citric acid (catalyst) and sorbitol (plasticiser). This type of product is a network of glucose molecules associated at random and containing several terminal groups formed from sorbitol and citric acid. These polymers are not absorbed in the large intestine, but are partly fermented in the colon and their caloric value can be estimated at around 8 kJ/g. With the exception of glucose polymers, this type of substitute is considered to be a food substitute since it is produced as a result of the normal processes used in the food industry applied to traditional raw materials.

13.6.2.3 Protein substitutes

Protein-based fat substitutes share the qualities of being composed of small-sized non-aggregated protein particles. These particles, dispersed in an aqueous phase, mimic the organoleptic properties of oil/water emulsions.

Most foods are found naturally in a particular form. Starch grains, the protein portions of grains, granules of egg yolk, casein micelles and fat globules in milk are some examples of this. However, there is a threshold below which particles are not distinguished by the palate as individual particles but rather as a continuous fluid, similar to a fat emulsion. This threshold of perception is close to 3 μm. A product that contains particles whose size exceeds this threshold will taste powdery or granular. Although particle size is an important factor, the texture of these particles is also extremely important. In fact, the impression of richness provided by chocolate is attributable to the presence of large particles; milk chocolate (particles whose size is less than 65 μm) possesses a texture that is richer on the palate than dark chocolate (particles whose size is less than 35 μm). So, particles that are large yet compressible can contribute to the rich or creamy texture of a food, especially if these particles can roll freely against each other.

'Micronisation' of proteins (see Section 5.2.3.4), that is to say, moderate thermal processing associated with intense shaking can, because of these small particles, produce the sensation of creaminess and smoothness in the mouth, normally associated with fats.

Although always made up of protein microparticles, protein-based fat substitutes differ in the way in which they are obtained and/or in the raw materials used.

As the substitute is often used with water, its caloric value will vary between 6 and 15 kJ/g. The protein nature of the product does, however, limit its use to food products which do not undergo heat processing at a temperature higher than 60 °C. In addition, the allergenic potential of these products is no different from that of the raw materials used in their manufacture (milk or egg proteins).

To sum up, the future of these substitutes is not yet – not by a long way – clearly established. In fact, substitutes of a lipid kind which offer the best advantages from a technological point of view are those that show quite a few disadvantages in terms of digestibility and nutritional effects. On the other hand, those that do not present these disadvantages (substitutes of a carbohydrate and protein kind) have technological limitations (resistance to heat treatments, organoleptic properties, compatibility with the other constituents) which restrict their field of application.

14

Amino acids and peptides

It is obvious that a large number of chemical products, such as ethylene, butadiene and propylene, will predominate for some time in the field of carbon chemistry and petrochemistry. It has also been clearly established that the processes of fermentation and bioconversion are the only economical ways – and in many cases the only technical means – of obtaining complex macromolecules such as vaccines, antibiotics and certain carbohydrate polymers.

One of the fields in which the problem of competition between the bioindustry and the chemical industry arises is that of the manufacture of molecules such as amino acids and peptides. So, depending on the amino acids involved, chemistry sometimes takes precedence over biotechnology (methionine) or vice versa (lysine); chemistry takes precedence over biotechnology for the small peptides whereas the opposite situation is true for long chain peptides: the equilibrium region between the two technologies currently lies around those peptides that possess between 20 and 30 residues.

Moreover, small peptides and amino acids greatly influence the flavour of most foods. It is generally accepted that proteins in their native state possess either no taste or very little, and that this develops during protein hydrolysis. This is what happens during the maturation of numerous food products: for example meat and cheese in the west, and soya and rice (sake) in the east.

With the discovery of sapid peptides with a wide range of tastes, the food industry has been able to take advantage of these new molecules which can be used as flavour enhancers, agents of sapidity (monosodium glutamate and aspartame, for example) and very good quality nutritional ingredients.

This aspect is also one of the definite advantages of amino acids and peptides, since they can be used in diets for people whose metabolism is deficient or induces health problems: diabetes, high blood pressure, difficulty in digesting proteins (cystic fibrosis). In fact these small molecules can be assimilated directly and their flavour can help to make following a strict diet less disagreeable. In addition, a number of peptides generated in the course of digestive transit possess biological activities (anti-hypertension, immunostimulant, opiate activity, etc.).

The importance of amino acids and peptides in the diet and their highly promising use within the agro-food industry will be discussed in this chapter. No doubt in the future it will be possible to obtain products 'à la carte', made up of peptides or peptide mixtures with specific flavours, depending on the intensity and specificity of the enzymatic hydrolysis of proteins.

Note Unless there are notifications to the contrary, all amino acids (AA) are taken to be in their natural L configuration.

14.1 Production and use of amino acids

14.1.1 Extraction and preparation

14.1.1.1 Protein hydrolysates

In the case of certain amino acids, their extraction from natural substances, generally waste matter (human hair from the Far East or India, residues from processing animal skins, poultry feathers, etc.) remains competitive. Thus, L-cystine and L-tyrosine, two amino acids which are not very soluble in water, can be separated from other water-soluble amino acids during the processing of protein hydrolysates of animal or plant origin. This method is still used today in order to obtain L-tyrosine and L-cystine, from which a mild reduction process provides L-cysteine. Hydrolysis of gelatin produces L-hydroxyproline, which is separated by extraction.

However, these methods of extraction are subject to increasing competition from microbiological methods and already L-arginine and L-histidine, formerly separated from protein hydrolysates, are produced more economically by fermentation. Certain plant seeds can also contain a sufficient concentration of amino acid to allow extraction to take place. This is the case with L-dopa (3-hydroxytyrosine) which can be obtained by extraction with methanol from faba beans (*Vicia faba*): 1 kg of faba beans can produce between 18 and 20 g of L-dopa (Table 14.1).

These extraction methods are not necessarily redundant in comparison with others. However, they are dependent on the price of the raw materials and their availability. In addition, in the case of extraction from plants it must be possible to improve the production yield of any kind of amino acid with the aid of genetic selection and methods of genetic engineering.

Table 14.1 Structures of some rare amino acids and their by-products

Dopa	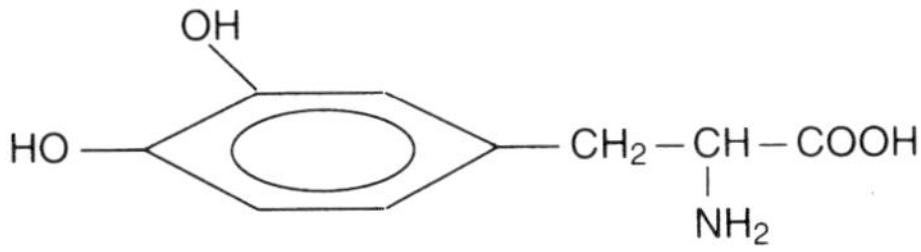
Cyclo-alline	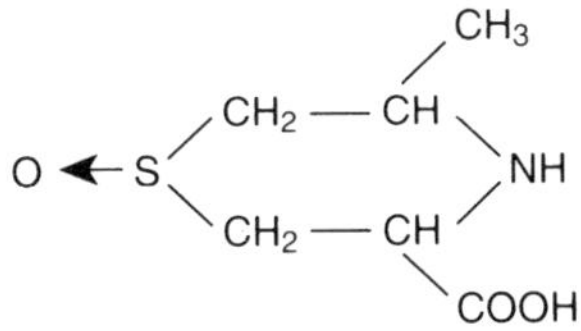
Lanthionine	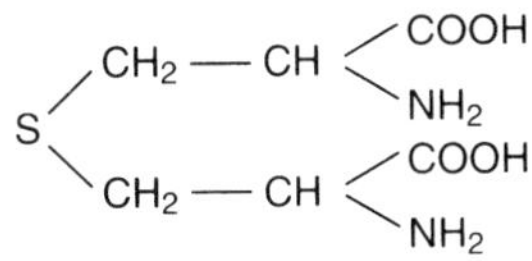
Betaine	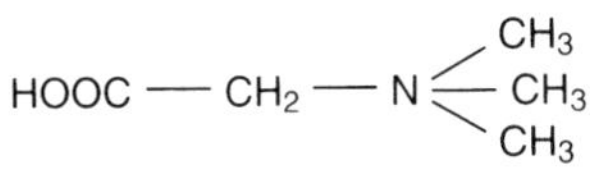
Morchelline	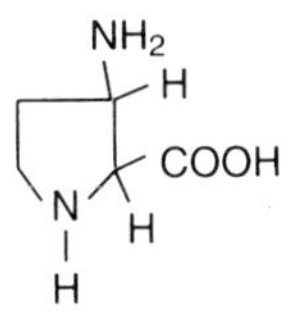
Cucurbitine	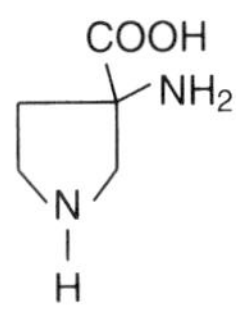
4-Methylproline	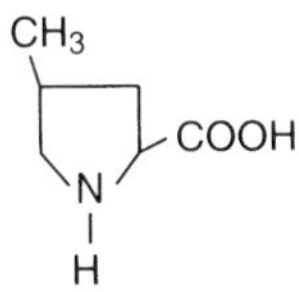
Azetidine carboxylic acid	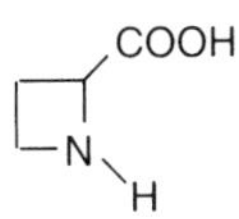
Guvacine	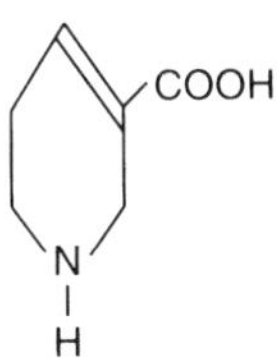

14.1.1.2 Enzymatic methods

Instead of using an entire micro-organism with the whole range of its enzymatic activities, it is possible to use only certain microbial enzymes which may or may not be immobilised on a support. Some examples of the use of enzymatic reactions for preparing amino acids are as follows: that of L-aspartic acid from fumaric acid, that of L-alanine from L-aspartic acid and that of L-lysine from cyclohexene.

The continuous production of aspartic acid can take place starting from a chemical raw material, ammonium fumarate, which uses a microbial aspartase bonded to polyacrylamide gel. A 95% yield of crystallised L-aspartic acid is obtained.

The disadvantage of these processes is the progressive loss of enzyme activity: active for one week, it loses three-quarters of its activity after 20 to 25 days. The aspartic acid obtained in this way can be converted into L-alanine by decarboxylation using L-aspartate-β-decarboxylase. Obviously the industrial profitability of a process is increased if it has a common core so that various amino acids can be prepared. These enzymatic methods are suitable for continuous production, so in Japan, L-aspartic acid is produced by passing the reaction solution through columns equipped with immobilised microbial cells. Tryptophan has been enzymatically synthesised from indole and pyruvate using a tryptophanase; the L-tryptophan formed is precipitated from the medium by forming a compound which is insoluble in inosine, which allows the reaction to continue.

The main disadvantage of these methods is the progressive loss of activity of these immobilised enzymes. Nevertheless, the potential of these methods is considerable and, from an industrial point of view, so is the possibility of carrying out a series of operations without having to separate at each stage. As this economises on labour and keeps equipment in constant use it seems very promising. In addition, many research projects have been developed all over the world in order to obtain more stable, high-

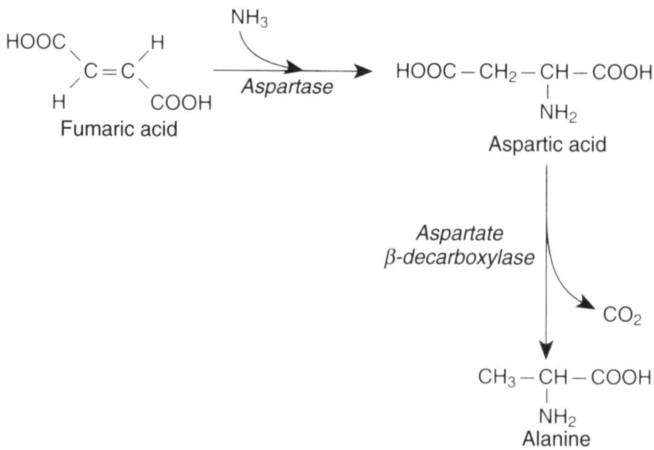

performance enzymes, both in terms of heat stability and duration of use. There is no doubt that this research will produce positive results in the near future, in the form of very competitive enzymatic processes which are very flexible to use.

Manufacturing amino acids by fermentation is based on a knowledge of metabolic pathways, the use of high-quality sources and mutants, the existence of retro-inhibition reactions (e.g. production of Ile by a mutant of *Serratia marcescens* which is no longer inhibited by the Ile hydroxamate).

The most spectacular example of production from an economic point of view concerns the production of L-glutamic acid from glucose using wild strains of *Corynebacterium glutamicum* or *Brevibacterium flavum*: the glutamic acid is excreted in the culture medium. The industrial technique uses sugar beet or sugar cane molasses. The vitamins and mineral salts necessary for the growth of the micro-organism are added to this aqueous solution in the fermenter. After inoculation of the fermentation medium maintained at 35 °C and shaken, air and ammonia are passed through. After about 40 hours, fermentation is halted and the glutamic acid which has been formed is separated. Approximately 500 kg of L-glutamic acid is obtained from one tonne of glucose. Sources of *C. glutamicum* have been modified by genetic engineering with a resulting production of high concentrations of Thr, Tyr and Phe.

Fermentation methods will benefit from progress in micro-organism genetics and genetic engineering. Fermentation processes have the advantage of not being particularly dependent on the availability of numerous raw materials, with sources of carbon and nitrogen generally being fairly inexpensive. Fermentation methods require labour and time to isolate and purify the amino acids, because their concentrations are generally low (especially when compared with the concentration from chemical synthesis). In addition, by-products which must be removed are formed (biomass and liquor), and these can cause environmental problems. Increasing the scale is not always easy, particularly because of the low concentrations.

14.1.1.3 Chemical methods

These methods remain dependent on the availability of raw materials or intermediate products at competitive prices. The traditional methods of synthesising amino acids should shortly give way to stereospecific methods of producing amino acids from the natural series.

Amino acids produced by chemical synthesis are as follows: Asn, Cys, Gly, Met, Phe, Ser, Thr, Trp.

14.1.2 'Rare' free amino acids

In addition to the 20 amino acids with a known transfer RNA, foods supply over a hundred other amino acids of varying chemical composition. These 'rare' amino acids can represent a considerable proportion (between 2 and 5%) of the nitrogen portion.

We have classified them – rather artificially – into several categories. When their formulae are not given in the text they are shown in Table 14.1.

14.1.2.1 β-Alanine: by-products and related substances

β-Alanine, with the formula $H_2N—CH_2—CH_2—COOH$, appears to be the product of the decarboxylation of aspartic acid; nevertheless, this reaction, which takes place in numerous micro-organisms, has not been observed in higher animals. β-Alanine is produced in higher animals during uracil degradation. The biological importance of β-alanine is demonstrated by its involvement in the creation of pantothenate (although not in humans, where pantothenic acid is a vitamin factor) and it is involved in the structure of two peptides found in abundance in muscle tissue, β-alanine-L-histidine (carnosine) and β-alanyl-L-N_{Π}-methyl-histidine (anserine) (Π = proximal; the methyl group is carried by the nearest nitrogen atom on the side chain). These two peptides are not extensively metabolised and are found partly in the urine; the content of the urine β-alanyl-peptide is an excellent marker of the amount of meat in the diet. However, it is also interesting to note the presence of a glutathione analogue in beans, in which the glycine is replaced by β-alanine: this is γ-L-glutamyl-L-cysteinyl-β-alanine. As for the isolated β-alanine, this can undergo oxidative degradation which produces a malonic semi-aldehyde followed by the formation of acetate.

Taurine is odd in two ways, since it acts both as a β-amino acid and a sulphonic amino acid, whose formula is $H_2N—CH_2—CH_2—SO_3H$. This analogue of β-alanine is synthesised from cysteine, using cysteic acid of which it is the product of decarboxylation. As a powerful neuron modulator and also a constituent of certain biliary acids (taurocholic acid), taurine is partially eliminated in the urine: one fraction undergoes deamination to produce isothionic acid, which is also found in the urine: $HO—CH_2—CH_2—SO_3H$.

Taurine is the most important of the rare β-amino acids in the animal kingdom. The highest concentrations are found in the growth and nerve tissues especially during periods of cell proliferation such as foetal and neonatal development. The brain, the retina, the heart, the liver, the platelets and lymphocytes possess active transport systems for taurine from the extracellular medium towards the intracellular medium.

The amino group of taurine can form a stable bond with the carboxyl group of the biliary acids. In addition to this well-characterised reaction, taurine can have other biological effects. Two of these are as follows:

- Reduction in platelet aggregation, probably by increasing calcium concentration.
- Increase in the concentration and availability of intracellular calcium for muscular contraction.

Ciliatine or 2-ethylphosphonic amino acid, whose formula is $H_2N—CH_2—CH_2—PO_3H_2$ represents a much more unusual molecule, since it is the first known biomolecule to possess the covalent C–P bond. Initially isolated in *Tetrahymena pyriformis* and in other ciliated organisms (from which it takes its everyday name of ciliatine) this aminophosphonate has a wider zoological distribution (in numerous marine invertebrates, especially molluscs). Ciliatine is found mainly in lipid structures similar to the cephalines or sphingomyelines; the presence of ciliatine in these 'phospholipids' can be explained by the great similarity of structure between amino-2-ethylphosphonic acid and phosphorylethanolamine, which only differs from this by its phosphoric ester function, that is to say, in the final analysis by one additional atom of oxygen.

Oysters and mussels supply ciliatine in very large quantities. One mussel contains several milligrams of this phosphonate which is poorly absorbed in the intestine. The fraction absorbed is partially eliminated in the urine, but is also found stored in the form of phospholipids in the liver, the brain and the kidneys. Dairy products and meat from polygastric animals constitute a significant source of ciliatine; the digestive tract of cows and sheep actually contains numerous saprophytic ciliates protozoarge whose lysis frees phosphonates.

14.1.2.2 Sulphur amino acids

Cysteine can be prepared from crushed feathers. It is used as an additive in the baking industry (conditioning bread dough) in the meat products industry (colourant for hams and sausages) or as an anti-oxidant in fruit juices.

The **S-substituted by-products of cysteine** are found in plants both in the free state and in the form of γ-glutamylpeptide, or even tripeptide. The alkyl substitute can be a saturated aliphatic chain (*S*-methyl- or *S*-*n*-propyl-cysteine) or ethenic chain (*S*-allyl- or *S*-propenyl-cysteine) or even support other functional groups (cystathionine); the sulphur atom, normally involved in a sulphur bond, can also exist in the oxidised form of sulphoxide. A more complex aroma compound has been identified in cycloaline, which exists in a free state in onions.

Products from the sea also provide us with other amino acids such as **lanthionine**, present in the muscle flesh of the lobster; this sulphur amino acid has been known for some time, but as a constituent of the proteins in wool, a constituent which is moreover disputed, since some consider it to be an artefact formed in the course of certain industrial treatments of this animal fibre. We can also draw attention to the presence of lanthionine in the sequence of certain bacteriocins and in the proteins processed by alkaline reagents.

Polysulphurous by-products of fungal origin, such as marasmine or lentinic acid are both found in the form of glutamylpeptide, in *Marasmus alliaceus* and *Lentinus elodes* respectively.

The nutritional advantages of sulphur amino acids come essentially from the fact that these compounds make a considerable contribution to the specific taste and odour of a particular plant food: garlic, onion, cabbage, bean, turnip or mushrooms.

14.1.2.3 Methylated amino acids
The monomethylated by-product of glycine is called **sarcosine**. The trimethylated by-product is **betaine** which is a product of oxidation of the corresponding alcohol, choline. The glycine betaine, is very abundant in beetroot (*Beta vulgaris*) which gave it its everyday chemical name. Betaine from proline (stachydrine) is of animal origin, and is a constituent of *Arca nove*, a bivalve eaten in Senegal. That from histidine is found in fungi widely eaten in France (*Agaricus campestris*, *Boletus edulis*).

The methylsulphonium salt from methionine represents another type of methylation: in this process *N*-methylation of betaines is replaced by *S*-methylation. This product is found in cabbage and asparagus.

14.1.2.4 Heterocyclic amino acids and cyclic iminoacids
Numerous amino acids of fungal origin possess a heterocyclic nucleus, and mainly come from the mushrooms eaten in the Far East. **Morchelline**, which gives the morel some of its gastronomical qualities can be compared to cucurbitane, an amino acid from the marrow responsible for the anti-helminthic properties of the seeds of this plant. For many years proline was the only known iminoacid. Nowadays natural iminoacids represent a very large family. **Azetidine carboxylic acid** was initially characterised in the rhizome of lily-of-the-valley and in other Liliacaea. It is also found in beet-root. Two products resulting from the polymerisation of azetidine car-boxylic acid are also known. Among the numerous methylated by-products of proline, **4-methylproline** is found in apples. **Pipecolic acid**, a product of the endogenous metabolism of lysine, is found in beans. **Guvaroline**, which only differs from baikiaine in the position of its acid group, is found in the betel (*Areca catechu*) whose leaves are chewed in the Far East.

14.1.3 Organoleptic properties and uses of amino acids

14.1.3.1 Taste of amino acids
The theory of the four tastes, sweet, salty, bitter and sour, has prevailed for many years. Each of these has an area of the tongue in which it is predominant. For some time this theory has been questioned, because of the discovery of a fifth taste which could not be incorporated in the 'traditional' concept: the *umami* taste. Research into the physiology of taste (electro-physiology, psychosociology, etc.) showed that glutamate possessed a taste independent of the four basic tastes. The taste of glutamate cannot be reproduced by combining these four fundamental tastes.

It now seems to be well established that the receptors of the sensory cells

Table 14.2 Flavour and threshold of detection of L-form amino acids

L-form amino acids	Chemical structure	Flavour	Threshold of detection (mg/100 ml)
His	Basic	Bitter	20
Met	Hydrophobic	Bitter	30
Val	Hydrophobic	Bitter	40
Arg	Basic	Bitter	50
Ile	Hydrophobic	Bitter	90
Trp	Hydrophobic	Bitter	90
Phe	Hydrophobic	Bitter	90
Leu	Hydrophobic	Bitter	190
Tyr	Hydrophobic	Bitter	Not detected
Ala	Hydrophobic	Sweet	60
Gly	Non-charged polar	Sweet	130
Ser	Non-charged polar	Sweet	150
Thr	Non-charged polar	Sweet	260
Lys	Basic	Sweet + Bitter	50
Pro	Hydrophobic	Sweet + Bitter	300
Asp	Acid	Acid	3
Glu	Acid	Acid	5
Asn	Non-charged polar	Acid	100
Gln	Non-charged polar	Vented	
Cys	Non-charged polar		Not detected
GluNa		*Umami*	30
AspNa		*Umami*	100

are specific to certain molecules. So, when we examine the flavour of sweet and bitter amino acids, we note that for most of the amino acids the L form is bitter whereas the D form is sweet. In addition, the hydrophobic groups of the side chains of the amino acids such as D-Val, D-Leu, D-Trp and D-Phe, are involved in the intensity of the sweet taste, which is greater than with D-Ala or D-Gly. Likewise the *umami* taste is eliminated after acetylation of the amino group or after esterification of the carboxylic group and when the hydrogen in α is replaced by a methyl group.

Table 14.2 lists the different amino acids, their taste, their detection threshold in mg/ml (the lower the value of the threshold the more pronounced the taste of the amino acid). Generally the intensity of the flavour increases with the concentration in amino acids.

Bitter amino acids produce disagreeable flavours in foods, which are often rejected by consumers. The hydrophobic amino acids are responsible for bitterness, with the main ones being L-Phe, L-Tyr, L-Leu, L-Val and L-Ile. The enantiomer forms do not produce the same sensation. L-Form amino acids are much more bitter than D-form amino acids which are very often sweet.

Sulphur amino acids are generally considered to be lacking in flavour, except for methionine which is said to have a certain amount of bitterness.

There are a maximum of 6 **sweet amino acids** out of the 20 amino acids of the L-form at known tRNA. Only L-alanine and glycine have a significant sweetening power. Their polarity does not seem essential since Thr, Ser and Gly are polar and Ala is non-polar. The sweetening power of D-form hydrophobic amino acids is higher than that of sucrose. Finally, certain sweet amino acids can provide the flavour characteristic of the flesh of certain animals, for example glycine reproduces the sweet flavour of crab and lobster.

No free **amino acid** is **salty** and its taste only appears at the peptide level of the structure. However, a salty taste has been revealed for proline and lysine chlorhydrate in particular.

Only **amino acids** of an **acid** form such as aspartic and glutamic acids have an acidic taste when they are in a dissociated form which is the case in the majority of foods with a slightly acid pH.

The ***umami*** **amino acids** are represented by the sodium salts of acid amino acids. Their taste is defined as a mixture which is both sweet yet has a meaty taste or a taste of chicken broth.

14.1.3.2 Some examples of the role of amino acids in food

Glycine is used in the pharmaceutical industry (buffer substance in an aspirin tablet for example) and in the food industry for domestic animals, but the food applications could become wider. Glycine could have anti-oxidant, bactericide, aroma-reinforcing and flavour-enhancing properties, all of which are widely used in Japan. In the EU matters are different. Although some countries authorise the use of glycine (in particular as an aroma enhancer), others do not permit its use. Glycine can only be used for its properties as a flavour enhancer in low doses, not sufficient to involve its other action. In addition, glycine can be involved in the Maillard reaction (browning agent) and has a certain amount of sweetening power. As a result it is used (in the USA, Germany and Switzerland) to mask the bitterness of aspartame in low-calorie drinks (Assugrin, for example, is a mixture of aspartame and glycine). Glycine, as a starting molecule of synthesis, may also find application in the cheese-making industry (bacteriocide effect) or in curing (the combination of glycine and its by-product sodium glycinate offer strong water-holding potential).

Glutamic acid and its various sodium, potassium, calcium, ammonium or magnesium by-products are generally designated by their generic name 'glutamate'. The most widespread is monosodium glutamate (MSG).

Glutamate is a natural compound which is found in practically all foods such as meat, fish, milk (including maternal milk) and numerous vegetables. The human organism also produces glutamate naturally and this is used in various metabolic reactions necessary to life.

In nature, glutamate is found in two forms, either in a 'bound' form when it is attached to other amino acids in proteins, or in a 'free' form when this is not the case. Only the 'free' form has gustatory qualities and is extensively involved in the palatability and acceptability of various foodstuffs. This explains why foods such as tomatoes, mushrooms and cheese, which naturally contain high levels of free glutamate, are very often used for their flavour in numerous culinary preparations.

Glutamate has the remarkable property of bringing out and enhancing the original flavour of various foodstuffs, thus increasing their appeal. Glutamate also contributes to the perception of a balanced and 'rounded' flavour in foods such as sauces and stews. Glutamate is used in the food industry in a wide variety of products:

- culinary broths;
- soups and broths (dehydrated, tinned, etc.);
- ready-made meals based on meat, fish, vegetables;
- sauces (dehydrated, tinned);
- delicatessen and salted foods;
- various products based on meat or fish.

The levels of use most frequently encountered are in the range 0.1–0.8% of the ready-to-eat product.

Lysine, an essential amino acid, is widely used in the animal feed industry. Supplement levels vary according to the animal species, from 500 to 4000 g/tonne of feed. The poultry and pork meat sectors constitute the main outlet for current uses of lysine but its utilisation in feed for rabbits and veal intended for slaughter is fairly extensive. Applications relating to fish farming and to dogs and cats are currently being developed. Lysine is not used for ruminants because the way in which their digestive system works does not allow satisfactory returns on investment.

Supplements of pure L-lysine can permit the reduction of the overall protein content of a food. The resulting saving in protein is used to reduce the cost of formula feed while maintaining equivalent metabolic performances. As an example, in pig and poultry feed, by using lysine, 5–6% of protein-rich soya cake can be replaced by cereals.

Incorporating pure amino acids in a food thus provides a rational response to the variability in the composition and quality of raw materials. For food manufacturers, this also means greater flexibility in formulation, especially by using protein sources low in lysine more economically (sunflower, colza, etc.) or those that are deficient in certain amino acids (peas).

Improving the nitrogenous balance in diets aimed at reducing excess protein also has the advantage of making better economic use of portions per animal, both from a nitrogenous and an energy point of view. In particular, this reduces the pathological effect associated with nutritional imbalance (digestive problems). These technical advantages have been used for many years by the industry producing foods prepared with lysine supplements, associated where appropriate with methionine. Threonine and tryptophan have followed lysine onto the market, and are used in the manufacture of competitive-priced animal feeds, allowing the dietary intake of amino acids to be adjusted more accurately to the requirements of the animals.

So, lysine supplements, a factor limiting the development of lean tissue, can make foods available to animals being reared for accelerated growth, which enable them to reach their full genetic potential and, in particular, encourage the depositing of muscle, thus satisfying the consumer's demand for less fatty meat products.

Finally, the developments in the use of amino acids in animal feed open up interesting prospects on the macro-economic scale:

- They provide a neat response to the problems of surplus cereal. In fact cereals are generally lacking in lysine, threonine (barley, wheat) and tryptophan (maize). Systematic recourse to these amino acids in formulating animal feed increases the potential of these cereals, thus increasing their consumption on farms.
- The use of amino acids offers a solution for the nitrogen pollution which occurs in intensive animal-rearing areas. So, a reduction of 2% in the protein content of the food, which can be achieved without altering performance simply by adding lysine supplements, can lower the quantity of nitrogen excreted by a pig farm by almost a quarter.

14.2 Peptides

14.2.1 Peptides used in foodstuffs

Over the past few years a number of synthetic peptides have proved to be very useful raw materials in the food industry. They can act as flavour enhancers when added, for example, to textured products to make them more tasty: *surimi* or substitute crab, reformed steak, for example, or they can be used even for the purposes of producing a more balanced food. We will look at the way in which the structure and properties of the peptides change according to whether they have a sweet, salty, *umami*, sour or bitter taste. At the end of the paragraph we will discuss the advantage of special peptides (amphiphilic, dicetopiperazine).

14.2.1.1 Sweet peptides

Five dipeptides can be considered to be sweet, and this sweetness is generally accompanied by bitterness. The peptide whose sequence is L-Asp-L-Phe-OMe, and which is better known under the commercial name of **aspartame**, has no bitter aftertaste (see Fig. 10.5). This peptide, whose sweetening power is between 180 and 200 times that of sucrose, can be used by people with diabetes because it does not induce hyperglycaemia. This sweetener supplies the same quantity of energy as the proteins (17 kJ/g).

The solubility of aspartame depends on pH and temperature. It is lowest at the isoelectric point (pH = 5.2). In solution, the ester bond is likely to be hydrolysed, with the resulting reduction in sweetening power. When dry, below 8% moisture content and at ambient temperature, stability is excellent over several years. Aspartame is very stable at pH = 4 but cannot endure acid (pH = 1) or alkaline pHs (pH = 7–8), or high temperatures. In the case of the latter, aspartame is modified by cleavage and/or cyclisation and the by-products obtained have no sweet taste.

Consequently, users are recommended to overdose culinary preparations, in order to retain, at the end of cooking, an appreciable sweet taste equivalent. Moreover, the aromatic residue of this peptide is responsible for the condition of phenylketonuria which affects some babies. This is the result of an excessive concentration of phenylalanine in the blood supplying the brain and because of an enzymatic deficiency in phenylalanine hydroxylase in these babies. A warning is therefore obligatory on the packaging of all foods containing this sweetener. In addition, the cyclo(Asp-Phe) encourages skin rashes. This problem should be reduced by limiting the duration of use (and therefore of storage), and by lowering the storage temperature of the sweetener contained in culinary preparations. Finally, an important point on which its success is based is its use in slimming diets. It has proved to be useful for those following strict diets. On the other hand, if sucrose is replaced by this sweetener in diets without 'taking care' then the opposite effect occurs: increased weight gain. It seems that the absorption of the sweet taste alone, that is to say without adding any calories, results in more food being consumed during subsequent meals, because if the sweet taste of the sweetener is eliminated by encapsulating it, this phenomenon is no longer observed. Phenylalanine actually stimulates the secretion of the hormone of satiety, cholecystokinin.

Aspartame is authorised in numerous countries (almost 100). The RDA is between 40 and 50 mg/kg.

Another sweet dipeptide now seems to be making a name for itself: **alitame** (Fig. 10.5). This is an amide of the dipeptide L-aspartyl-D-alanine. Its sweetening power is approximately 2000 times that of sucrose, and 12 times that of aspartame. Its solubility in water is excellent and its stability at high temperatures and when there are variations in pH is greater than that of aspartame.

14.2.1.2 Salty peptides

Salty peptides are not very numerous as such, since they are often accompanied by a bitter aftertaste. Recently a dipeptide L-ornithyl-taurine has been discovered. This has a salty taste but no sodium ions, so it can be used in so-called 'salt-free' diets. However, other studies have observed that the taste of the dipeptide Orn-Tau-chlorhydrate is probably not salty and that the flavour noted could be due to the contamination of the sample with NaCl during the stages of purifying and/or synthesising the dipeptide.

14.2.1.3 Umami *peptides*

Initial investigations on the subject of *umami* peptides have established that a peptide containing the residue L-glutamyl in the N terminal position possesses a *umami* taste, and more specifically the dipeptides in which the glutamyl residue has been added to an acidic amino acid. On the other hand, when combined with Gly, Val, Pro or Ala, it produces peptides without any flavour. When associated with hydrophobic amino acids, the glutamyl residue produces bitter peptides.

However, the presence of a glutamyl residue at the N terminal extremity does not seem to be a prerequisite for the *umami* taste. The *umami* taste can also be found in the absence of the glutamyl residue. A *umami* dipeptide with a Ser-Phe sequence has been discovered. The amino acid, serine, is also involved in a tripeptide, Gly-Glu-Ser, whose *umami* taste is very pronounced and appears during the enzymatic hydrolysis of soya proteins. The flavour of this peptide is, however, less intense than that of MSG (the detection threshold of this peptide is 200 times higher).

These peptides could be used in culinary preparations in order to give them a new flavour, by supplementing the use of MSG.

Finally, a peptide that releases a *umami* taste, and can be used as a flavour enhancer, has been purified from beef gravy. The octapeptide Lys-Gly-Asp-Glu-Glu-Ser-Leu-Ala possesses a *umami* flavour as strong as MSG, accompanied by a slight acid taste due to the acid fragment of the central portion of the peptide. Eliminating the two residues Lys and Gly in the N-terminal position leads to a loss of *umami* flavour.

14.2.1.4 Sour peptides

Acid peptides always contain in their sequence one or several acid amino acids. Within this category, the following oligopeptides have been identified: Asp-Phe, Glu-Phe, L-Ala-L-Asp, L-Glu-L-Glu, L-Gly-L-Asp and L-Ser-L-Glu. The acidity is probably due to the releasing of the proton when the free carboxylic groups of the dipeptide dissociate.

14.2.1.5 Bitter peptides

Our knowledge of the taste of the free amino acids might lead us to think that the taste of peptides depends solely on their amino acid sequence. This is exactly the case with the peptides containing hydrophobic amino acids, that is to say those that contain Phe, Leu, Trp, Ile, Pro, Tyr and Val, and are generally bitter.

Generally speaking, bitterness increases when the hydrophobic amino acid is placed in C terminal position in the peptide sequence. So, by adding Gly-Gly to Phe-Phe in C terminal position, the bitterness of the peptide drops by a third of its initial value. Likewise, when Gly-Gly is added to $(Val)_3$ in C terminal position the bitterness completely disappears.

As for the best-known dipeptide, Arg-Pro (more bitter than Pro-Arg), its bitterness disappears by adding two glycyl residues to each end. The dipeptide Pro-Gly appears sweet or tasteless, whereas Gly-Pro seems bitter (as Pro is in the C terminal position).

The most bitter peptide obtained to date (detection threshold value lower than that of strychnine) is an octapeptide Arg-Arg-Pro-Pro-Pro-Phe-Phe-Phe. The bitterness seems to result from two main conditions: a hydrophobic amino acid in C terminal position and an alkaline amino acid in N terminal position. In addition, an effect of synergy takes place when there is a structure of a $(Arg)_l$-$(Pro)_m$-$(Phe)_n$ type with $l = 1–2$, $m = n = 1–3$. Finally, bitterness is more intense in the peptides than in the amino acids. It actually seems that the NH_2 groups and more particularly the free carboxylic groups contribute to lowering the bitterness of the amino acids.

14.2.1.6 Special peptides

The abundance of recent data on **amphiphilic peptides**, preferentially in an α-helix form (Fig. 14.1), active in translocation, anchorage (signal peptides) and membrane lysis (cytolic peptides of the mellitin type) can be exploited in the food sector.

The interfacial properties of this type of oligopeptide as well as its capacity for interacting with the phospholipids can be useful in formulating foams, emulsions and films. Some of these peptides are also suitable for use in the manufacture of liposomes for the agro-food and pharmaceutical industries.

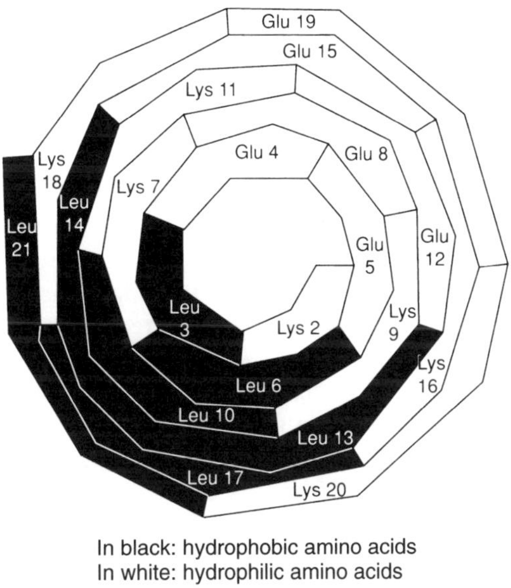

Fig. 14.1 Axial projection of an amphiphilic peptide forming an α helix.

Following the formation of a peptide link between the NH_2 and COOH terminal groups of the peptide chain, we can obtain a **dicetopiperazine** with the following formula:

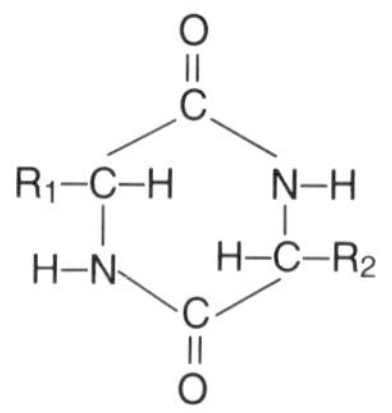

R_1 and R_2 correspond to the side chains of the two cyclic amino acids. A cyclic peptide in a casein hydrolysate with a cyclic formula (Leu-Trp) has been discovered. This cyclisation reinforces a fault: the intensity of bitterness of this dipeptide increases by blocking the two end-groups, and especially that of the C terminal one. Heat is also an important factor in the formation of dicetopiperazine. For example, the bitter taste of chocolate probably comes from the mixture of theobromine and cyclo (Gly-Phe) in the quantity of 2/1 during the roasting of cocoa beans. A strong interaction probably takes place between the two compounds, via the hydrogen bonds between the amide groups of the two molecules.

Storing sake at relatively high temperatures also leads to the formation of dicetopiperazine, a cyclo (Pro-Leu) which makes this drink bitter when it ages. This cyclic dipeptide also intervenes during the fermentation of other drinks such as sherry, *raoshu* and even during the fermentation of soy sauce.

14.2.2 Biologically active peptides from the hydrolysis of food proteins

For some time the products from the digestion of proteins contained in foods have been considered solely as a supply of essential and non-essential amino acids and of the nitrogen indispensable to the biosynthesis of the proteins and nucleic acids in the organism. However, it has now been clearly established that the proteins from extremely varied media (milk, cereals, blood, flesh from fish and seafood) contain chains of amino acids in their primary sequence which, once freed, are capable of influencing the physiology of the organism.

So, several opiate, anti-opiate, anti-hypertensive, mineral transporter, immunomodifier, platelet aggregation inhibitor, mitogenous and bombesic peptides have been clearly shown to further the digestion of various proteins, in particular by enzymes from the digestive tract.

These principal activities, as well as most of the peptides that are responsible for them, are summarised in Table 14.3. We can review the properties of these peptides.

Table 14.3 Principal biological activities revealed in peptides resulting from the hydrolysis of proteins of food origin

Activity	Protein source	Type of hydrolysis	Method of revelation	Sequence
Anti-opiate	Human lactoferrin	Pepsic	Competition for binding on the opiate receptor with the [^{3}H] naloxone	$_{318}$Tyr-Leu-Gly-Ser-Gly-Tyr$_{323}$-OCH_3 IC_{50} = 22.5 μM $_{536}$Arg-Tyr-Tyr-Gly-Tyr$_{540}$-OCH_3 IC_{50} = 10.0 μM $_{673}$Lys-Tyr-Leu-Gly-Pro-Gln-Tyr$_{679}$-OCH_3 IC_{50} = 15.3 μM
	Bovine κ casein	Pepsic	Raised inhibition of contraction of ileum in guinea pigs	Casoxin A: $_{35}$Tyr-Pro-Ser-Tyr-Gly-Leu-Asn$_{41}$ Casoxin B: $_{58}$Tyr-Pro-Tyr-Tyr$_{61}$
	Bovine κ casein	Trypsic	Raised inhibition of contraction of ileum in guinea pigs	Casoxin C: $_{25}$Tyr-Ile-Pro-Ile-Gln-Tyr-Val-Leu-Ser-Arg$_{34}$
Opiate	β chain haemoglobin	Pepsin and pancreatin	Raised inhibition of contraction of ileum in guinea pigs Binding on the opiate receptor	Haemorphin 4 (H4): $_{34}$Tyr-Pro-Trp-Thr$_{37}$ Haemorphin 5 (H5): $_{34}$Tyr-Pro-Trp-Thr-Gln$_{38}$
	Mitochondrial cytochrome b	Gastro-intestinal enzymes	Raised inhibition of contraction of ileum in guinea pigs Binding on the opiate receptor	Cytochrophin-5: $_{345}$Tyr-Pro-Phe-Thr-Ile$_{349}$ IC = 290.4 μM Cytochrophin-4: $_{345}$Tyr-Pro-Phe-Thr$_{348}$ IC_{50} = 120.1 μM
	Bovine α_{s1} casein	Pepsic	Inhibition of adenylate cyclase Inhibition of contractions of the vas deferens in mice Competition with [$_3$H] dihydromorphine and [^{3}H] met-encephaline-amide	Exorphins: $_{90}$Arg-Tyr-Leu-Gly-Tyr-Leu-Glu$_{96}$ $_{91}$Tyr-Leu-Gly-Tyr-Leu-Glu$_{96}$ $_{90}$Arg-Tyr-Leu-Gly-Tyr-Leu$_{95}$ $_{91}$Tyr-Leu-Gly-Tyr-Leu$_{95}$

Table 14.3 *Continued*

Activity	Protein source	Type of hydrolysis	Method of revelation	Sequence
	Bovine β casein	Multiple	Inhibition of contraction of ileum in guinea pigs *In vivo* analgaesia in rats Binding on the opiate receptor	β-casomorphin-7: ${}_{60}$Tyr-Pro-Phe-Pro-Gly-Pro-Ile${}_{66}$ β-casomorphin-5: ${}_{60}$Tyr-Pro-Phe-Pro-Gly${}_{64}$
Inhibition of platelet aggregation	κ casein	Trypsic	Measurement of platelet aggregation Inhibition of binding of fibrinogen I^{125}	${}_{106}$Met-Ala-Ile-Pro-Pro-Lys-Lys-Asn-Gln-Asp-Lys${}_{116}$
	Human lacto-transferrin	Choice in sequence	Inhibition of binding of fibrinogen I^{125}	${}_{39}$Lys-Arg-Asp-Ser${}_{42}$
Bombesic	Bovine and human milk	None	Marked rabbit antibody against the Gly-Asn-Gin-Trp sequence	
Mitogenous	Human β casein	Trypsic	Incorporation of 3H-thymidine by the cells BALB/c3T3	${}_{1}$Arg-Glu-Thr-Ile-Glu-Ser-Leu-Ser-Ser-Ser-Glu-Glu-Ser-Ile-Pro-Glu-Tyr-Lys${}_{18}$ ${}_{105}$Gln-Pro-Thr-Ile-Pro-Phe-Phe-Asp-Pro-Gln-Ile-Pro-Lys${}_{117}$
Immunostimulant	Human β casein	Trypsic	Increase of mice resistance to infection by *Klebsiella pneumoniae*	${}_{54}$Val-Glu-Pro-Ile-Pro-Tyr${}_{59}$
	Bovine β casein	Trypsic and chromotrypsic	Increase of mice resistance to infection by *Klebsiella pneumoniae*	${}_{63}$Pro-Gly-Pro-Ile-Pro-Asn${}_{68}$
ACE inhibitor	Bovine α_{s1} casein	Trypsic	*In vitro* inactivation of hydrolysis of a Hip-His-Leu substrate by ACE	CEI_{12}: ${}_{23}$Phe-Phe-Val-Ala-Pro-Phe-Pro-Glu-Val-Phe-Gly-Lys${}_{34}$ IC_{50} = 77 μM
	Bovine α_{s1} casein	Trypsic	Measurement of blood pressure in rats anaesthetised after injecting fractions	$CEI\alpha_{c6}$: ${}_{193}$Thr-Thr-Met-Pro-Leu-Trp${}_{199}$ IC_{50} = 16 μM

CEI_{12} (α_{s1} casein)	Endopeptidase protein from flavobacterium aeningosepticum	Inhibition of the contractions of the uterus and ileum in guinea pigs through inactivation of bradykinin	CEI_5: $_{23}$Phe-Phe-Val-Ala-Pro$_{27}$ $IC_{50} = 6$ µM
Bovine β casein	Trypsic	Inhibition of the contractions of the uterus and ileum in guinea pigs through inactivation of bradykinin	$CEI\beta_7$: $_{177}$Ala-Val-Pro-Tyr-Pro-Gln-Arg$_{183}$ $IC_{50} = 15$ µM
Latex from *Ficus carica*	Heat	Inactivation of hydrolysis of Hip-His-Leu by ACE *in vitro*	FLP-1: Ala-Val-Asn-Pro-Ile-Arg $IC_{50} = 14$ µM FLP-2: Leu-Tyr-Pro-Val-Leu $IC_{50} = 5$ µM FLP-3: Leu-Val-Arg $IC_{50} = 14$ µM
Maize endosperm	Thermo-lysin	Inactivation of hydrolysis of Hip-His-Leu Measurement of blood pressure *in vivo* in anaesthetised rats	Val-His-Leu-Pro-Pro $IC_{50} = 200$ µM Val-His-Leu-Pro $IC_{50} = 15$ µM Leu-Pro-Pro $IC_{50} = 9.6$ µM
Flesh from fish and seafood	Pepsin	Reduction of ACE activity	Sardine: $ID_{50} = 27.1$ mg Pearl oyster: $ID_{50} = 28.8$ mg

ACE: enzyme-converting angiotensin I.
CEI: inhibitor of the enzyme-converting angiotensin I.
IC_{50}: concentration of the molecule tested for which 50% of the maximum effect is attained.
ID_{50}: mass of the molecule tested for which 50% of the maximum effect is attained.

14.2.2.1 *Opiate and anti-opiate peptides*

The link between schizophrenic psychosis and the diet in persons predisposed towards schizophrenia was the starting point for the research into peptides from food proteins which have an opiate activity. In fact there is a close correlation between the consumption of milk or cereals and mental disorders or the coeliac disease syndrome. These problems disappear when gliadin is eliminated from the diet, but reappear if milk or soya is added to this intake. These results have been confirmed in several animal species in which gluten hydrolysates, injected into the blood stream or taken by mouth lead to modifications in behaviour.

The morphino-mimetic action of these peptides has led researchers to name them exorphins, given their exogenous origin in comparison with the endorphins. Studies of the sequence of the various exorphins produced by the hydrolysis of several proteins underline the importance of a Tyr-X-Phe or Tyr-X_1-X_2-Phe sequence in the C-terminal side of the peptide.

Exorphins also play an important role in appetite control and food intake. If naloxone (a specific opiate inhibitor) or an endorphin (Met or Leu-enkephaline) is injected intravenously this effect disappears.

Other proteins are sources of opiate peptides; for example peptide fragments 34–37 and 34–38 from the β chain of haemoglobin – known as haemorphins 4 and 5 – and a fragment of mitochondrial cytochrome b, cytochrophine-5, display an opiate activity.

On the other hand, peptides produced from the digestion of human lactoferrin which have an affinity for the morphine receptor, but do not transmit opiate activity, have been discovered. These peptides have been shown to be opiate antagonists and will be referred to as lactoferroxine A, B and C as a reminder of casoxine, the anti-opiate peptide in casein.

14.2.2.2 *Anti-hypertensive peptides*

The biological system responsible for vasoconstriction (constriction of blood vessels) involves the kidneys, the blood and the adrenal glands through the intermediary rennin, angiotensin and aldosterone. The enzyme which converts angiotensin (ACE) is a dipeptidylcarboxypeptidase (EC 3.4.15.1) responsible for the conversion of inactive angiotensin I into angiotensin II, the most powerful of the known vasoconstrictors.

An examination of the sequence of peptides inhibiting ACE reveals Pro-Pro, Ala-Pro or Ala-Hyp sequences on the C-terminal side. This observation leads us to look for peptides capable of inhibiting ACE in protein sources (caseins, fish flesh, cereals, latex) whose constituents contain regions rich in proline.

14.2.2.3 *Mineral transporter peptides*

α_{s1}-, α_{s2}- and β-caseins contain in their sequences numerous phosphoseryl residues which give the corresponding peptides, named casein phospho-

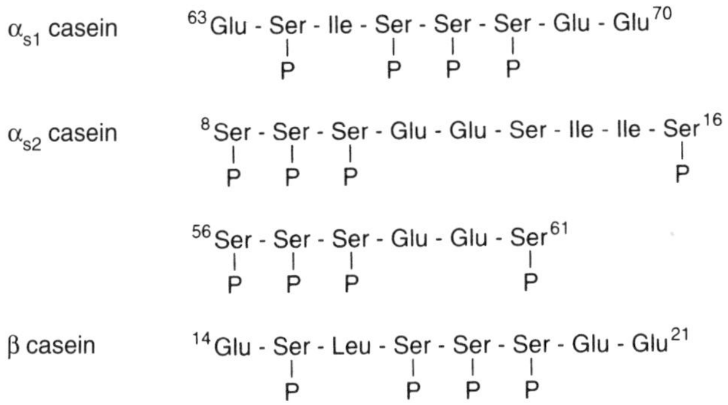

Fig. 14.2 Sequences of phosphopeptides of bovine caseins.

peptides, a strong chelating power towards certain divalent cations (Ca^{2+}, Mg^{2+}, Fe^{2+}, Zn^{2+}, Cu^{2+}) (Fig. 14.2).

Figure 14.3 shows one of the preparation diagrams suggested by INRA (the French National Institute for Agricultural Research) researchers in Rennes, starting with a solution of Na or K caseinate. This involves the following stages:

- Pancreatic hydrolysis of a caseinate solution adjusted to pH 8.0 by means of a membrane enzymatic reactor.
- Complexation of phosphopeptidic sequences by adding calcium chloride and disodium phosphate, which results in polymerisation up to a high molecular size.
- Separation and purification of these phosphopeptidic sequences by using ultrafiltration.

The phosphopeptides obtained using this ingenious process are characterised by a Ser/total amino acid ratio of between 8 and 20% depending on the working methods used and the starting product. At pH 7.2, 90% of the Ca and the P are combined with the peptides. In spite of having a high calcium phosphate content these peptides have excellent solubility, in the order of 25%.

14.2.2.4 Immunomodifier peptides

Observations have shown that new-born breast-fed babies are more resistant to bacterial infection than those who are not breast-fed. In fact, milk, and especially colostrum, contains numerous immune factors such as the immunoglobulins IgA, IgM, IgG, as well as enzymes (lysozyme and lactoperoxidase) or iron proteins (lactoferrin and transferrin).

By analogy with the immunomodifying activities of peptides produced by bacteria, these activities have been discovered at the level of the peptides produced by the hydrolysis of human casein.

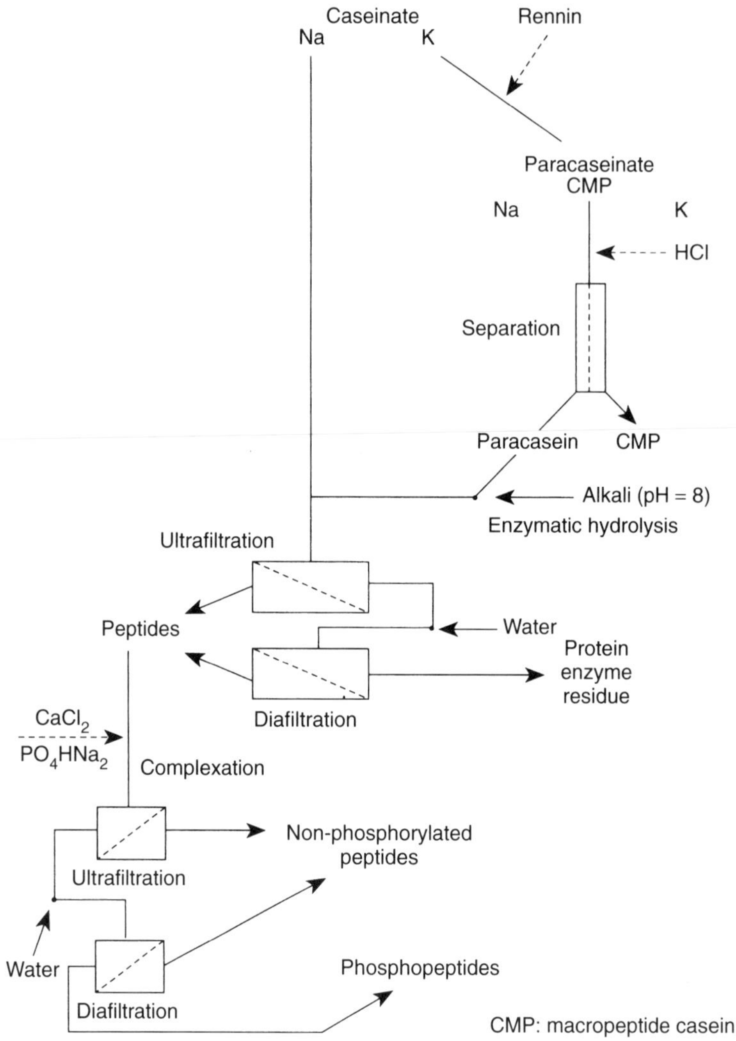

Fig. 14.3 Preparation of phosphopeptides from caseins.

14.2.2.5 Antithrombotic peptides

The positive results of two tests have shown that the 106–116 fragment of κ-casein (homologue from the 400–411 region of fibrinogen) inhibits the binding of human fibrinogen on the platelet receptors; however, peptide 106–116 from the trypsic hydrolysate of caseinomacropeptide from mice has been shown to be more active. In addition, the sequence Lys_{39}-Ser_{42} (KRDS) of human lactoferrin presents the same activity.

Other biological activities (bombesic, mitogenous) are shown in Table 14.3. Studies to confirm these activities are under way.

15

Pigments and aromas

The marketing of more and more elaborately formulated products, manufactured from fractionated and reconstituted raw materials, which bring together constituents of various origins, is making additives extremely important, in particular pigments and aromas. In keeping with the approach of the rest of the book, this chapter will deal only with natural food colourants and aromas.

The meaning of the term 'natural' can vary from one country to another, but the generally accepted definition is that provided by the Code of Federal Regulations of the USA (21 CFR 101-22.a.3). In addition to natural extracts from plants and animals, natural substrates modified by microorganisms or enzymes and cultures of plant tissues meet the currently accepted definition of natural products.

15.1 Natural pigments

The majority of natural colourants are of plant origin. They can be pure compounds or products of extraction. The latter come from edible materials and can be combined with other molecules.

The principal natural pigments belong to three broad categories:

1 Porphyrin pigments, including chlorophylls, haem pigments (for example, myoglobin and haemoglobin, dealt with in Chapter 7).
2 Carotenoids, including β-carotene, a precursor of vitamin A, lycopene, and the xanthophylls.
3 Flavonoids and their by-products.

We should also mention the following: tannins, betalaines, quinones, xanthones, etc.

15.1.1 Chlorophylls (E 140)

Chlorophyll is the green pigment in plants, but comprises a group of at least four fairly similar derivatives whose proportions vary according to the plants in which they are found. Commercial chlorophyll is soluble in water, ethanol and oil, and produces a dark green liquid. From a chemical point of view, chlorophylls have a tetrapyrrolic nucleus similar to that of haem, but with an atom of Mg^{2+} in the centre. A long hydrocarbon side chain gives the molecule its lipid-soluble character.

The magnesium can be replaced by copper, thus obtaining copper compounds, with saponified ester functions which improve water solubility. These soluble copper forms are more stable. However, as they do not exist in nature, they are not recognised as being natural.

15.1.2 Carotenoids (E 160) and xanthophylls (E 161)

Carotenoids are natural pigments, extremely widespread in nature and are the source of brilliant colours: yellow, orange and red in numerous edible fruits (lemons, peaches, apricots, oranges, strawberries, cherries, etc.) and in vegetables (carrots, tomatoes, etc.) in fungi (chanterelle) and flowers. They are also present in animal products such as eggs, lobsters, greyfish and various types of fish (Fig. 15.1). Over four hundred have been listed.

From a chemical point of view, these are chains of isoprene units which sometimes have, at the end of the chain, a cyclic nucleus which carries various functional groups (alcohol, acid, etc.). The best known are β-carotene and lycopene but others are also used as food colourants: α-carotene, γ-carotene, bixin, norbixin, capsantein, lycopene, β-apo-8′-carotenal, the ethyl ester of β-apo-8-carotenic acid. These are fat-soluble compounds, but the industry manufactures water-dispersible preparations by formulating colloid suspension, by emulsifying the carotenoids or by dispersing them in the appropriate colloids. Their colour varies from yellow to red.

β-Carotene is found in the form of a crystalline powder. It is insoluble in water and ethanol, and not very soluble in vegetable fats. In chloroform the maximum spectrometric absorption is found between 466 and 496 nm. It has a vitamin A activity: 1 g of β-carotene corresponds to 1.67 million international units (IUs) of vitamin A and the vitamin activity of 0.6 μg of β-carotene is almost equivalent to 0.3 μg of vitamin A. β-Carotene is sensitive to oxygen (air), heat, light and humidity.

The anti-oxidant properties of β-carotene are currently the subject of special attention because they may be involved in the mechanisms of preventing certain types of cancer. So, certain companies are interested in the

Name	Structure	Max. absorption colour (nm)
β-Carotene	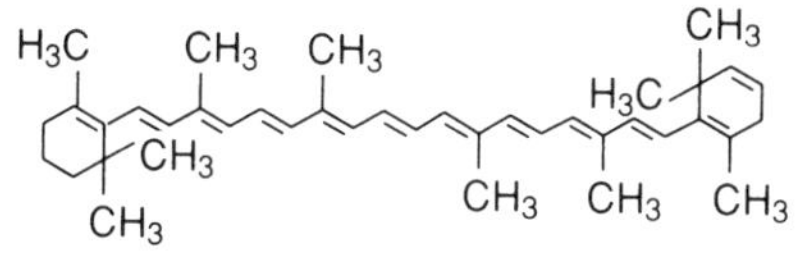	466 and 496 orange
Bixin	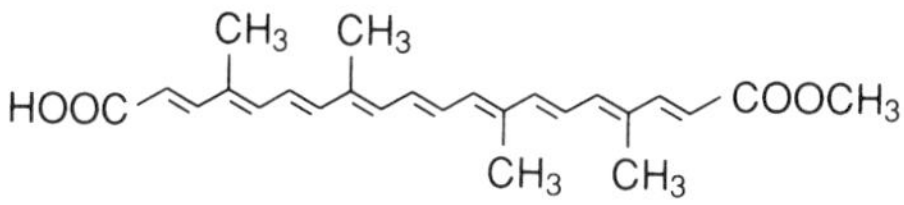	475 and 509 orange
Capsantein (capsorubin or capsanthin)	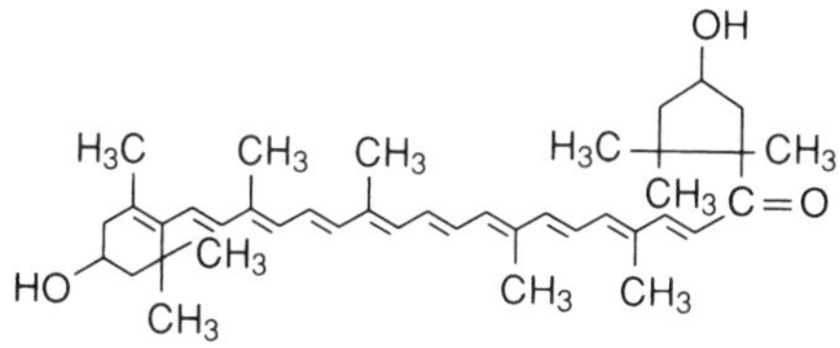	483 red
Lycopene	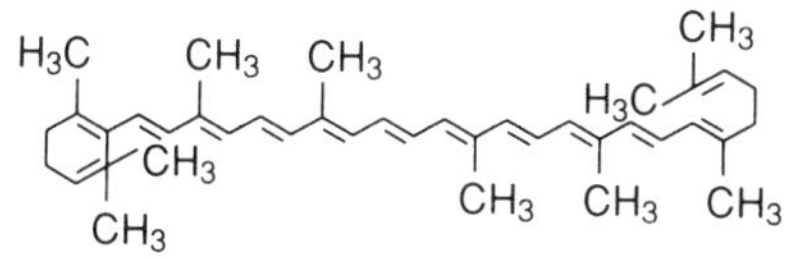	446, 472 and 505 red
Lutein	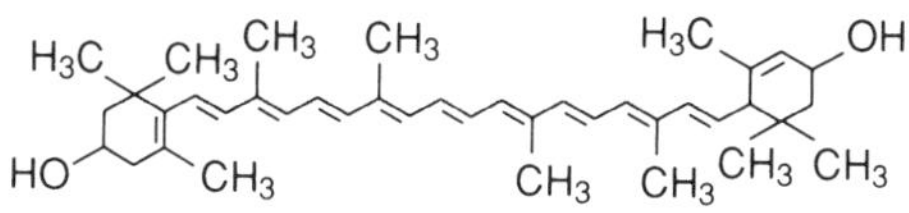	453 and 481 yellow
Cryptoxanthin	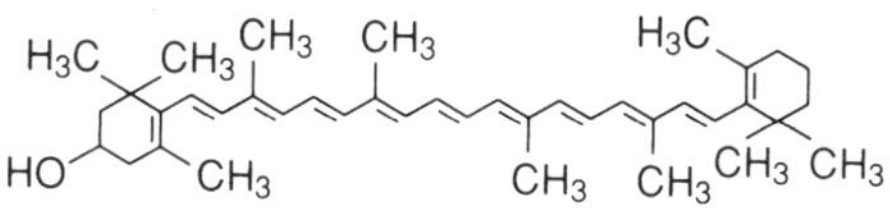	452 and 480 yellow
Cantaxanthin	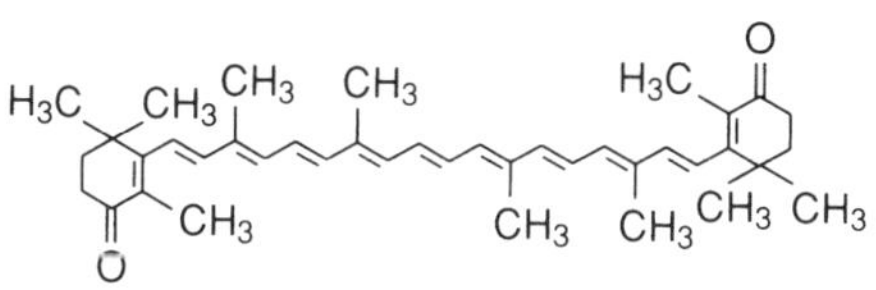	464 and 467 violet

Fig. 15.1 Structure of the principal carotenoids and xanthophylls.

large-scale manufacture of β-carotene of natural origin. In the spirullina, photosynthetic micro-algae rich in proteins, the level of β-carotene can reach 10% of the dry matter. Currently, 300 hectares of salt lakes are devoted to growing a variety of spirullina (*Dunaliella salina*) and the β-carotene extracted from algae is crystallised and dispersed in soya oil at a 30% concentration.

α- and γ-carotene are distributed in much lower quantities, exhibiting physicochemical and colourant characteristics very similar to those of β-carotene.

Bixin is one of the coloured constituents of the annatto seed envelope. The annatto tree (*Bixa orellana*) is grown in the tropics. Bixin is soluble in oils and fats; this solubility increases with the degree of unsaturation of the oil, until it stabilises at around 0.1%. It is soluble in chloroform, pyridine and glacial acetic acid. On the other hand, norbixin is less soluble in fats than bixin, and has good solubility in alcohols and alkaline solutions.

Heat stable up to 100 °C, bixin possesses the basic 'skeleton' of carotene, a carboxyl group makes up one of its extremities, with the other one being an esterified group. The carbonyl group and the double bonds explain the very strong resonance of its structure, thus giving it good stability and a very high colourant power.

In this category we can also mention capsantein (capsorubine, capsanthine) which is a red pigment, extracted from paprika *Capsicum annuum,* a mild edible pigment.

Lycopene is the red colourant in ripe fruits, especially tomatoes. It has maxima of absorption at 446, 472 and 505 nm (for the *trans* form). It is soluble in chloroform and benzene, and virtually insoluble in methanol and ethanol.

β-Apo-8′-carotenal (*trans* form) is widespread in nature (citrus fruits, vegetables, grass) but often a synthetic carotenoid is used, in the form of a fine purple crystalline powder, insoluble in water, slightly soluble in ethanol and vegetable oils and very soluble in chloroform. This pigment is heat sensitive.

Xanthophylls are pigments very similar to carotenoids, generally with hydroxyl or ketone replacements on the nucleus (Fig. 15.1). They are distributed in the leaves and petals, where they represent almost 10% of the colouring matter. Their solubility in ethanol is greater than that of the carotenoids.

The most frequently used food colorants are lutein (yellow), cryptoxanthin (yellow), flavoxanthin (yellow), violaxanthin (orange), rubixanthin (orange), rhodoxanthin (red) and cantaxanthin (violet).

15.1.3 Flavonoids and by-products

This category includes the pigments known as **anthocyanins** which are soluble pigments, extremely widespread in nature and responsible for the blue, purple, red and sometimes even orange coloration of numerous fruits and vegetables. Over two hundred anthocyanins have been identified. Red anthocyanins take the form of a flavylium cation. This flavylium is generally represented as an oxonium ion carrying a positive charge on the oxygen atom of the heterocycle. This characteristic property gives them an amphoteric character which brings about changes in colour depending on the pH

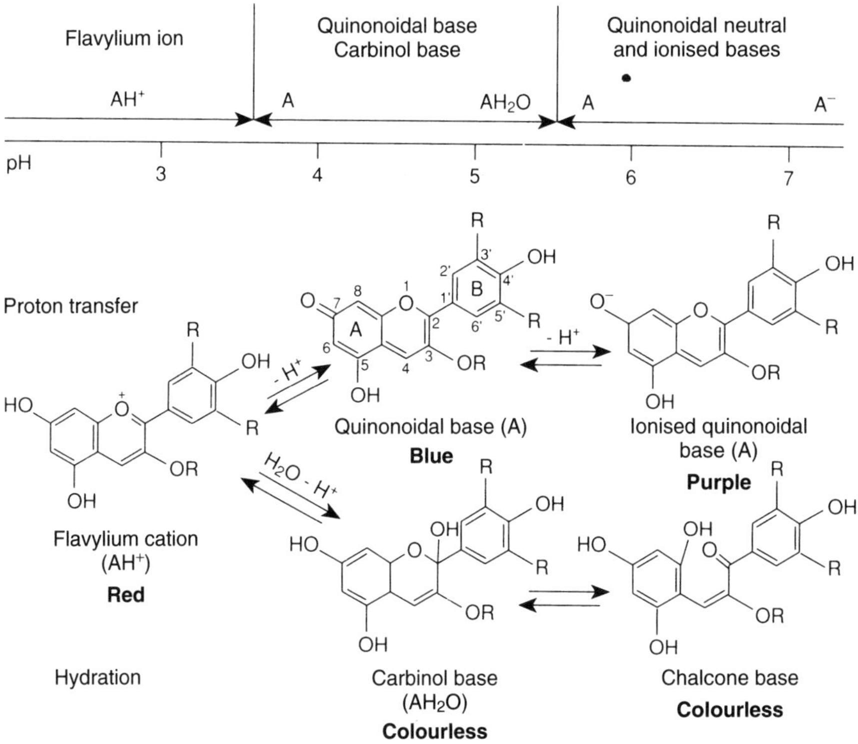

Fig. 15.2 Chemical reactivity of the anthocyanins depending on the pH.

(Fig. 15.2). In practice, the anthocyanins used as food colourants supply colours going from purplish red to cherry red and can be used at a pH of between 3.5 and 5.5. Anthocyanins have the advantage of being relatively insensitive to heat and light.

The anthocyanins used in food (E 163) can only be obtained from edible fruits and vegetables such as strawberries, blackberries, cherries, plums, raspberries, blackcurrants, redcurrants, red cabbage, red onions, bilberries, aubergines, grapes, elderberries, etc.

Only six **anthocyanidins** are important within food technology. In addition to the presence of an OH group on carbons 3, 5 and 7, they can be differentiated from each other by the replacement of hydroxyl or methoxyl groups on the 3′, 4′ and 5′ carbons of the β nucleus.

These compounds, which are red, blue or purple depending on their structure, are generally found in the form of anthocyanosids, that is to say with one or several carbohydrates. These pigments change colour either according to the pH or when the carbohydrate group is separated by hydrolysis. Certain flavonoids are colourless, for example leucocyanidol. When oxidised (by heating in an acid medium), these colourless compounds undergo a conversion into anthocyanidins and the colour changes from pink to red (this is what

Table 15.1 Flavonoid composition of a bitter orange* extract (expressed as a percentage after integration of an HPLC chromatogram)

Flavonones		Flavones		Flavonols
Neoeriocitrin	2.0	Rhoifolin	1.5	0.5
Isonaringin	1.1	Neodiosmin	0.7	
Naringin	21.8	Diosmin	0.2	
Hesperidin	1.7			
Neohesperidin	12.0			
Naringenin	0.4			
Poncirin	0.6			
Hesperitin	0.5			

* Approximately 2.5 times more bitter than caffeine.

happens in certain varieties of apples, pears, cabbages, beans, etc.). This is why their preferred name is proanthocyanidins. These molecules, still known as pycnogenols (pharmaceutical term), are attracting the interest of researchers. They are concentrated substances, found in significant proportions in grape seeds, and are powerful captors of oxygenated free radicals. They appear to have anti-inflammatory and anti-viral properties.

Flavonols of a yellow colour are generally characterised by the presence of a carbonyl group in position 4 and a hydroxyl group in position 3. Unlike the anthocyanidins, the carbohydrate group is usually combined in position 7. Among the most widespread flavonols are quercetol and myricetol, in particular.

Flavonones, which do not contain an OH group in position 3, have strong structural similarities to the flavonols. This category includes the flavonoids responsible for the bitter flavour of some grapefruits, lemons, oranges: naringin (naringenol combined with glucose and rhamnose) and hesperidin.

In an extract of bitter oranges, naringin and neohesperidin are in dominant concentrations (Table 15.1). In addition, hydrogenation of the neohesperidin can produce neohesperidin dihydrochalcone (neohesperidin DC). This has a sweetening power which is 1500–1800 times higher than that of sucrose (Fig. 15.3).

15.1.4 Other compounds

Alongside these large categories of pigments, foods of plant origin contain numerous phenolic compounds which, by enzymatic conversion, can produce coloured polymers, usually brown or black.

15.1.4.1 Tannins

Tannins are a good example of this type of compound, because they can be split into two large groups: condensed tannins (still called cathecol tannins)

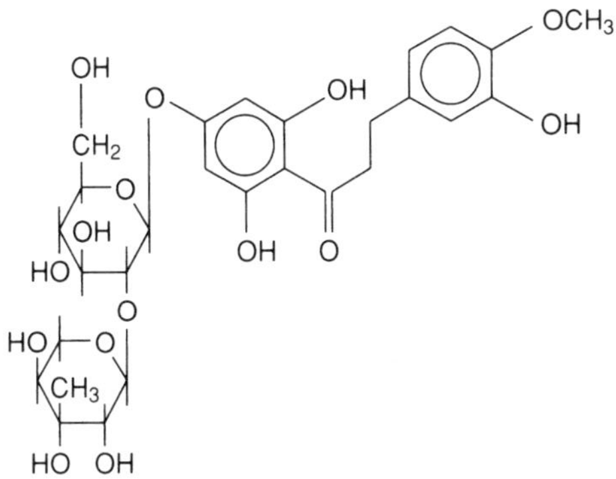

Fig. 15.3 Structure of neohesperidin dihydrochalcone.

whose chemical structure is very close to that of the anthocyanidins and the hydrolysable or pyrogallic tannins which are produced from the esterification of the five alcohol functions of glucose by various polyphenolic acids.

15.1.4.2 Betalains

The appearance of the betalains seems to resemble that of the anthocyanidins, which is how they came to be given the incorrect name of nitrogenous anthocyanidins. They all have the same basic structure and their colour comes from the resonance between the different mesomeric structures (Fig. 15.4).

The root of edible red beetroot (*Beta vulgaris*) contains numerous pigments among which are red betalains (betanin, betanidin, prebetanidin) and yellow betaxanthins (vulgaxanthin I and II). Maceration and aqueous extraction produces a dark red syrup and then a very water-soluble powder. The red of beetroot (E 162) is very tolerant to variations in pH (from 3.5 to 7).

15.1.4.3 Quinones and xanthones

Quinones and xanthones also produce a red pigment which is very widespread in flowers, fungi, algae and bacteria. Among the two hundred compounds that have been identified, the naphtoquinones are the most interesting as colourants or purgatives.

15.1.4.4 Pigments resulting from a browning reaction

The **pigments** that form by enzymatic browning are given the general name of **melanins**. Their final colour is brownish-black, but intermediate shades are observed: pink, red, bluish-black. The formation of these polymers from quinones takes place without the intervention of enzymes.

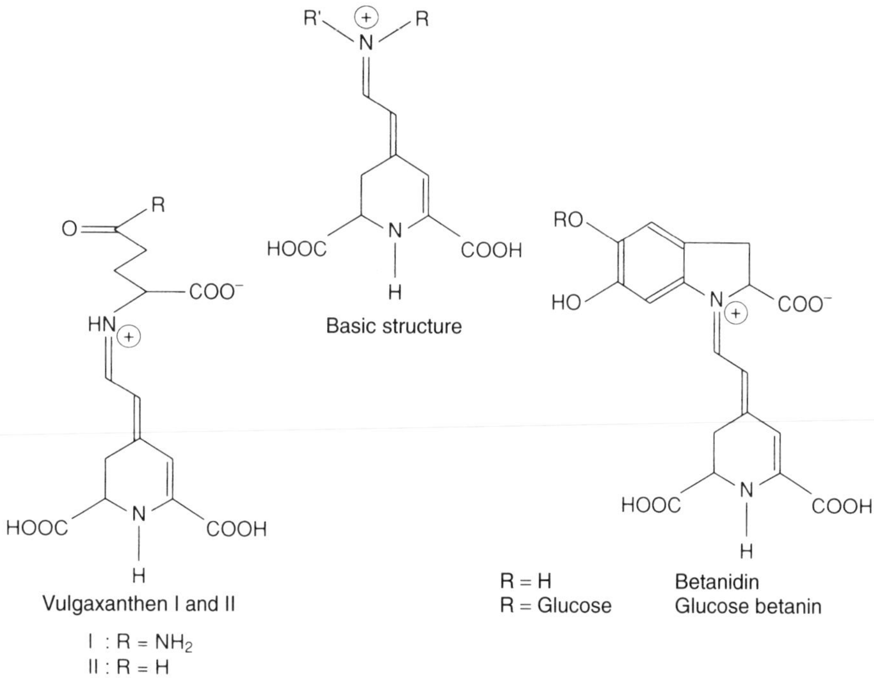

Fig. 15.4 Structure of betalains.

Caramel (E 150) is a brown culinary colourant which has been known for a great many years and is traditionally obtained by heating sucrose (critical temperature: 170 °C). This browning of sucrose or other food carbohydrates is regulated by adding small quantities of ammonia (ammoniacal caramel), carbonate or traces of mineral acid during heating. Caramel is soluble in water and dilute ethanol solutions, and insoluble in organic solvents.

15.1.4.5 Miscellaneous pigments

Turmeric (E 100)
Turmeric is one of the ingredients in curry. It is extracted from *Curcuma longa* or saffron, a plant from the Zingiberacaea family grown in the Far East and Madagascar. It is a brownish yellowy-orange powder with purplish highlights, and can be crystallised in methanol. It is insoluble in water and ether, weakly soluble in ethanol and glacial acetic acid (light yellow solution), soluble in alkalis (brownish-red solution). The purity of the preparations is higher than 99%.

Cochineal – carminic acid (E 120)
This colourant is obtained from the dried bodies of the females of the insect *Coccus cacti* containing most importantly eggs and young larvae (15 000

insects are needed to obtain 100 g!). This insect lives in Central America, in the Canary Islands, in North Africa and in southern Spain. The product obtained contains a certain proportion of carminic acid (between 10 and 15%) which is a bright red pigment.

Carminic acid is soluble in water (red solution in a neutral medium with absorption maximum at 500 nm, yellow solution below pH 4), ethanol and sulphuric acid, and is practically insoluble in organic solvents. In order to remedy this instability, mixtures consisting of carminic acid, alum (aluminium and potassium sulphate) and a pH-regulating agent have been suggested.

Riboflavin (E 101)

Riboflavin (or vitamin B2) is obtained from natural sources (yeasts, wheat germ, eggs, animal liver, etc.) or by synthesis. It is soluble in water, but insoluble in oils and organic solvents. Aqueous solutions are yellow with optimum green fluorescence at pH 6–7 and can be successfully stored in cool conditions away from light, except in an alkaline solution.

Riboflavin forms part of the structure of the nucleotides FMN and FAD and as such is involved in the phenomena of cellular respiration and oxidative phosphorylation. The daily requirement for humans is approximately 1.5 mg.

15.1.4.6 New sources of food colourants of natural origin

The lack of a stable red colourant has led to the interest in sandalwood. Three pigments (**Santalin A**, **B** and **C**) have been isolated. These colourants are soluble in ethanol, light sensitive and stable under heat. The physicochemical characteristics of these pigments are close to those of the anthocyanins and share a flavonoid structure. In practice, santalins often take an ionic form with a positive charge located on the oxygen, which is why their behaviour resembles that of the anthocyanins towards changes in pH (Fig. 15.5).

They are used mainly with concentrated fruit juices, sorbets, liqueurs, cocktails, tomato sauce, jams, fruit purées, vegetable fats and smoked fish. The colourants in sandalwood produce an orangey-red colour up to pH 5.0 and beyond this they move on to purplish shades.

The **rose hip** (*Hibiscus sabdariffa*) is a bush whose fruits are used to make jams and jellies, drinks or aromatic extracts. The fibres can be woven to make fabric or fishing nets, the leaves can be eaten in salads and the seeds produce oil. The flowers contain large quantities of anthocyanins produced from cyanidin and delphinidin: 1.5 g/100 g of dry matter. The clear red colour of these pigments is used to colour fruit-based products. Their thermal stability is comparable to that of grape anthocyanins.

Monascus is a micro-organism which produces several types of orange or red pigments: monascin, and monascorubin in particular. The colouring appears when the molecule reacts with an amino group. These pigments are

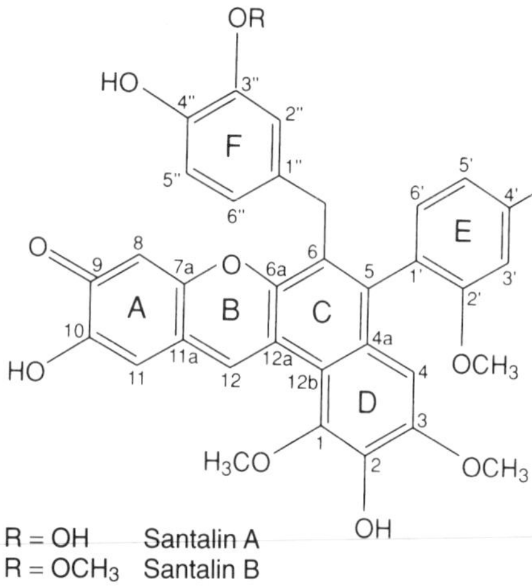

Fig. 15.5 Structure of santalins.

stable between pH 2 and 10. Meat or sausages can be coloured easily and the colour is resistant to cooking, and has no tendency to migrate. However, this type of pigment has not yet received authorisation for use in Europe.

15.2 Aromas

15.2.1 General information and definitions

Aromas are very complex mixtures of raw materials and semi-manufactured products, of natural and/or synthetic origin. There are often hundreds of constituents which make up an aroma and give it its very special organoleptic characteristics. Each basic constituent generally has variants which reveal its origin. Aromas have clearly defined olfactory and gustatory properties. When incorporated in foods they can give them their characteristic and agreeable smells and flavours. Aromas therefore constitute an important and essential factor in the manufacture of food products.

In fact, the technologies used to manufacture foods generally cause all or some of the most characteristic aroma molecules to disappear. In addition, adding aromas in the form of additives allows manufacturers to combat problems of supply to a certain extent. Industrial aromas are available all year round whereas fruits and vegetables produce crops only in season.

The word 'flavouring' has been introduced by specialists to describe this step. This can be defined as follows: 'Substances or preparations added to a food (or a drink) in order to give it a new aroma or to modify the existing one.' This fairly broad definition establishes that the aim of the flavouring is to give the food an aroma, so its role is therefore the same as that of the actual aroma itself.

As the flavour of a food seems to be so important to our experience of food, many specialists consider that the substances added for this purpose should not be considered to be additives but actual ingredients. These ingredients can take the form of powders or liquids of varying viscosity which are easy to store; in this way their aromatic power can be controlled more easily.

In addition, the composition of these aromas can be adapted to the particular technological process that the food product is to undergo, that is to say they can be prepared so as to withstand a method of manufacture involving particular constraints, such as extrusion-cooking, or special preparation before consumption, such as heating up in a microwave oven.

To sum up, an aroma is an essential element in our experience of a food. These substances have a very direct action on the assimilation capacity of the other ingredients: encouraging the individual to eat certain foods, secretion of saliva or gastric juice. Aroma therefore has a direct nutritional role; flavourings have the same role which cannot be considered as a subordinate one.

15.2.2 Different classes of aroma – physico-chemical characteristics

To date, over 10000 odorant molecules have been identified. This number alone is sufficient to illustrate the complexity of this family of substances. Consequently, there have been several attempts to classify them into broad families. This classification is based on either the type of preparation or on the raw material or materials used, or on the aromatic note.

15.2.2.1 Aromatic preparations

Three large families are usually distinguished:

- Spices and aromatics: plants or parts of plants used fresh or dried.
- Condiments and processed aromatic preparations: substances prepared from the previous ones.
- Aroma formulations obtained by mixing extremely varied raw materials.

15.2.2.2 Aromatising raw materials

Depending on their method of manufacture, aromas can take the form of natural extracts, products of reaction or products of synthesis.

Natural extracts – essentially of plant origin – include the following:

- Essential oils or essences obtained either by cold, using mechanical processes (pressing citrus fruit skins, for example), or by steam distillation (distilled essences).

- Oleoresins obtained from tissues using solvent extraction. These are also known as resinoids or consistents. After extraction, fractional distillation is carried out in order to recover the solvent and collect the aroma compounds which contain very few of the most volatile compounds.
- Infusions, percolates, macerations, concentrates, etc.
- Distillates from alcohol preparations.
- Elementary compounds collected by crystallisation (for example: menthol).

The **products of reaction** obtained by subjecting certain plant or animal raw materials to simple processes: enzymatic, microbiological, physical. The use of enzymes in the manufacture of aromas is justified because of at least three advantages possessed by these biocatalysts:

- Specificity of the reaction, as in the example of L-menthol manufacture. DL-menthol is synthesised by hydrogenating the thymol which produces a racemic mixture of four isomers: menthol, iso-menthol, neo-menthol and isoneo-menthol (Fig. 15.6). After esterification of this mixture of terpenic alcohols and specific hydrolysis of the esters of the L-menthol by microbial esterases, fractional extraction using organic solvents can isolate the L-menthol in the pure state.
- Providing conditions for growth reaction.
- Reducing the formation of waste.

One of the most noteworthy of the various applications of enzymes in the development of aromas is their action on the aroma precursors.

Proteases and amylases, for example, hydrolyse proteins and starch. The resulting amino acids and simple carbohydrates can be used in new reactions involving non-enzymatic browning in order to create special aromatic nuances. This technology is used to produce certain culinary aromas, in particular those for meat:

- The extraction of the aromatic ingredients contained in fruits and plants.
- The production of aromas found in enzymatically modified products. Some examples of these are the esterases which can synthesise esters, the molecules making up numerous fruit aromas, and lipases which are responsible for cheese aromas, etc.

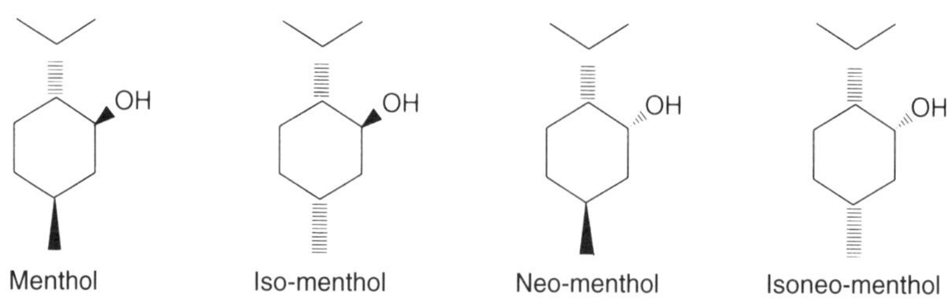

Fig. 15.6 Structure of menthol isomers.

For over sixty years we have known that micro-organisms can produce chemical structures with specific aromas. The list of all the abilities of the micro-organisms is still far from complete. There are two ways of manufacturing natural aromas using micro-organisms:

- The synthesis of aromas by specific organisms, cultivated on the appropriate and/or selective media: this is the fermentation method.
- The microbiological conversion of a natural precursor into an aroma: this is the method of bioconversion.

One illustration of this is the synthesis of lactones by yeasts. *Saccharomyces cerevisiae* produces hydroxy-3-dimethyl-5, 5 furanone-2 (HDMF) which is one of the ingredients of the aroma of sake and yellow wine from the Jura mountains in France.

The manufacture of complex aromatic mixtures using micro-organisms (for example aromas from bread, meat, dairy produce or fruit) is expanding rapidly and is of major interest to the food industry.

The techniques referred to above seem, together with cultures of plant tissues, to be opening the way to important developments within the field of aromatic preparations, since European legislation recognises aroma substances obtained biochemically as being natural.

In fact, one of the main problems that arises in the case of aromas of plant origin is that of permanence of supply. Thus, several research and development projects concern the formation of food aromas in cultures of plant cells.

Theoretically there are four methods of using plant tissue culture in order to manufacture aroma ingredients:

- Direct extraction of the ingredients from the cells or the medium.
- Enzymatic synthesis.
- Bioconversion.
- Use of the cellular mass for subsequent conversion.

This distinction between conversions of enzymatic and microbiological origin is artificial. We can look at the example of producing cheese aromas obtained in the course of the following reactions:

- Fermentation of lactose in order to obtain – depending on the micro-organisms present – lactic acid, acetoin, diacetyl, 2-butanol.
- Hydrolysis of lipids effected by esterases and lipases of bacterial origin with the release of fatty acids, the most volatile of which are involved in the aroma.
- Enzymatic degradation of the caseins by endogenous proteases (plasmin) and those of bacterial origin into amino acids which in turn undergo a certain number of enzymatic conversions.
- The appearance of secondary reactions between the volatile products resulting from the previous reactions.

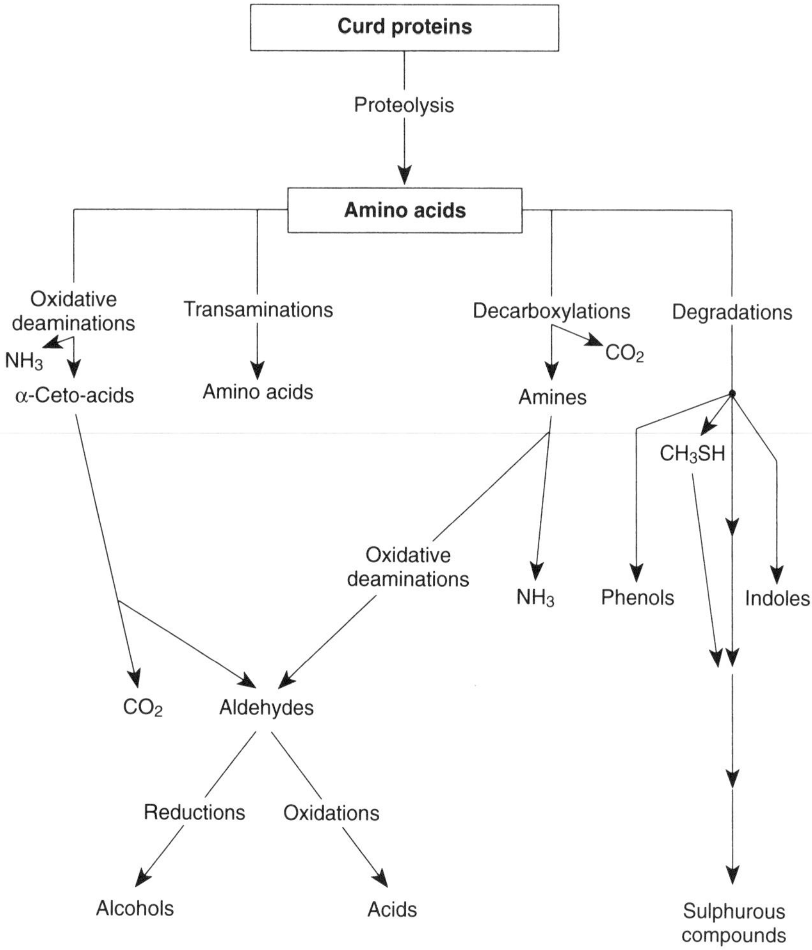

Fig. 15.7 Catabolism of amino acids during cheese ripening.

The aroma of a cheese is the result of all these reactions and the 'industrial reproduction' of cheese aromas requires perfect mastery of their chains of reaction (Fig. 15.7). So, given the still unidentified role of a large number of reactions, it has not yet been possible to obtain any noteworthy success in terms of the 'artificial' production of a cheese aroma.

In addition, processes which have been established for many years, such as, for example, roasting and cooking, do, however, retain indisputable advantages. These processes produce, in particular, volatile ingredients such as, for example, nitrogenous and/or sulphur heterocycles.

An illustration of this is the sulphur product dimethyl-3,5-trithiolane-1,2,4 which has been identified in beef stock, roasted hazelnuts, mushrooms, pork, kidney beans and potatoes. Its presence in aromas of cooked or roasted prod-

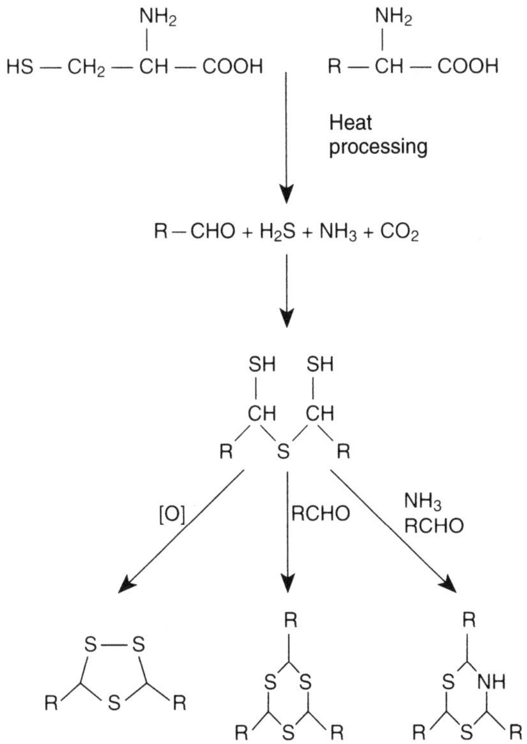

Fig. 15.8 Formation of nitrogenous and sulphurous heterocycles by degradation of cysteine and other amino acids.

ucts that are rich in cysteine can easily be explained, as can that of its homologues, the trithianes-1,2,4 and the dithiazines-1,2,5 by the recombination of the products of thermal degradation of amino acids (Fig. 15.8).

15.2.2.3 Aromatic notes

Aromatic notes are very varied. They can be classified into three large categories:

- **Sweet aromas** which include vanilla, caramel, coffee, cocoa, coconut and kola nut, honey as well as all the fruit aromas.
- **Salty aromas** can be of plant origin (spices, herbs, fruits, etc.) or those of animal origin (meat products, dairy products, etc.).
- Finally a '**miscellaneous**' category which includes in particular alcohols (rum, whisky, etc.).

15.2.2.4 Physico-chemical characteristics

Aromas that take a liquid or powder form can be classified according to their **solubility**. Water-soluble aromas include the true water solubles and the pseudo-water solubles whose relative solubility is due either to their wide dispersion (essence pastes) or to their low concentration. Fat-soluble aromas are generally essential oils or products extracted by organic solvents (oleoresins, resinoids, etc.).

The concentration of aromas used most frequently is between 1 part per 30 and 1 part per 1500. The power reference supplied is arbitrarily the dosage to be used in order to obtain the desired aroma intensity; so, depending on the use required we will select aromas of a particular power.

The **viscosity** of solutions of aromas is also very variable: it embraces virtually the whole range from alcohol to that of a pasty concentrate. However, this property has a crucial role on the exchange surface which is itself a determining parameter within the operation for converting aromatic substances into the gaseous phase.

The **chemical functional groups** most often found are, in descending order: esters, ketones, alcohols, aldehydes, phenol ethers, nitrogenous heterocycles, oxygen heterocycles and carboxylic acids.

Figure 15.9 shows the groups of molecular structures which are very important in the perception of the flavour of foods. These compounds are involved in a great number of olfactory notes. Most hydrocarbons of a terpenic nature make only a modest contribution to the aromatic note of a product; nevertheless it would be wrong to underestimate their importance. Moreover, the thresholds' detection greatly depends on the medium in which they are measured and it would be very misleading to believe that a chemically pure compound possesses a privileged position within a mixture of aromatic ingredients simply because its threshold of detection measured in an aqueous solution at 20 °C is less than those of the other ingredients in the mixture. The part played by the various ingredients of a food in the phenomena of holding and salting-out considerably modifies these thresholds. In addition, an aroma compound is never found in the pure state in a food, but always forms part of a complex mixture of volatile molecules which interact. As a consequence, caution must be exercised in any consideration of thresholds of detection.

The majority of aromatic molecules have the following characteristics:

- Molecular mass of between 17 and 300 Da.
- Number of carbons lower than 18.
- Boiling point below 380 °C.

Many aromatic mixtures also have anti-oxidant properties. Among the spices, rosemary, sage, thyme and nutmeg flower are the most active. For example, four substances that possess anti-oxidant properties have been isolated in rosemary: carnisol, rosmanol, rosmaquinone and rosmasdiphenol. These molecules have an anti-oxidant activity equal to, or greater than, that of butylhydroxytoluene (BHT-E 321) or butyl-tertiary hydroxyanisol (BHA-E 320). In addition, these compounds are stable at high tempera-

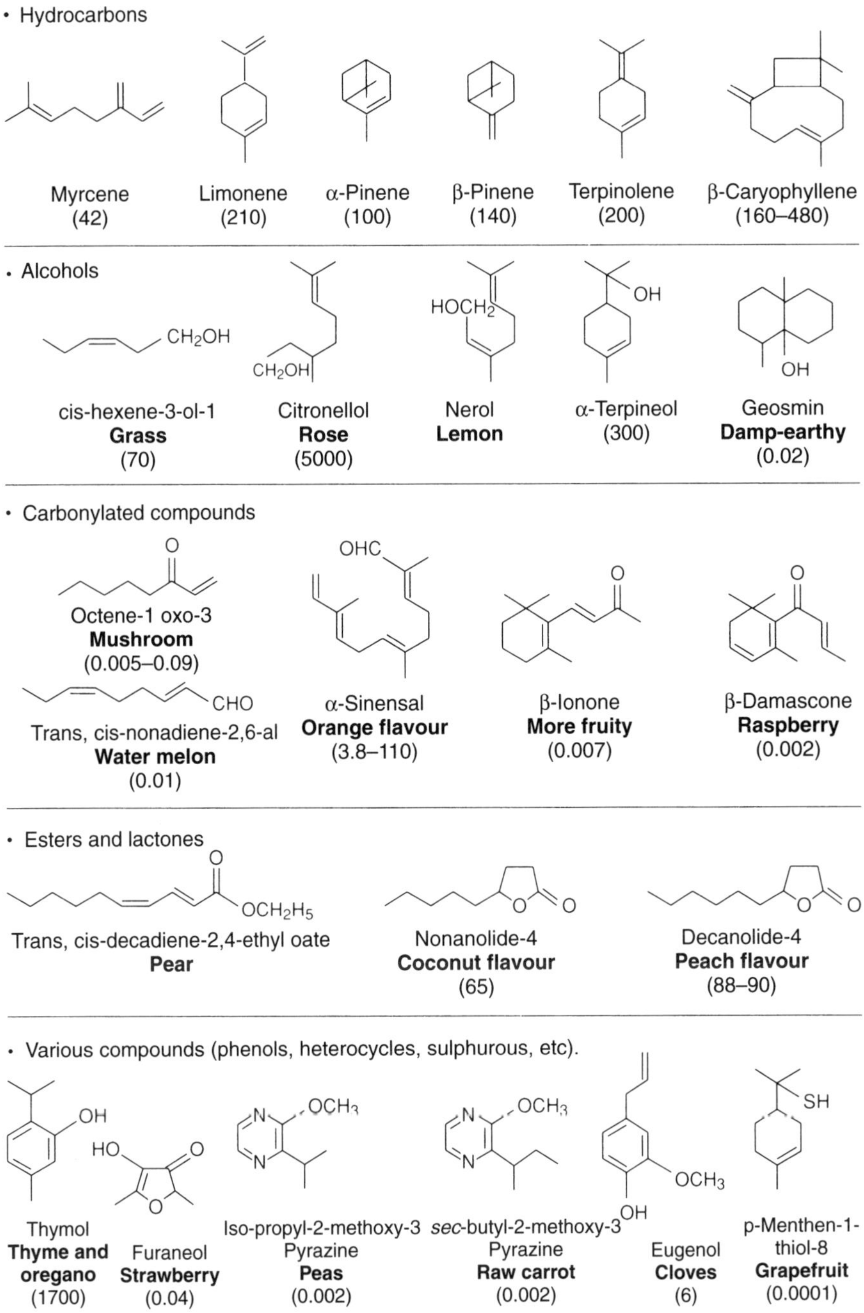

Fig. 15.9 Structure of characteristic flavour constituents. The olfactory note appears in bold. The thresholds of detection determined in water at 20 °C and expressed in mg/l are indicated in brackets.

tures and are very resistant to steam distillation during cooking or drying operations.

15.2.3 Formulation and manufacture of formulations

If, in the case of additives, one single molecule is generally sufficient to obtain the desired effect, within the field of aromas one single molecule can never re-create a flavour since one flavour actually has several facets. Their nature is similar to that of other food ingredients in terms of their great complexity; we need to know, for example, that several hundred compounds can frequently be identified in the discharge from a food product.

A satisfactory formulation must restore the image of the reference aroma at different times, which is the most difficult aspect: initial odour, first perception by the retronasal route, residual body and notes, etc. In addition the formulation must take into account not only the way in which it is to be industrially manufactured but also the processes it will have to undergo when it is incorporated in the food. We need to understand that any modification relating to one single compound in the aroma, even in proportions that cannot be measured, can easily be perceived by the consumer; the aroma of a dish corresponds to a balance of volatile constituents, and any change in the concentration of a single element can change the aroma in terms of quality, even if the overall intensity remains the same. The modification can just as easily entail the reduction, or even the disappearance, of a substance, as it can its appearance or development of new compounds. The essential point in the manufacture of flavourings lies in one or several blending operations of varying degrees of complexity. Generally, liquid forms are blended, but the industry also offers powders.

15.2.3.1 Factors intervening in the manufacture of a formulation

In the first instance, we must retain the relationship between head notes, body notes and tail notes and the relationship between odour and taste. The **head notes** are provided by the very volatile compounds of the aroma, those which are the first to diffuse into the atmosphere surrounding the food product and which are responsible for the very first olfactory impression. The **base notes** or body notes are provided by compounds of average volatility. They are the most important because they must be maintained with the same intensity over the whole time the food is being consumed. The **tail notes**, or persistence notes, result from ingredients that are scarcely or not at all volatile, which generally delay the release of the more volatile ingredients as well. These compounds often possess properties of sapidity.

The **physical characteristics of the food** are involved for various reasons. Among these, the properties of density, degree of aeration, adsorbent capacity and dispersion of the aroma are the most important. So, an aerated product is generally of low density, but it also holds the aromatic ingredients directly in the gaseous state; the head notes will then appear directly in fairly intense forms.

Likewise, the adsorbent capacities of the medium are extremely important; the more an aroma is adsorbed, the less it is released, and the entrapment of the aroma in its medium increases the duration of the response period. Lipids are known for their binding capacity, but it is also true that many proteins have a higher binding capacity.

The physical state of the food determines its surface area (fluid or viscous liquids, crumbly or hard solids, etc.) and the aromatic compounds have varying degrees of mobility within these. So the starting point for the factors making up flavour depends directly on this surface area. In the same way the temperature at which the food is eaten encourages the release of the aromatic constituents to a greater or lesser degree. Finally, certain molecules can enhance the aroma, which is therefore required in lower amounts.

15.2.3.2 Flavour modifiers

A number of substances of very different types and whose methods of action are still poorly understood have the ability to modify the taste of foods.

MSG undoubtedly occupies the most important place among the ingredients habitually grouped together under the term 'flavour enhancer'. It does not modify the nature of the aroma, but sensory measurements show an overall increase of olfactogustatory perceptions for foods containing between 0.1 and 2% of glutamic residue (see Section 14.1.2.2).

Other molecules, many of which derive from amino acids, also have flavour-enhancing properties, especially in the stock/meat notes. Some of these are as follows:

- sodium aspartate;
- isovaline;
- sodium DL-threo-β-hydroxyglutamate;
- sodium DL-homocysteinate;
- sodium L-α-amino-adipate;
- L-tricholomic acid;
- L-ibotenic acid;
- sodium inosinate;
- sodium guanylate.

Diethyl glutamate can develop both sweet and bitter flavours and methylthio-propylic alcohol, although not an amino acid, has an action similar to that of glutamate.

The 'fruity', 'jam' and 'caramel' notes also have their enhancers in molecules with a pyrone nucleus: maltol, ethylmaltol (artificial aroma) and furaneol. Likewise, red fruit aromas have a quality and quantity impact which is higher in the presence of fructose than of sucrose.

In addition, substances that act at extremely low doses (in the order of 10^{-4} p.p.m.) have a special action on the 'dairy' sensation. Sodium dioctyl

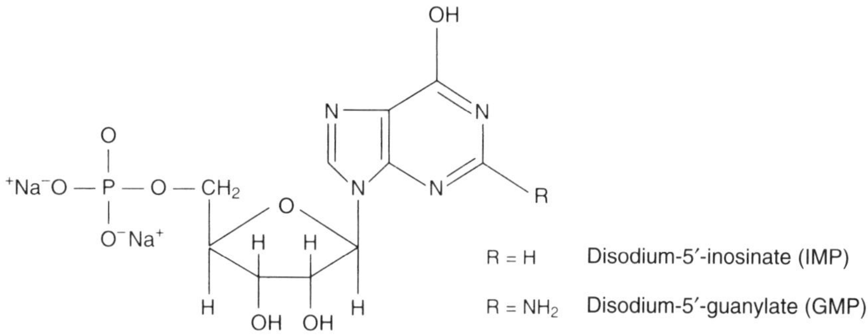

Fig. 15.10 Molecular structure of ribonucleotides.

sulphosuccinate is involved in the fresh milk note and *N*, *N*-diorthotolylethylene diamine and cyclamic acid in the butter note. There are also natural flavour enhancers such as the nucleotides, for example.

Ribonucleotides are present in most natural, plant or animal products. Of all the ribonucleotides found in nature it is the 5′-inosinate (IMP) or the 5′-guanylate (GMP) produced from the breakdown of adenosine triphosphate which have the most intense aromatic effects (Fig. 15.10). Fungi have a high GMP content, whereas IMP is widespread in the tissues of fish and animals. Although these ribonucleotides and MSG have many points in common, it is important to note that their uses are different. IMP and GMP can enhance a large number of flavours and modify both sweet and salty tastes either alone or in combination, whereas MSG – in a higher quantity – is essentially used to bring out the flavour of meat.

There are numerous substances acting on other senses. The principal acidulants (citric acid, lactic acid, tartaric acid, orthophosphoric acid, etc.) have different effects depending on which flavouring is used and they also reinforce the anti-oxidant action of other substances. Bitter compounds are numerous and differ widely in their chemical formulae; each substance produces one original note. Depending on its environment this occurs sooner or later, with varying degrees of persistence and varying degrees of sharpness.

Some compounds are also capable of producing the sensation of a change in temperature in the mouth: various volatile or sapid compounds are responsible for impressions of coolness or burning. These compounds are widespread within the field of aromatic herbs and spices: eucalyptol, menthol, sinalbin and sinigrin in mustard, piperine in pepper, gingerol in ginger.

Finally, certain substances also have a direct action on the flavouring. This is the case, for example, with thaumatin (see Section 10.7.4.1). The principal industrial application of this sweetener is to mask the bitterness or astringency of various food products. For example, in calcium-rich foods or in dietary food containing potassium chloride, thaumatin added in the ratio of 1 p.p.m. considerably reduces the sensation of bitterness. It can also mask the specific taste of pastes made from freshwater fish.

To conclude, we must remember that all the senses contribute to identifying a food. Aroma is the most important factor, but if our other senses have also allowed us to identify a food, then its aromatic power is clearly higher.

15.2.3.3 Composition of extracts of spices and aromatic herbs

In order to illustrate the fact that an aroma is a complex product, we can look at the composition of some spice and aromatic herb extracts. For example, the essential oils of **pepper** are rich in mono- and sesquiterpenic hydrocarbons, which represent, on average, 90% of the essence and show significant quantitative variations according to variety. Among the most abundant of these are: δ-3-carene (from traces to 20.2%), limonene and β-phellandrene (16.4–24.4%), α-phellandrene (0.5–27.4%), α-pinene (1.1–16.2%), β-pinene (4.9–14.3%), sabinene (0.2–19.4%), β-bisabolene (0.1–5.2%), β-caryophyllene (9.4–30.9%), α-copaene (from traces to 3.9%), α-humulene (1.0–2.1%), α-selinene and β-selinene (0.2–7.9%).

Depending on their origins there are also significant differences in the composition of essential oils from **cinnamon** peel: β-caryophyllene (1.4–3.3%), cinnamic alcohol (0.7–4.9%), linalol (2.3–4.9%), benzaldehyde (1.0–1.2%), cinnamaldehyde (58–78.1%), cineole-1,8 (1.7–3.3%), benzyl benzoate (0.7–1%), cinnamyl acetate (2.3–6%), eugenol (from traces to 8.8%) and eugenyl acetate (0.1–1.5%).

Likewise, depending on the origins of the plant, there is a wide diversity in the composition of **basilic** essences. Some of the variations observed in terms of principal constituents are as follows: linalol (0.2–75.4%), methyl chavicol (0.3–88.6%), methyl cinnamate (from traces to 15.5%), eugenol (from traces to 11.2%), *cis*-ocimene and cineole-1,8 (from traces to 13.6%).

One reference work deserves mention when considering any analytical research on the subject of aromas: the manual published by the TNO-CIVO Food Analysis Institute. This brings together, plant by plant, the qualitative and quantitative data relating to the volatile compounds present in foods.

Bibliography

General books

Alais, C. and Linden, G., *Abrégé de Biochimie Alimentaire*, Masson, Paris, 2nd edn, 1991.

Birch, G.G. and Lindley, M.G., *Interactions of Food Components*, Elsevier Applied Science Publishers Ltd, London, 1986.

Belitz, H.D., *Food Chemistry*, Springer-Verlag, London, 2nd edn, 1990.

Cheftel, J.C., Cheftel, H. and Besancon, P., *Introduction à la Biochimie et à la Technologie des Aliments*, Lavoisier, Paris, vol. 1, 1976, vol. 2, 1977.

Cheftel, J.C., Cuq, J.L. and Lorient, D., *Protéines Alimentaires*, Lavoisier, Paris, 1985.

Dickinson, E. and Stainsby, G., *Colloids in Food*, Applied Science Publishers Ltd, London, 1982.

Dickinson, E., *An Introduction to Food Colloids*, Oxford Science Publications, Oxford, 1992.

Dupin, H., Cuq, J.L., Malewiak, M.I., Leynaud-Rouaud, C. and Berthier, A.M., *Alimentation et Nutrition Humaine*, E.S.F., Paris, 1992.

Everett, D.H., *Basic Principles of Colloid Science*, Royal Society of Chemistry, London, 1991.

Fennema, O.R., *Food Chemistry*, Marcel Dekker New York, 2nd ed., 1985.

Hunter, R.J., *Foundations of Colloid Science*, Oxford University Press, Oxford, vol. 1, 1990, vol. 2, 1991.

Robinson, D.S., *Food Biochemistry and Nutritional Value*, Longman Scientific and Technical, Harlow, 1987.

Specialised books

Aspinal, G.O. (ed.), *The Polysaccharides*, Academic Press, New York, 1986.

Bourgeois, C.M. and Le Roux, P. (eds), *Protéines animales*, Lavoisier, Paris, 1982.

Charambalous, G. and Inglett, G.E., *Flavor of Food and Beverages*, Academic Press, New York, 1978.

Dickinson, E., *Food Emulsions and Foams*, Royal Society of Chemistry, London, 1987.
Dickinson, E. and Stainsby, G., *Advances in Food Emulsions and Foams*, Elsevier Applied Science, London, 1988.
Fox, P.F. (ed.), *Developments in Dairy Chemistry Proteins*, Applied Science Publishers Ltd, London, 1982.
Girard, J.P. (ed.), *Technologie de la viande et des produits carnés*, Lavoisier, Paris, 1988.
Godon, B. (ed.), *Protéines végétales*, Lavoisier, Paris, 1985.
Godon, B. (ed.), *Biotransformation des produits céréaliers*, Lavoisier, Paris, 1991.
Karleskind, A. (ed.), *Manuel des corps gras*, Lavoisier, Paris, 2 vols, 1992.
Larsson, K. and Friberg, K. (eds), *Food Emulsions*, Marcel Dekker, New York, 1990.
Law, R. (ed.), *Developments in Meat Science*, Applied Science Publishers, London, 1981.
Le Meste, M. and Colas, B. (ed.), *L'eau dans les procédés de transformation et de conservation des aliments*, Cahiers ENSBANA, Lavoisier, Paris, 1990.
Lorient, D., Colas, B. and Le Meste, M. (eds), *Propriétés fonctionnelles des macromolécules alimentaires*, Cahiers ENSBANA, Lavoisier, Paris, 1988.
Mitchell, J.R. and Ledward, D.A. (eds), *Functional Properties of Food Macromolecules*, Elsevier Applied Science Publishers Ltd, London, 1986.
Multon, J.L. (ed.), *Additifs et auxiliaires de fabrication dans les industries agroalimentaires*, Lavoisier, Paris, 2nd ed., 1992.
Multon, J.L. (ed.), *Sucres, édulcorants et glucides de charges dans les industries agroalimentaires*, Lavoisier, Paris, 1992.
Myers, D., *Surfaces, Interfaces and Colloids: Principles and Applications*, VCH Verlagsgesellschaft, Berlin, 1991.
Pomeranz, Y., *Functional Properties of Food Components*, Academic Press, London, 1991.
Richards, H. and Multon, J.L. (eds), *Les arômes alimentaires*, Lavoisier, Paris, 1992.
Richards, H. (ed.), *Epices et aromates*, Lavoisier, Paris, 1992.
Stadelman, W.J. and Cotterill, O.S. (eds), *Egg Science and Technology*, The Avi Publishing Co, Connecticut, 3rd ed., 1986.
Whistler, R.J., Bemiller, J.N., Paschall, E.F. (eds), *Starch*, 2nd ed., Academic Press, New York, 1984.
Wilson, A.J., *Foams: Physics, Chemistry and Structure*, Springer Verlag, London, 1989.

Conferences and symposia

Fishman, M. and Jen, K.J., *Chemistry and Function of Pectins*, ACS Symposium Series, 1986.
Martin, R. (ed.), 'International Symposium on Engineered Seafood Including Surimi', Seattle, 1985.
'Produits Alimentaires Intermédiaires', International Symposium, Paris, 1984.
'Colloque APRIA', Paris, 1991.
'La Restructuration des viandes', Colloque CNRS-CNERNA, Versailles, 1985.
'Food Ingredients Europe (F.I.E.)', Conference Proceedings, Expoconsult Publishers, Maarssen, The Netherlands, 1989, Düsseldorf, 1990, London, 1991, Paris, 1992.

Technical documentation

Sociétés Ferruzi, Kelco, National Starch, Roquette Frères, Sanofi.

Index

Page numbers in bold, e.g. **232**, denote references to figures or entries in tables.